BLOCKCHAIN TECHNOLOGY

Unlocking the Value of 3Ts in the Business World

Electronic Commerce Management for Business

BLOCKCHAIN TECHNOLOGY

Unlocking the Value of 3Ts in the Business World

Richa Chauhan, PhD
Vidhi Kaul, PhD
Editors

AAP | APPLE ACADEMIC PRESS

First edition published 2026

Apple Academic Press Inc.
1265 Goldenrod Circle, NE,
Palm Bay, FL 32905 USA

760 Laurentian Drive, Unit 19,
Burlington, ON L7N 0A4, Canada

CRC Press
2385 NW Executive Center Drive,
Suite 320, Boca Raton FL 33431

4 Park Square, Milton Park,
Abingdon, Oxon, OX14 4RN, UK

Library and Archives Canada Cataloguing in Publication

Title: Blockchain technology : unlocking the value of 3Ts in the business world / Richa Chauhan, PhD, Vidhi Kaul, PhD, editors.
Other titles: Blockchain technology (Palm Bay, Fla.)
Names: Chauhan, Richa (Lecturer in business), editor. | Kaul, Vidhi, editor.
Description: First edition. | Series statement: Electronic commerce management for business | Includes bibliographical references and index.
Identifiers: Canadiana (print) 20250208377 | Canadiana (ebook) 20250208431 | ISBN 9781998511907 (hardcover) | ISBN 9781998511914 (ebook)
Subjects: LCSH: Blockchains (Databases)—Industrial applications. | LCSH: Blockchains (Databases)—Economic aspects.
Classification: LCC QA76.9.B56 B5672 2026 | DDC 005.75/88—dc23

Library of Congress Cataloging-in-Publication Data

CIP data on file with US Library of Congress

ISBN: 978-1-99851-190-7 (hbk)
ISBN: 978-1-99851-191-4 (ebk)
DOI: 10.1201/9781998511914

About the Book Series Electronic Commerce Management for Business

Editors-in-Chief:

Dr. Vikas Garg
Symbiosis Institute of Business Management, Noida
Symbiosis International (Deemed University), Pune, India

Dr. Richa Goel
Symbiosis Centre for Management Studies, Noida, India

Electronic trading, typically called e-commerce, involves procurement and distribution of electronic systems and other digital networks of goods or services. E-commerce stands for electronic commerce and concerns the electronic medium trade in goods and services. As the Internet grew, the volume of online trade has increased significantly. B2B, B2C, C2C, and related resources help to build online technology for future customer needs and markets. E-commerce has revolutionized industry, disrupted trade with the Internet, and created an e-commerce market for customers and enterprises through the electronic communications system.

The advances in internet and web-based technology have steadily eroded boundaries between traditional economies and the new online economy, including the amount of corporate money. The growth in e-commerce is enormous. Another explanation for this is the low cost of the PC and the rising use of the Internet. The corporate community is rapidly conscious of the possibilities provided by e-commerce.

This series mainly aims to discuss the role of e-commerce in today's business. It further highlights some critical issues in e-commerce, suggestions, and future strategies for e-commerce in years to come.

Topics of Books in the Series:

The primary endeavor of this series is to introduce and explore contemporary research developments in a variety of rapidly growing research areas. The volumes will deal with the following topics but not limited to:

- E-Business and Mobile Commerce

- Big Data Analytics
- Business and Data Analytics
- Implications of Technological Disruption
- Technology and the War for Talent
- Impact of New Technologies on Business
- Artificial Intelligence and Its Impact on Business
- Managing Global Competitiveness Using Technology
- Impact of New Technology on Global Labor Markets
- Blockchain in International Business
- IoT and Smart Cities
- IoT Strategies and Business models
- Industrial Internet of Things (IoT)
- AR (Augmented Reality) in Product Design and Manufacturing
- AR (Augmented Reality) and Training
- Use of AR (Augmented Reality) in Sales and Marketing
- Smart Manufacturing
- Cryptocurrency
- Strategies of Emerging Countries in the Era of New Technology
- The Roadmap to Thriving in the New Skills Economy
- Future of Mobile Apps
- Social Media – A Powerful Tool
- Big Data Powers Organizations
- Integration of Technology in HR Practices
- Skillset for Industry 4.0
- Managing Virtual Teams and Environments for Innovation
- Innovation in Business
- Effects of Shifts in Globalization on Emerging Economies
- Smart Global Supply Chain Management
- Corporate Governance in the New Global Technology-Driven Economy
- Business Intelligence Tools & Applications
- Cyber Defense & Information Technology
- Cloud Computing
- Data & Intelligent System Modelling
- Digital Marketing & Internet Marketing
- Smart Connected Products
- Impact of Technology on Financial Decisions
- Finance and Risk Management for Using New Technology

- Digital Transformation and Financial Services
- Banking services in the digital era
- Virtual Reality
- Robotics and Its Use in Business
- Industry 4.0 and Labor Issues

The series aims to serve as a valuable resource for researchers, academicians, and industry professionals and undergraduate, postgraduate, doctoral students in electronic commerce, Internet marketing, business innovation, digitization, technology in product design, management, and more.

Features of the volumes under this series will explore recent trends, model extensions, developments, solutions to real-time problems, and case studies.

BOOKS IN THE SERIES

Applications of Artificial Intelligence in Business and Finance 5.0
Editors: Richa Goel, PhD, Vikas Garg, PhD, and Michela Floris, PhD

Blockchain Technology: Unlocking the Value of 3Ts in Business World
Editors: Richa Chauhan, PhD, and Vidhi Kaul, PhD

Entrepreneurship and Sustainable Business Development
Editors: Ruchi Khandelwal, PhD, Deepa Kapoor, PhD, Nidhi Gupta, PhD, and Neha Gupta, PhD

Smart Technology for a Smart City and Industry 4.0
Editors: Vinay Kandpal, PhD, Manoj Nallapaneni, PhD, and
Deep Chandra, PhD

About the Editors

Richa Chauhan, PhD
Assistant Professor, Jaypee Business School, Jaypee Institute of Information Technology, Noida, India

Richa Chauhan, PhD, MBA, is Assistant Professor (Senior Grade) at the Jaypee Business School at the Jaypee Institute of Information Technology, Noida, India. With more than 14 years as an academician, she has contributed to many research endeavors on emotional intelligence, spirituality, indigenous management, blockchain, and consumer behavior. She has authored many national and international journals and published in ABDC- and Scopus-indexed journals. She strongly believes the role of an academician is not only to nurture young minds to bring in a fundamental change in approach to problem-solving but also to embark upon constant research endeavors for the development of the industry and society at large. Dr. Chauhan earned her MBA from ICFAI University, Dehradun, and PhD in Management from Banasthali Vidyapith, India.

Vidhi Kaul, PhD
Visiting Faculty, Banasthali Vidyapith, Rajasthan, India

Vidhi Kaul, PhD, is currently Visiting Faculty at Banasthali Vidyapith, Rajasthan, India. She was formerly Assistant Professor at S. S. Jain Subodh P. G. Mahila Mahavidhyalaya, Jaipur, India, as well as at St. Xavier's College, Jaipur, India, where she headed the Department of Business Administration from 2012–2014. With teaching experience of more than 12 years, she was also actively involved in the career counseling of undergraduate students in professional programs and has contributed to a book on personality development. She has published a number of research papers in peer-reviewed journals and has contributed chapters in books published by Taylor and Francis and other publishers in the field of commerce and management. Dr. Kaul earned her bachelor's degree with honors in Business Administration (Commerce) from the University

Maharani College, Jaipur, and completed her master's and doctorate degrees in Commerce at the P.G. School of Commerce, University of Rajasthan, India.

Contents

Contributors

Sandhya Avasthi
ABES Engineering College, Ghaziabad, Uttar Pradesh, India

Satya Banerjee
London College of Fashion, University of the Arts London, United Kingdom

Divya Bansal
Amity University, Noida, Uttar Pradesh, India

Mona Chaudhary
Amity University, Noida, Uttar Pradesh, India

Richa Chauhan
Jaypee Business School, Jaypee Institute of Information Technology, Noida, Uttar Pradesh, India

Monica Choudhary
SP Jain School of Global Management, Australia

Aditi Dhama
Amity School of Fashion Technology, Amity University, Greater Noida, Uttar Pradesh, India

Ruchi Jain Garg
Sharda School of Business Studies, Sharda University, Greater Noida, Uttar Pradesh, India

Anju Gupta
Khandelwal Vaish Girls Institute of Technology, Jaipur, Rajasthan, India

Shashi Gupta
UPID, Noida, Uttar Pradesh, India

Anshul Jain
WISDOM, Banasthali Vidyapith, Tonk, Rajasthan, India

Sweta Jain
IManage, San Francisco, CA, USA

Prateek Jangid
Zudio Tata-Trent, Mumbai, Maharashtra, India

Vidhi Kaul
University of Rajasthan, India

Ebru Kemer
Niğde Ömer Halisdemir University Vocational School of Social Sciences, Department of Hotel, Restaurant, and Catering Services, Dr. Öğretim Üyesi Mustafa Kemal Üniversitesi, Turkey

Khyati Kochhar
WISDOM, Banasthali Vidyapith, Tonk, Rajasthan, India

Vipin Gupta
Wipro LTD, Edmonton, Canada

Alka Kumawat
Khandelwal Vaish Girls Institute of Technology, Jaipur, Rajasthan, India

Leena
Banasthali Vidyapith, Jaipur, Rajasthan, India

Nidhi Maheshwari
Delhi Technological University, University School of Management and Entrepreneurship, New Delhi, India

Edward Miller
University of Greenland, NuuK, Greenland

Divya Mishra
Assistant Professor, Vivekananda School of Business Studies, Vivekananda Institute of Professional Studies-Technical Campus, AU-Block (Outer Ring Road), Pitampura, Delhi, India

Shikha Mishra
Amity Business School, Noida, U.P., India

Nandita Mittal
Barnet and Southgate College, Wood Street Campus, London

Vipin Nadda
University of Sunderland, London, UK

Sujit S. Nair
Amity Business School, Noida U.P., India

Raghu Vamsi Potukuchi
Department of Computer Science and Engineering, Jaypee Institute of Information Technology, Noida, Uttar Pradesh, India

Tanushree Sanwal
KIET School of Management, KIET Group of Institutions Delhi-NCR, Ghaziabad, Uttar Pradesh, India

Parul Saxena
Sharda School of Business Studies, Sharda University, Greater Noida, Uttar Pradesh, India

Rajeev Semwal
Amity Institute of Psychology and Allied Sciences, Amity University, Greater Noida, Uttar Pradesh, India

Mohd. Salman Shamsi
Department of Marketing, Faculty of Business Administration, University of Tabuk, Tabuk, Saudi Arabia

Shobhna Sirohi
Tata Consultancy Services, Deutschland GmbH, Germany

Pooja Tiwari
Sharda School of Business Studies, Sharda University, Greater Noida, Uttar Pradesh, India

Nandita Tripathi
Amity Institute of Psychology and Allied Sciences, Amity University, Greater Noida, Uttar Pradesh, India

Meenakshi Tyagi
KIET School of Management, KIET Group of Institutions Delhi-NCR, Ghaziabad, Uttar Pradesh, India

Deep Sagar Verma
Department of Fashion Management Studies, NIFT Panchkula, Haryana, India

Priyanka Yadav
WISDOM, Banasthali Vidyapith, Tonk, Rajasthan, India

Sapna Yadav
Jaypee Business School, Jaypee Institute of Information Technology, Noida, Uttar Pradesh, India

Acknowledgments

This academic endeavor stands as a testament to the collaborative efforts and unwavering support of numerous individuals who have contributed their expertise, guidance, and encouragement throughout this journey into the intricate realm of blockchain technology for master's students.

First and foremost, we express our deepest gratitude to the esteemed mentor and book advisor, whose invaluable guidance, profound insights, and unwavering encouragement propelled this work forward. Their mentorship has been an indispensable compass, steering this research toward clarity and depth.

We extend heartfelt appreciation to the participants and experts in the field of blockchain technology who generously shared their time, knowledge, and experiences during interviews, surveys, and discussions. Their contributions formed the bedrock upon which this academic pursuit was built.

This work would not have been possible without the support of our family and friends, whose unwavering belief in my capabilities served as a constant source of motivation. Their understanding, patience, and encouragement sustained me through the highs and lows of this academic pursuit.

Lastly, we express our gratitude to the academic community, whose collective body of knowledge served as a guiding light, shaping and informing the ideas presented within this book.

Preface

What if the logistics world is free from intermediaries? What if data and transactions can be recorded, autonomously coordinated, and verified without the need for third parties? If such assumptions can be achieved, then the complications of the supply chain and logistics industry could be eliminated globally. Blockchain technology is efficient enough to facilitate this assurance to the logistics industry.

Presently, blockchain technology is still very far from maturity, along with various challenges and complexities to be overcome before its effective deployment at scale in the world of business. Probably, the substantial challenge would be accomplishing successful acceptability by the industry via collaboration and even partnership between manifold supply chain and logistics, insurance and banking, marketing, and finance stakeholders, as well as diversified interests. However, premature applications of this emerging disruptive technology across various industries–from energy to insurance, marketing to finance, and production to retail–propose that blockchain technology has an approving possibility of accomplishing its comprehensive potential shortly. Blockchain technology is structured enough to facilitate higher effectiveness in new business modeling, incorporating leaner, transparent, traceable, and trustworthy mechanisms not only in logistics but also in other sections of business processes.

This present academic pursuit is an endeavor to explore and bring about detailed and elaborated data and information on the successful acceptability and approachability of blockchain technology not only in logistics but also in other sections of the business world. This book tends to present a holistic 360° acceptability of blockchain mechanisms for the business world and intends to serve as an effective instrument to professionals, experts, academicians, and students who want to understand the practical implementation of blockchain technology in various sections of the business world.

Introduction

New technological advancements and their applicability often simplify our social, business, and economic lives. One of the technologies that is fundamentally changing our numerous antiquated and conventional business models is blockchain. The global valuation of blockchain technology in 2023 was around 17.46 billion and its CAGR (compound annual growth rate) is anticipated to be 87.7% from 2023 to 2030. Although most people believe that blockchain technology can only revolutionize the finance industry, this technology has the power to revolutionize all business models, structures, and types, regardless of their nature, ownership, location, or even size. Blockchain technology has many benefits for businesses that rely heavily on middlemen because of its special ability to eliminate middlemen.

With the advent of Industry 4.0, the focus has shifted from massive automation through cyber-physical systems (CPS) to a more customer-driven approach. The development of hyper-cognitive systems, the fusion of virtual and extended reality, digital machinery prototyping and twin designs, trusted machine boundaries, cooperative robots, and supply chains driven by artificial intelligence (AI) have all been cited as contributing factors to the change. Industry 5.0, the new wave that emerges beyond Industry 4.0 ecosystems, is predicted to combine mass production with user-centric customization. Various verticals of industries, including the tourism industry, advertising and fashion industry, supply chain production ecosystems, smart farming and grids, drones, and healthcare, are anticipated to benefit from Industry 5.0. Nonetheless, information is exchanged across several diverse networks that span various authoritative domains. Thus, to secure the industrial perimeters and promote synergy, reliable and secure data transfer is essential.

As blockchain (BC) has the inherent qualities of immutability, chronology, and auditability in industrial systems, it is a preferred option for enabling security in Industry 5.0 ecosystems. The present book is an attempt to provide and enhance the understanding of Blockchain Technology via an in-depth study of its usability and precedence over contemporary business systems, as well as exploring its desideratum in the new-fangled business world.

CHAPTER 1

Blockchain: A Key Enabler of Futuristic Industry 5.0 Solutions

RICHA CHAUHAN,[1] VIDHI KAUL,[2] and SWETA JAIN[3]

[1]*Jaypee Business School, Jaypee Institute of Information Technology, Noida, Uttar Pradesh, India*

[2]*University of Rajasthan, India*

[3]*IManage, San Francisco, CA, USA*

ABSTRACT

The introduction of Industry 4.0 emphasized the transition from extensive automation via cyber-physical structures to a strategy that is close to customer requirements and assumptions. The evolution of hyper-cognitive nodes, the combination of virtual and extended reality, digital machinery, robots, trusted machine boundaries, artificial intelligence-driven supply chains, and prototyping and twin designing have all been reported to contribute to the transition. This new wave, called Industry 5.0, is expected to leverage massive development and production with client customization, outside the scope of Industry 4.0 ecosystems. Various industrial verticals, including smart grids, farming, supply chain production ecosystems, drones, and healthcare, are anticipated to benefit from Industry 5.0. Nonetheless, information is exchanged across several heterogeneous and diverse networks that span numerous authoritative sections. Thus, to secure the industrial perimeters and promote synergy, reliable and protected data sharing is essential. As blockchain (BC) has the inherent qualities of chronology, audibility in industrial systems, and immutability,

Blockchain Technology: Unlocking the Value of 3Ts in the Business World.
Richa Chauhan & Vidhi Kaul (Eds.)

DOI: 10.1201/9781998511914-1

it is a preferred option for enabling security in Industry 5.0 ecosystems. The proposed chapter is an attempt to showcase the overall picture and vision of Industry 5.0 applications supported by Blockchain Technology.

1.1 INTRODUCTION

The introduction of Industry 4.0 emphasized the transition from extensive automation via cyber-physical structures to a strategy that is close to customer requirements and assumptions. The evolution of hyper-cognitive nodes, the combination of virtual and extended reality, digital machinery, robots, trusted machine boundaries, artificial intelligence-driven supply chains, and prototyping and twin designing have all been reported to contribute to the transition. This new wave, called Industry 5.0, is expected to leverage massive development and production with client customization, outside the scope of Industry 4.0 ecosystems. Various industrial verticals, including smart grids, farming, supply chain production ecosystems, drones, and healthcare, are anticipated to benefit from Industry 5.0. Nonetheless, information is exchanged across several heterogeneous and diverse networks that span numerous authoritative sections. Thus, to secure the industrial perimeters and promote synergy, reliable and protected data sharing is essential. As blockchain (BC) has the inherent qualities of chronology, audibility in industrial systems, and immutability, it is a preferred option for enabling security in Industry 5.0 ecosystems. The proposed chapter is an attempt to showcase the overall picture and vision of Industry 5.0 applications supported by Blockchain Technology.

The Fourth Industrial Revolution is currently underway. It aims to leverage the vision of smart factories and enhanced production by automating industrial systems and processes, integrating critical technologies such as edge and cloud computing, artificial intelligence, and the Internet of Things. Industry 4.0 has completely changed earlier iterations, where the main goals were to increase output and achieve massive production. The primary driving forces of Industry 1.0, which began in the 1700s, were the production of steam and water energy. Assembly line production, which Henry Ford introduced in 1870 and turned into a low-cost mass production unit for electricity generation and production, became the hallmark of Industry 2.0. The transition from mechanical to digital production occurred in Industry 3.0, and some automation was incorporated into the industry's procedures. Large-scale computers and memory-programmable controller logic are integrated into industrial designs in this generation, which minimizes the need for human intervention and labor. The world underwent a dramatic transition to Industry 4.0 in 2011. Communication and information technologies are used in this generation to support remote production lines that are coordinated by assisted networks. Industry 4.0 also saw the emergence of the Internet of Things-driven cyber-physical systems, which sparked the idea of the smart factory and led to self-organized production, connected logistics, and autonomous monitoring to boost corporate profits. Industry 4.0 is focused on smart manufacturing, with automation serving as the primary engine. Industry 5.0, which introduces the cognitive control process from Industry 4.0 with increased engagement in human-machine interactions, is the next revolution. It would therefore be value-driven as compared to process-driven. Artificial intelligence would be the driving engine in Industry 5.0, combining human knowledge to create cognitive capabilities and precise benchmarks (Longo, Padovano, and Umbrello, 2020). The Industry 5.0 concept has been established by experts and is based on intelligent behavior, control, and manufacturing fundamentals. The primary definitions provided by various researchers are shown in Table 1.1.

TABLE 1.1 Major Definition of Industry 5.0 by Industry Specialists

Advised By	Major Definitions
Longo et al. (2020)	According to the European Economic and Social Committee, Industry 5.0 combines the strength of human intelligence with CPPR to build cooperative factories. Authorities are trying to develop ethical and innovative solutions to overcome labor shortage problems in Industry 4.0.
Show et al. (2021)	Industry 5.0 is an incongruous innovation and the next generation of governance, and it attempts to isolate hyper-connected manufacturing and industrial automation systems.
Friedman et al. (2019)	Industry 5.0 necessitates that industry practitioners and information technologists prioritize the consideration of human factors in industrial system technology.
Rada et al. (2018)	Industry 5.0 is the first human-led industrial evolution, based on the 6R principles of industrial upcycling, such as revolutionize, reconsider, realize, reduce, reuse, and recycle. It's a method of systematic waste prevention and logistics efficiency design used to assess life standards and inventive ideas, and it manufactures high-quality custom items.
Nahvandi et al. (2019)	To enhance process efficiency, Industry 5.0 combines human brain power and ingenuity with intelligent systems. It integrates the capabilities of the cyber-physical production system (CPPS) and human intelligence to address the labor shortfall present in Industry 4.0. Industries, in collaboration with researchers, develop creative and human-centered design solutions.

TABLE 1.1 *(Continued)*

Advised By	Major Definitions
Koch et al. (2017)	Industry 5.0 is a society of smart factories where robots, intended for direct communication with humans, interact. The smart factories use social and corporate networks to connect humans and cyber-physical systems.
Maddikunta et al. (2022)	Industry 5.0 is a human-centered design approach where humans and cobots collaboratively work for multiple resources to customize autonomous production. Cobots are not programmable but can detect and understand human presence. This enables cobots to handle repetitive activities and labor-intensive work while humans are in charge of overall management.

The production and manufacturing methods would be adjusted to meet the specific needs of the final user. The goal of Industry 5.0 is to improve production quality by assigning repetitive and mundane jobs and duties to robots to perform while humans maneuver critical and intelligent jobs. This will give rise to a specialized market for skilled laborers. In Industry 5.0, humans will direct robots with an emphasis on a multitude of customizations. Industry 5.0 and CPS are closely related, as both envision a collaboratively imagined procedural relationship between humans and robots. Collaborative robots, as they are now called, would serve as the cornerstone of process management. In addition, it would be more environmentally friendly, with an emphasis on sustainable development and production activities. Automation is a key necessity and an efficient structure for reinforcement learning, machine learning, and deep learning, and would contribute to Artificial Intelligence-driven systems. Furthermore, the human aspect must be included to manage the processes effectively. The procedures in Industry 5.0, driven by cyber-physical systems, would use assisted wireless networking channels to transfer enormous amounts of data over the Internet. As a result, the data process needs to be secured against malevolent attacks that might compromise the confidentiality and security of sensitive information (Show, Chew, and Ling, 2021). Additionally, with customized requirements, client data needs to be kept safe across communication nodes. A foundational approach to addressing privacy concerns is the use of user-defined norms for data access and fair information practices with authorization. As explicit information sections would be hidden due to privacy protection structures, Artificial Intelligence machinery will not be capable of customizing itself properly, though the best possible balance of trade-offs between privacy and personalization is very much needed. Moreover, the security and privacy-based approaches fall short because they fail to account for the unusual behavior of sensor nodes. In Industry 5.0, trust in control data and sharing is essential as heterogeneous and autonomous networks perform together. Blockchain presents a viable solution that enables the creation of transparent ledgers that make controlling and managing industrial operational data simple. Within a Peer-to-Peer structure network, Blockchain is a shared, immutable, and distributed ledger that makes it easier to record transactions and track assets. It creates reliable review platforms that support auditing and compliance needs (Friedman and Hendry, 2019). Briefly, blockchain is a digital data ledger that consistently and sequentially gathers required information. A block header in the blockchain receives a hash that is connected to the block that came before it. After the data is added to the chain, it cannot be changed. The hash of the block is altered if any of the transactions within it are changed (Figure 1.1).

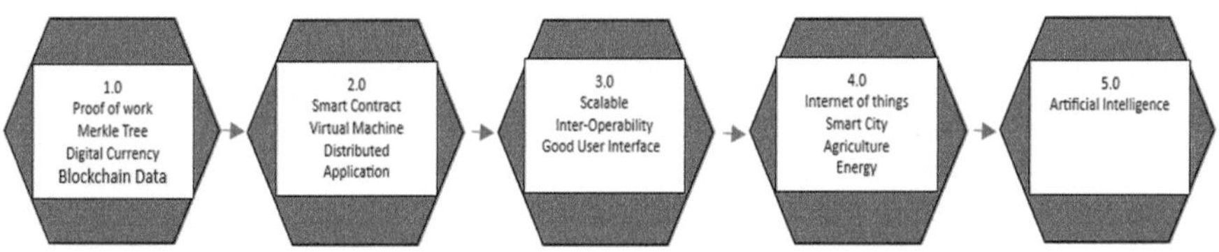

FIGURE 1.1 Evolution of blockchain.

Smart contracts are essential to the implementation of Industry 5.0 supply chains because they guarantee security enforcement, authentication, automated service-oriented behaviors, and access control. Assets, services, commodities, and items are managed via smart contracts backed by digital entities (Rada, 2018). Through a low-level contract interface, which automates the stakeholder agreements, smart contracts communicate through an underlying network of blockchain technology. Consensus protocols are crucial to the efficient management of the blockchain network as they regulate node throughput, mining latency, and network scalability. As Proof of Stake and Proof of Work require a lot of resources, consensus is not appropriate for responsive data allocation in industrial operations. While low-powered consensus techniques like Omni Ledger, Tangle, RAFT, and IOTA are mostly used in ecosystems of Industry 5.0, a permissioned blockchain is the recommended method for managing and orchestrating real-time data (Nahavandi, 2019). Developing authentic structures to thwart attacks such as Sybil, 51% attacks, and distributed denial-of-service (DDoS) is another crucial component of blockchain's fourth generation. Code injection, out-of-order, reentrancy, and gas attacks are just a few of the attack setups that commonly cause smart contracts to malfunction and should be prevented by secure blockchain technology and design. Different business-related applications and Hashgraph technology ensure that the suggested consensus of Industry 5.0 is equitable and scalable in its methodology.

1.2 MOTIVATION

The automation structure of Industry 5.0 is expected to incorporate human elements using accurate control and precise modeling structures and systems. Since open wireless networks are used for data trade-offs, security has become an essential element in protecting industrial structures. Therefore, blockchain serves as a foundational solution for Industry 5.0, utilizing time-stamped ledgers that are chronological and auditable to establish a trusted structure and control. Industry 5.0 has been the subject of preliminary research and studies that discuss pivotal technologies, process-driven methodologies, perspectives, and vision. However, the security and safety factors in Industry 5.0 are just as significant. The automation aspect of Industry 5.0 is expected to incorporate a human-centric approach via accurate control and precise modeling systems. Therefore, blockchain serves as a foundational solution for Industry 5.0, utilizing time-stamped ledgers that are auditable and chronological to establish a trusted mechanism of control. Industry 5.0 has been the subject of preliminary research and studies that discuss fundamental processes and technologies, process-based methodologies, and vision. Nevertheless, in Industry 5.0, the security perspective is just as significant.

1.3 INDUSTRY 5.0 AND INDUSTRY 4.0

The foundations of Industry 4.0, its potential constraints, and the rise of Industry 5.0 are covered in this section. To fully realize the potential of industrial structures and processes, the chapter conceptualizes the vision, perspective, and enablers of consensus protocols (Koch et al., 2017).

While Industry 5.0 is concentrated on solutions that are influenced by humans, Industry 4.0 focuses on sensor-based connections. Industry 5.0 creates unique solutions and modules that are highly required in the upcoming 10 years by fusing the accuracy of robots with human inventiveness. Robots that connect communication between the industrial unit, transportation, supply chain, and 5.0 architecture satisfy the needs of various industry verticals, justifying a close relationship between humans and machines.

People can learn about the preservation of the ecosystem, environmental stewardship, and the effective economic usability of readily available resources for future generations from Industry 5.0. Industry 5.0 offers personalization, the amalgamation of cognitive structures, user-driven customization, and the intelligence of humans, and above all, a return to a real environment. While Industry 5.0 is focused on creating smart societies, factors such as cloud computing, robotics, the Internet of Things, artificial intelligence, big data, and large-scale production navigate Industry 4.0. Moreover, the major technological setup of Industry 5.0 is robot-human cooperation and collaboration with renewable energy. Industry 4.0 is focused on improving structural innovation, operational study and research, and business administration as its applicability and research areas (Maddikunta et al., 2022). However, Industry 5.0 also pertains to a smart environment that encompasses the economy, agriculture, biology, and waste management.

1.4 UNDERSTANDING INDUSTRY 5.0 AND ITS INTRICACIES

Industry 5.0 seeks to increase the efficiency of industrial procedures by fusing human intervention and automation. The autonomous procedural structure of Industry 5.0 is acquainted with human desires and intentions. As robots now understand and can communicate with humans, it is expected that humans too will be able to perform closely and fearlessly with them. As a result, the procedures will also show synergy and align setups with numerous business models, resulting in lower waste of resources and reduced operating and capital costs. In addition, Industry 5.0 would promote green production by being self-sustaining and self-healing.

The autonomy of robots will decline in Industry 5.0 to the point where they will be used only for laborious, mundane, and repetitive assignments.

It will be designed to comprehend human instructions and use machine learning and data learning algorithms to process them into the appropriate outputs. As a result, the risk within production procedures will be reduced by this next generation of robots. Robots will be able to sense and perceive not only the human expectations in their immediate environment but also their future goals. Robots will operate on a reward-penalty-based system in which they will follow guidelines or set fundamental structures and will learn from the environment with which they are associated using deep reinforcement learning algorithms. Industry 5.0, for all its ambiguity, is built upon simple but powerful techniques like the logic efficiency design (LED) principle and the 6R (recognize, realize, reconsider, reuse, reduce, and recycle) methodology. The 6R technique seeks to incorporate the quasi-approach into the waste treatment layout. It is a superstructure of the 3R (reduce, reuse, and recycle) strategies. Although 6R techniques are used in waste management, the idea behind them is universal and can be applied in a wide range of industrial processes. The LED concept seeks to optimize operations and maintenance efficiency by creating logical process blueprints for industrial applications. The primary goal of the principle is to increase supply chain profitability. Appropriate methodologies are created to promote potential revenue among many stakeholders, operational transparency in supply chain projects, and functional connectivity. There have been disagreements and difficulties related to the working population, despite technological progressions like 6R and logic efficiency design in Industry 5.0. Worker conditions, geographical and networked locations, and social consequences are the primary causes of these challenges. The human workforce in any industry is gradually aging, meaning that middle-aged skilled individuals typically hold managerial positions. This makes it difficult to have candid conversations with the younger workforce and improperly documents accountability. Furthermore, many industrial workplaces continue to operate with antiquated equipment and outdated setups.

As a result, the total output decreases because these workplaces are not connected to other manufacturing sites via CPS. Adopting Industry 4.0 artificial intelligence technologies is hesitant because many industrial stakeholders are set in their ways and resistant to change. The issues influence yearly turnover rates, artificial intelligence-based profit forecasts, and business models. Ultimately, they force the cancellation of numerous promising projects. Consequently, Industry 5.0 specifically addresses these issues. It permits important stakeholder training in which the development plan, project vision, feasibility, and dependability are discussed before the project's start. In addition, they will be maximizing the value chain, which will, in turn, foster systemic confidence in robotic systems. Additionally, the topic of technology and automation is covered, with specific regulations (Lu, Xu, and Wang, 2020). The important design technologies that underpin the Industry 5.0 visions are covered in subsections.

1.4.1 LARGE-SCALE CUSTOMIZATION

The digital user experience for large customization will be at the core of Industry 5.0 procedures. Initially, a market segment-based strategy is developed in which a particular product is created depending solely on the unique inputs and information of that target market segment. The recommendations obtained from customers via targeted interviews, which represent the unique selling proposition they hope to see in the developed product, make up the input parameters. Thus, the stakeholders designed the desired parameters of mandatory functionalities for the look and feel of the new product through interviews. To collect the data, a large group of users and customers is contacted and interviewed. User recommendations, competitive design of the product, and demographic elements all play a significant role in the process of selecting parameters. Target users, notable similarities, and dissimilarities are identified through intelligent artificial intelligence systems, and then customized products are created to meet end-buyer requirements. Large-scale personalization enables businesses to target decision-makers across industries and improve customer engagement.

1.4.2 HYPER PERSONALIZATION

The idea behind hyper-personalization is to utilize actual customer behavioral information to tailor services, goods, experiences, and preferences of the individual consumer. To accomplish hyper-personalization, the industrial trial process necessitates a thorough comprehension of the associated technology costs, the product customer base, and the designed product. A customized marketing plan is established using the gained and collected knowledge. Digital technologies, such as machine learning pipelines, computer vision, artificial intelligence, and high cognitive systems, help achieve this goal by enabling industrial systems and procedures to choose a particular good or item for each customer. Human inputs and robots collaborate closely to produce goods in large quantities while allowing for good customization. The functional behavior of products is crucial, with a prime focus on cost reduction. To scale the manufacturing/production procedures, the supply chain makes the shift from archetype design to an agile production system. For instance, an e-commerce website can assess and design a product based on consumers' past click streams and interactions and can suggest the ideal option to a new consumer. The algorithm of a click stream recommender is required to structure and design a comparative analysis of the most liked products in the market for quick comparison (Demir, Döven, and Sezen, 2019; Tanwar et al., 2022).

1.4.3 COLLABORATIVE ROBOTS

Robots, or collaborative robots, are made to work closely with people; this close collaboration enables the robots to bring a humanistic approach into the production process. In 1996, professors at Northwestern University created the first robots. Modern robots are driven by sensors and can develop and structure actuation depending on human signals, in contrast to first-generation robots, which were mechanical and typically engaged in mechanical procedures. In Industry 5.0, the complete made-to-engine of a product is created through tailor-made robots. In the context of Healthcare 5.0, for instance, surgical robots help doctors and surgeons with tomography and 3D-dimensional vision. Telesurgery can also be added in addition to the above example, where a robot operates under the guidance of a distant surgeon via a responsive channel network, like the tactile web (Lee, Ahn, and Bang, 2011). Robots are a vital workforce in contemporary industries, contributing to increased productivity and boosting growth with the possibility of automating repetitive tasks.

1.4.4 DIGITAL TWINS

Digital twins are computerized depictions of actual industrial procedures or configurations, designed to precisely duplicate the actions of real-world items (Gupta et al., 2021; Jiang et al., 2018). For instance, the actual object being studied, a turbine motor, has several sensors installed to calculate various aspects of its power kinematics, motion, and operations. Such sensors produce information in several ways, such as temperature, meteorological situations and conditions, and energy output, as well as the operation of the actual solid object or process. After that, a processing structure receives this information and incorporates it to create a digital replica of the system. After gathering this data, a virtual model is made and utilized to simulate the actual physical object simultaneously. With predetermined error, the performance can be assessed and analyzed by using the generated digital twin. The software iterates received inputs several times, collecting the most apt inputs and information to be sent to the actual object in the real world.

Real object simulators function similarly to each other, but the primary distinction between digital twins and a simulator is the industrial subsystem scale. Whereas a simulator might only be able to operate with a limited set of inputs, digital twins can create different simulations to investigate various procedures. In contrast to digital twins, which depend on a two-way information flow where sensor nodes feed real-world data into the system, this is rarely helpful in simulations.

With input sets, the information is analyzed utilizing machine learning and data learning models, and reviews with regression problems are returned to the control system. Compared to standard simulations, dependencies and complex issues can be better investigated by digital twins because they have access to high-quality and much more recent statistics (Zhang et al., 2022).

1.5 CHALLENGES OF BLOCKCHAIN STRUCTURE

In the context of Industry 5.0, blockchain (BC) is crucial for safeguarding assets and facilitating the flow of information within various industrial methods and components. It establishes a reliable and traceable record to support the varied needs of endpoints in the application. Therefore, in terms of security, BC's role as a facilitator for Industry 5.0 is contingent upon the specific internal details and underlying structures of significant architectures, structures, and schemes that are compatible with BC ledgers.

Figure 1.2 illustrates the primary obstacles in implementing blockchain solutions within the Industry 5.0 structure.

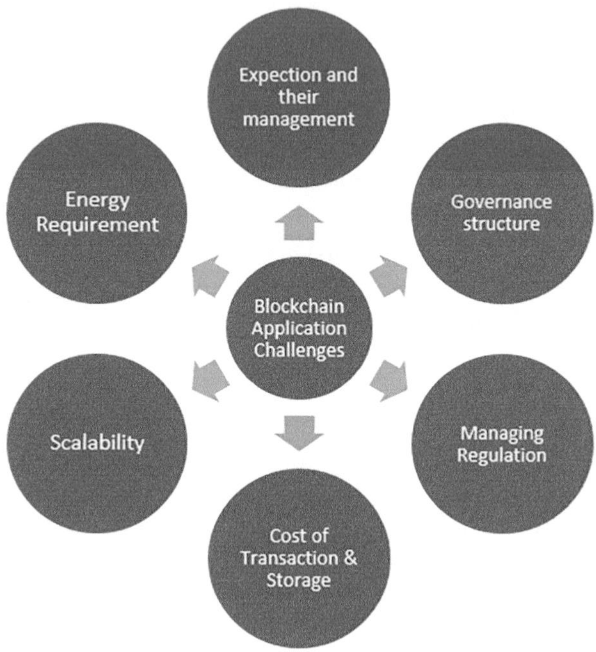

FIGURE 1.2 Complexities of blockchain application.

In the upcoming years, security tokens, smart contracts (SCs), and evolving legal regulations will define BC technology by recording legal information on the blockchain. A dependable and scalable BC implementation is imperative in achieving these objectives. The underlying consensus supports the importance of the BC type in harnessing the full potential of advanced technologies such as Cyber-Physical Systems, big data, Artificial Intelligence, and the Internet of Things. Therefore, this section of the chapter delves into the design and implementation challenges that traditional systems face and explores the potential advantages of utilizing BC technology. The chapter aims to address and examine the major probable obstacles and risks involved in accepting blockchain as the primary structure in Industry 5.0 applications (Bhattacharya et al., 2021; Kebande et al., 2021). The projected future leads are presented as in subsections.

1.5.1 EXPECTATIONS AND THEIR MANAGEMENT

In the business world, trust is crucial. Unfortunately, there are times when an attacker may cause chaos with a data leak; nobody is easily trustworthy in the network. To combat this, blockchain technology can be used to prevent tampering and establish trust. This is achieved using cryptographic hashes and a permanent ledger. However, the challenge lies in accurately measuring trust in real-time system management, as statistical analysis of trust computation becomes necessary if any nodes exhibit malicious behavior. Recent research proposes a reputation-based design for blockchain that assigns a reputation score to transactions and reaches a consensus based on the miner's reputation (Dev, Tsang, and Corchado, 2022; Singh et al., 2022).

1.5.2 GOVERNANCE STRUCTURE

The top priority for businesses is safeguarding crucial information, which can have varying security needs depending on the application. For instance, in cloud monitoring, preventing unauthorized access and data manipulation requires strict authorization and authentication processes, such as evaluating attack patterns. Similarly, generating trustworthy certificates is a crucial concern in vehicular systems, where blockchain technology can provide reliable solutions. In healthcare, maintaining privacy is paramount to protect sensitive patient data from unauthorized parties. Consequently, a one-size-fits-all blockchain architecture is inadequate for addressing the diverse demands of Industry 5.0. Instead, custom blockchain networks can be developed to meet the specific requirements of various industries.

1.5.3 MANAGING REGULATIONS

The current industrial control systems face a significant hurdle due to their centralized procedure. Their size and complexity make it challenging to adapt to business processes. If the central system is compromised, all data must be redirected, resulting in increased network latency and computing demands. In such a centralized system, authorities have complete verification rights over user transactions. Conversely, in a decentralized system where devices connect directly, peer-to-peer verification takes place.

However, peer-to-peer communication standards should be uniformly regulated to ensure consistency across applications. One potential solution is to establish a unified web framework that utilizes application programming interfaces for data exchange, providing greater control between applications and service endpoints.

1.5.4 COST OF TRANSACTION AND STORAGE

While blockchain is reliable for maintaining trust and chronology in industrial processes, it does come at a high cost of resources. This cost is further increased by the communication expenses when sensor nodes connect with other nodes. Unfortunately, there are currently no established standards for addressing these costs, with most solutions only focused on optimization. Potential areas of research include creating a uniform set of protocols for the BC network to minimize communication expenses. However, this may not align with specific application requirements. Another option is developing low-cost smart contracts that can be stored within the sensor node's memory. This would allow for self-deployment once data is collected, leading to increased automation and reduced costs in industrial applications.

1.5.5 FLEXIBILITY OF SCALE

In today's world, the blockchain (BC) network stores and verifies all transactions. This results in a significantly large ledger size, causing delays in navigating through all the blocks in the chain. Each block has a fixed size and can only handle a limited amount of transactional data. The block size and addition time vary depending on the consensus mechanism used to verify new blocks. For example, public blockchain ledgers may only process seven to eight transactions per second, while real-time industrial situations require millions of transactions. This makes it challenging to implement blockchain and presents scalability issues. One solution is creating side chain ledgers that transfer transactional data between chains according to the specific use case. Another option is implementing sharding in blockchain, which forms smaller controlled ledgers managed by a sharded authority. The main concern is ensuring fairness in the consensus process for sharded blockchains so that transactions are ordered fairly and selective applications are not given priority.

1.5.6 ENERGY REQUIREMENT

With the increasing reliance on sensor-driven industrial processes, energy usage has become a major concern. To address this challenge, permissioned and private blockchain networks have been suggested as potential solutions. However, these private networks utilize their consensus methods, which may compromise security to reduce energy consumption for sensor operation. This often involves shifting towards a centralized or semi-distributed approach where a main validator node is responsible for addressing mining requirements. In turn, maintaining an overall level of trust within the network becomes crucial. As a result, researchers have turned their attention towards developing verifiable and trusted blockchain solutions for permissioned and private networks to find an optimal balance between energy cost and security for practical applications (Chen et al., 2021).

1.6 SECURITY CHALLENGES AND ATTACK SCENARIOS WITHIN BLOCKCHAIN-SUPPORTED INDUSTRY 5.0

In blockchain, transactions can be digitized and distributed throughout the network, eliminating the need for a central authority's validation. This system is made possible by its immutability through cryptographic hashes and its identity and access control through smart contracts. However, despite its strong security features, blockchain does have some limitations. These limitations primarily stem from social engineering attacks, where unsuspecting users may accidentally reveal their wallet security phrase through deceptive websites designed to steal information. Other common forms of attack include mining pool attacks, privacy theft, and smart contract-based attacks. A recent study by Chen et al. analyzed these different types of attacks on blockchain-assisted ecosystems and identified three critical areas: mining pool attacks, communication-based attacks, and smart contract-based attacks. The chapter also attempts to discuss implementation challenges and future directions in this area. Therefore, it is essential to address these vulnerabilities in the Industry 5.0 ecosystem that relies on blockchain technology. In this chapter, we will explore some of the key attacks on blockchain. The details are presented as in subsections.

1.6.1 MINING ATTACK

Beginning the discourse with mining-related blockchain assaults, these specific attacks aim to acquire authority over the mining network. The perpetrator's objective is to tamper with the network's hash rate, conceal a block from other miners in a private and separate chain, or manipulate the blockchain by elevating their alternate chain as the dominant one. These attacks can be further categorized as in subsections.

1.6.1.1 SELFISH MINING

This approach involves a deceptive miner who secretly mines a block without informing other miners. The block is then shared on a separate chain that is not recognized by the main network. Meanwhile, a trustworthy miner also mines the same block and broadcasts it to all other miners. The deceitful miner continues to mine and add blocks to the side chain, while honest miners continue to add the same blocks on the main chain. Once the side chain surpasses the length of the main chain, the attacker reveals their side chain to all nodes. According to the longest chain rule, the side chain will be treated as genuine, and honest miners are not rewarded. To prevent the addition of more blocks to the side chain instead of the main network, one method is to use sequencing for block confirmation and publish the block height beforehand. This enables nodes to catch selfish mining behavior in the main network. Another option is to include indelible timestamps during block validation in the main network, allowing any lower timestamped blocks to be further examined by other nodes.

1.6.1.2 BLOCK WITHHOLDING ATTACK

A miner who attacks the system by submitting a partial proof of work (PoW) can significantly decrease the overall profitability of the pool. This delay in submitting blocks from the victim's mempool to the mining pool's administrator ultimately reduces each miner's pay per share and diminishes their incentives. To address this issue, potential solutions involve implementing a random mining strategy where a random miner submits proof of work, using game theory among miners to monitor each other's activities, or optimizing strategies such as swarm-based optimization to increase pay-per-share for all miners. These solutions aim to create a fair ecosystem that prioritizes both incentives and control for all miners involved.

1.6.1.3 51% ATTACK

This attack is aimed at the mining pool's hash rate, within which a category of miners conspires to reach a combined hash rate greater than 51% of the entire network. This allows them to hold the majority and potentially reverse transaction validations or create an invalid side chain that would still be considered legitimate due to their majority consensus (Johng et al., 2018; Tian, 2017). Various methods have been proposed to prevent this category of attack. The initial solution is to avoid using standard proof of work consensus, which can be exploited by miners with a high hash rate. Another approach suggests implementing checkpoints after a set of blocks has been added to the chain. As a result, an irreversible chain will be created once a checkpoint is developed or when a block is mined successfully by various miners on the network. Many researchers have forwarded a method for determining the total hashing power of mining pools through a historical weighting system. Such a structure facilitates honest miners to settle conflicts among branches on the main chain. By calculating the historical weighted difficulty of both the main chain and side chain, honest miners can choose the branch with the higher difficulty level.

1.6.1.4 CRYPTO JACKING

This form of cybercrime involves crypto jackers attempting to gain unauthorized access to a user's device to mine cryptocurrency without the user's knowledge. The malicious application installs mining code on the targeted machine, running in the background and contributing to a mining pool by computing hashes. While this generates hash power, the actual recipient of any incentives remains unknown. To combat these issues, it is important to implement effective security measures on devices and inform law enforcement agencies as well (Angrish et al., 2018; King and Nadal, 2012).

1.6.2 NETWORK COMMUNICATION ATTACK

The decentralized blockchain application utilizes a peer-to-peer network to facilitate communication among nodes. The potential attack on this network, as well as on the smart contracts and supported applications, will now be addressed.

1.6.2.1 SYBIL

Sybil is a peer-to-peer attack model that allows a node to function under various identities. The major objective of the attack is to obtain control of the network and execute illegal commands. To achieve this, a malicious node sends a message to other peer-to-peer systems in the consensus process, gathering important details regarding the connection pools within the blockchain network while misleading the information. The attacker creates a false identity to appear legal and authorized and deceives neighboring nodes. Studies suggest credit-driven block consensus techniques and methods that ensure that only those blocks with the maximum credit scores are selected for joining the block consensus group. An alternative technique involves using a network flow algorithm to identify an agent node for block consensus, which facilitates credibility-driven data transactions by evaluating its credibility based on transaction data.

1.6.2.2 IDENTITY PRIVACY

The attacker achieves control of private data by utilizing the communicating wallet address and registration details on the chain. The user then observes the transaction on a public ledger to examine the exchange between parties. This type of privacy attack can involve key and replay, or impersonation techniques, within which the attacker attempts to obtain the user's key phrase or intercept their transaction data to demand authentication. In addition, a common tactic used by attackers is to distribute airdrop links to vulnerable contracts that can access the user's key phrase from their wallet. One potential solution to prevent this is to bind Internet Protocol addresses with wallet addresses, ensuring that every transaction includes the Internet Protocol and Media Access Control addresses of both the sender and receiver.

1.6.2.3 FLASH LOAN

In DeFi transactions, loans are instantly granted through smart contract (SC) execution without any security verification. These smart contracts have three stages: initial approval of the loan amount, specification of the loan purpose, and fixed monthly payments, followed by repayment timelines. If any stage fails, the blockchain will revert to its previous state, resulting in no loan disbursal. To take advantage of this weakness, borrowers must act quickly and transfer funds before the block is finalized. To address this issue, developers can carefully construct their smart contracts.

1.6.2.4 DISTRIBUTED DENIAL OF SERVICE

During a distributed denial of service attack, a set of managed machines focuses on a specific node by gathering details about its network and manipulating the devices connected to it. The assault is initiated through a botnet swarm, which inundates the targeted node with false data, rendering it incapable of completing block mining tasks. The attacks incorporate transferring cryptocurrency to a malevolent structure, where numerous small-scale transactions are executed in bulk. The network experiences a high volume of micro-transactions, known as dust transactions, which can strain the mining power. Consequently, the cost of these transactions exceeds what is feasible for the chain. To prevent distributed denial of service attacks on the chain, intrusion detection systems are recommended. These systems utilize deep learning and machine learning anomaly algorithms to recognize and monitor illicit transactions. The necessary computational power can be obtained through a cloud-based ecosystem. A hybrid ensemble learning approach has been proposed to enhance the accuracy of detecting such attacks on different blockchains. In their study, the authors suggest a data learning-dependent model that employs principal component analysis to identify attack features and utilizes real-world data for training and testing purposes.

1.6.2.5 ECLIPSE

During this attack, a malicious intruder targets an individual user in a peer-to-peer network by isolating them. The intruder then redirects connections from legitimate nodes to their node, manipulating the targeted node and causing issues with transactions and block mining. To address this issue, a potential solution could involve implementing a predetermined random eviction process that allows for changes in communication lists. Additionally, a random selection and testing method before evicting any nodes could ensure that the number of malicious nodes does not continually increase.

1.6.2.6 RE-ENTRANT

This type of attack falls under the smart contracts category and involves an external call from a smart contract function to a potentially malevolent contract on the main chain. The unauthorized contract then uses a looping call to drain funds from the authorized function. This materialized when the contract failed to update its state before transferring funds. To execute this attack, the perpetrator utilizes an incorrect address as the recipient's contract address at the time of performing the withdrawal function. To prevent such attacks, slight modifications must be made to the contract, such as updating the user's balance before transferring funds. Alternatively, employing both dynamic and static analysis in application binary coding using smart contracts can enhance detection capabilities and mitigate the risk of such attacks (Shrestha, Vassileva, and Deters, 2020; Verma et al., 2021).

1.6.2.7 RUG PULLS

Such an attack involves luring investors into projects that are relatively new for recently popularized cryptocurrency. Once the user has gained the investor's trust, they transfer all funds from their open wallet to a new one and then disappear, leaving the investor with nothing. These attacks are carried out with malicious intent and involve using intentionally crafted smart contracts with multiple wallet addresses.

1.6.2.8 PRIVACY ATTACK

Blockchain is a transparent public ledger that allows for the tracking of transaction information by attackers. This type of attack is commonly known as a privacy breach.

1.6.2.9 TRANSACTION INFORMATION

By exploiting the transaction, the attacker gains access to potentially sensitive information. This includes analyzing historical transactions to track privacy, provide false information, and leak information. Once acquiring the user's data, the attacker locates and comprehends the associated transaction. For registered users on a private chain, their registration details are safeguarded. Alternatively, hardware-based digital wallets can be used, utilizing a pseudo-random function to prevent identification of the wallet owner (Tenopir et al., 2011).

1.7 LEARNING ACQUIRED

The main takeaway from the chapter is potential future orientation. The chapter presents and discusses the challenges of current Industry 4.0 standards in today's technical scenario and proposes emerging Industry 5.0 stipulations as a potential solution. Specifically, it emphasizes the role of blockchain in ensuring trustworthy control of industrial methods to secure structures. This includes addressing confidence and faith management in blockchain-based smart manufacturing, security, and privacy, where organizations can use blockchain technology to observe their various methods and foster trusteeship among numerous business components exchanging data through unsecured channels. In support of this idea, the chapter introduces a suggested architecture that incorporates Industry 5.0 standard practices into business methods and structures, which is demonstrated within the chapter.

In the future, the integration of Internet of Things (IoT), blockchain-supported technology, and artificial intelligence will personalize service offerings for users, aligning with the Industry 5.0 vision. As we progress toward digital and smart industries, diverse data moves between various business processes, potentially creating security loopholes and privacy risks. Therefore, it is essential to prioritize security and privacy before providing services. Data accumulation poses respective challenges for privacy preservation as information is transferred between processes. To adapt to the industry's manufacturing process, organizations must train their employees in compliance with machine intelligence and prepare them to work alongside robots. Machine and data learning, automation, and artificial intelligence play a critical role in modern industries but also introduce new threats that need to be addressed. To effectively train our model, we need reliable datasets that are both secure and transparent in their decision-making across varying data inputs. To employ comprehensive international information for making more meticulous decisions, our structure leverages local data from every client and combines it to train the original model. This typically requires sharing sensitive data with competitors but can be avoided through federated learning. With this approach, organizations can share model updates to improve each other's models without compromising the privacy of their sensitive information. The use of a centralized node allows for the computation of an internationally recognized structure by aggregating local and ground-level up-to-date information from numerous organizations. Furthermore, in different business processes, there is a need for storing and sharing sensitive monitoring and control information through trusted cloud services. As we embrace the Industry 5.0 revolution, it is crucial to properly address the ethical and societal impact of artificial intelligence (Bhattacharya et al., 2021; Guo and Yu, 2022). Although automation may create job opportunities for humans, there must be a balance between human and robot interaction within the system.

1.8 REAL LIFE MANAGERIAL COMPLEXITIES

Industry 5.0 represents a convergence of smart technologies, all interconnected through the Internet of Things. With such extensive connectivity, attack vulnerabilities are numerous and cannot be fully anticipated. Hence, there is no one-size-fits-all security plan that can cover all possible attack vectors. This makes it crucial for organizations to be vigilant during the deployment phase, as attacks on authentication and integrity are quite frequent. Additionally, managing the massive machine-to-human communication that occurs over nodes requires secure trust management protocols and device registration before network communication (Leng et al., 2018; Ron and Shamir, 2013). As the system process engages with external parties in an open blockchain peer-to-peer network, there is a risk of exposing private information or making the system susceptible. This chapter considers the various attacks observed in blockchain-based smart manufacturing and their impact on system properties (Samreen and Alalfi, 2020). These attacks are classified into three categories: privacy, network communication, and mining. The table highlights the specific property of blockchain that is affected by each attack, emphasizing the need for thorough analysis to mitigate any negative consequences. To achieve this, incorporating artificial intelligence models into blockchain-aided Industry 5.0 ecosystems proves to be an effective solution (Bae and Lim, 2018; Zhang and Lee, 2019).

The emergence of Industry 5.0 has led to increased interconnectivity among organizations and departments, creating a need for efficient scaling of both manpower and organizational processes. However, determining the optimal balance between automation and human control is a challenging task, particularly when considering specific service level agreements. This highlights the importance of providing comprehensive training to skilled personnel who will be responsible for operating machinery in uncertain scenarios or potential failures within the organization. Furthermore, it is crucial to adhere to legal regulations and policies set by policymakers to fully embrace Industry 5.0 without encountering any legal complications.

1.9 CONCLUSION

This chapter offers valuable insights on how blockchain can enhance trust in Industry 5.0. Additionally, it covers various technologies that contribute to the Industry 5.0 concept and explains why integrating blockchain is crucial for ensuring security. Our research focused on a recommended reference architecture that utilizes blockchain to support different aspects of Industry 5.0, such as robotics, smart healthcare, digital twins, and manufacturing. In summary, implementing blockchain in Industry 5.0 will not only secure and reliable cyber-physical systems boundaries but also streamline transactional payments through smart contracts, leading to improved end-user satisfaction. A solution taxonomy for aided verticals in Industry 5.0 was entrusted, along with a study of the open matters and complexities that arise in practical deployments. It was also noted that blockchain may not be a one-size-fits-all solution, and open attacks on it were discussed, along with potential ways to mitigate them in the future. This was followed by a unified study showcasing the use of blockchain-assisted digital twin production in Industry 5.0. A security analysis of the study was provided, supporting its feasibility in real-world scenarios. Looking forward into the future, our goal is to create a trusted architecture for large-scale human-robot interaction utilizing blockchain, in which data transmitted to artificial intelligence models can be verified on distributed ledgers.

KEYWORDS

- **automation**
- **blockchain technology**
- **client customization**
- **cyber-physical structures**
- **hyper-cognitive nodes**
- **industry 5.0**

REFERENCES

Angrish, A., Craver, B., Hasan, M., & Starly, B. (2018). A case study for Blockchain in manufacturing: 'FabRec': A prototype for a peer-to-peer network of manufacturing nodes. *Procedia Manufacturing, 26*, 1180–1192.

Bae, J., & Lim, H. (2018). Random mining group selection to prevent 51% attacks on bitcoin. In *Proceedings of the 48th Annual IEEE/IFIP International Conference on Dependable Systems and Networks Workshops (DSN-W)*, 81–82.

Bhattacharya, P., Saraswat, D., Dave, A., Acharya, M., Tanwar, S., Sharma, G., & Davidson, I. E. (2021). Coalition of 6G and blockchain in AR/VR space: Challenges and future directions. *IEEE Access, 9*, 168455–168484. https://doi.org/10.1109/ACCESS.2021.3078409.

Bhattacharya, P., Tanwar, S., Bodkhe, U., Kumar, A., & Kumar, N. (2021). EVBlocks: A blockchain-based secure energy trading scheme for electric vehicles underlying 5G-V2X ecosystems. *Wireless Personal Communications*, 1–41.

Chen, M., Malook, T., Rehman, A. U., Muhammad, Y., Alshehri, M. D., Akbar, A., Bilal, M., & Khan, M. A. (2021). Blockchain-enabled healthcare system for detection of diabetes. *Journal of Information Security and Applications, 58*, 102771. https://doi.org/10.1016/j.jisa.2021.102771.

Demir, K. A., Döven, G., & Sezen, B. (2019). Industry 5.0 and human-robot co-working. *Procedia Computer Science, 158*, 688–695. https://doi.org/10.1016/j.procs.2019.09.101.

Dev, K., Tsang, K. F., & Corchado, J. M. (2022). The era of Industry 5.0—Technologies from no recognizable H-M interface to hearty touch personal products. *IEEE Transactions on Industrial Informatics, 18*(8), 5432–5434. https://doi.org/10.1109/TII.2022.3154651.

Friedman, B., & Hendry, D. G. (2019). *Value Sensitive Design: Shaping Technology with Moral Imagination*. MIT Press. https://doi.org/10.7551/mitpress/7585.001.0001.

Guo, H., & Yu, X. (2022). A survey on blockchain technology and its security. *Blockchain Research and Applications, 3*(2), 100067.

Gupta, R., Tanwar, S., Obaidat, M. S., Tyagi, S., & Kumar, N. (2021). Capsule: All you need to know about tactile internet in a nutshell. In *Proceedings of the International Conference on Communications, Computers, Cybersecurity, and Informatics (CCCI)*, 1–5. https://doi.org/10.1109/CCCI52658.2021.9686174.

Jiang, S., Cao, J., Wu, H., Yang, Y., Ma, M., & He, J. (2018). BlocHIE: A blockchain-based platform for healthcare information exchange. In *Proceedings of the IEEE International Conference on Smart Computing (SMARTCOMP)*, Taormina, Italy, 49–56. https://doi.org/10.1109/SMARTCOMP.2018.00016.

Johng, H., Kim, D., Hill, T., & Chung, L. (2018). Using blockchain to enhance the trustworthiness of business processes: A goal-oriented approach. In *Proceedings of the IEEE International Conference on Services Computing (SCC)*, San Francisco, CA, USA, 249–252.

Kebande, V., Awaysheh, F., Adeyemi, I., Alawadi, S., & Alshehri, M. (2021). A blockchain-based multi-factor authentication model for a cloud-enabled Internet of Vehicles. *Sensors, 21*(18), 6018. https://doi.org/10.3390/s21186018.

King, S., & Nadal, S. (2012). Ppcoin: Peer-to-peer crypto-currency with proof-of-stake. *Technical Report*, 1–6.

Koch, P. J., van Amstel, M. K., Dębska, P., Thormann, M. A., Tetzlaff, A. J., Bøgh, S., & Chrysostomou, D. (2017). A skill-based robot co-worker for industrial maintenance tasks. *Procedia Manufacturing, 11*, 83–90. https://doi.org/10.1016/j.promfg.2017.07.053.

Lee, D. J., Ahn, J. H., & Bang, Y. (2011). Managing consumer privacy concerns in personalization: A strategic analysis of privacy protection. *MIS Quarterly, 35*, 423–444. https://doi.org/10.2307/23042793.

Leng, K., Bi, Y., Jing, L., Fu, H. C., & Van Nieuwenhuyse, I. (2018). Research on agricultural supply chain systems with double chain architecture based on blockchain technology. *Future Generation Computer Systems, 86*, 641–649.

Longo, F., Padovano, A., & Umbrello, S. (2020). Value-oriented and ethical technology engineering in Industry 5.0: A human-centric perspective for the design of the factory of the future. *Applied Sciences, 10*(12), 4182. https://doi.org/10.3390/app10124182.

Lu, Y., Xu, X., & Wang, L. (2020). Smart manufacturing process and system automation—A critical review of the standards and envisioned scenarios. *Journal of Manufacturing Systems, 56*, 312–325. https://doi.org/10.1016/j.jmsy.2020.07.006.

Maddikunta, P. K. R., Pham, Q. V., Prabadevi, B., Deepa, N., Dev, K., Gadekallu, R., Ruby, R., & Liyanage, M. (2022). Industry 5.0: A survey on enabling technologies and potential applications. *Journal of Industrial Information Integration, 26*, 100257. https://doi.org/10.1016/j.jii.2022.100257.

Nahavandi, S. (2019). Industry 5.0—A human-centric solution. *Sustainability, 11*(16), 4371. https://doi.org/10.3390/su11164371.

Rada, M. (2018). Industry 5.0 definition. *Medium*. https://michael-rada.medium.com/industry-5-0-definition-6a2f9922dc48 (accessed on 14 December 2024).

Ron, D., & Shamir, A. (2013). Quantitative analysis of the full bitcoin transaction graph. In A. R. Sadeghi (Ed.), *Financial Cryptography and Data Security* (pp. 6–24). Springer.

Samreen, N. F., & Alalfi, M. H. (2020). Re-entrancy vulnerability identification in Ethereum smart contracts. In *Proceedings of the IEEE International Workshop on Blockchain-Oriented Software Engineering (IWBOSE)*, 22–29.

Show, P. L., Chew, K. W., & Ling, T. C. (2021). *The Prospect of Industry 5.0 in Biomanufacturing*. CRC Press. 326 pp.

Shrestha, A. K., Vassileva, J., & Deters, R. (2020). A blockchain platform for user data sharing ensuring user control and incentives. *Frontiers in Blockchain, 3*(1), 48.

Singh, S. R., Mithaiwala, H., Chauhan, N., Shah, P., Trivedi, C., & Rao, U. P. (2022). Decentralized blockchain-based framework for securing review system. In U. P. Rao, S. J. Patel, P. Raj, & A. Visconti (Eds.), *Security, Privacy and Data Analytics* (pp. 239–255). Springer. https://doi.org/10.1007/978-981-16-5333-9_17.

Tanwar, S., Popat, A., Bhattacharya, P., Gupta, R., & Kumar, N. (2022). A taxonomy of energy optimization techniques for smart cities: Architecture and future directions. *Expert Systems, 39*(5), e12703. https://doi.org/10.1111/exsy.12703.

Tenopir, C., Palmer, C. L., Metzer, L., van der Hoeven, J., & Malone, J. (2011). Sharing data: Practices, barriers, and incentives. *Proceedings of the American Society for Information Science and Technology, 48*(1), 1–4.

Tian, F. (2017). A supply chain traceability system for food safety based on HACCP, blockchain & Internet of Things. In *Proceedings of the International Conference on Service Systems and Service Management (ICSSSM)*, Dalian, China, 1–6. https://doi.org/10.1109/ICSSSM.2017.7996115.

Verma, A., Bhattacharya, P., Saraswat, D., & Tanwar, S. (2021). NyaYa: Blockchain-based electronic law record management scheme for judicial investigations. *Journal of Information Security and Applications, 63*, 103025.

Zhang, P., Hong, Y., Kumar, N., Alazab, M., Alshehri, M. D., & Jiang, C. (2022). BC-EdgeFL: A defensive transmission model based on blockchain-assisted reinforced federated learning in IIoT environment. *IEEE Transactions on Industrial Informatics, 18*(5), 3551–3561. https://doi.org/10.1109/TII.2021.3067575.

Zhang, S., & Lee, J. H. (2019). Double-spending with a Sybil attack in the bitcoin decentralized network. *IEEE Transactions on Industrial Informatics, 15*(10), 5715–5722.

CHAPTER 2

Blockchain Technology: Transforming the Future Outlook of the Banking Sector

MONA CHAUDHARY,[1,*] DIVYA BANSAL,[1] and EDWARD MILLER[2]

[1]*Amity University, Noida, Uttar Pradesh, India*

[2]*University of Greenland, Manutooq 1, Nuuk, Greenland*

*Corresponding author

ABSTRACT

In the 21st century, technology reigns supreme, and people are enthusiastically embracing new technologies in response to the increasing demand for modernization. Blockchain is one of these inventions that has emerged as a transformative force that might have a big impact on a lot of different businesses and industries. Blockchain technology has the ability to completely transform the finance, banking, and fintech industries by lowering costs, enhancing the digital user experience, and reducing the risks associated with data in a secure environment. Service providers who invest in blockchain capabilities are positioned for long-term success. This technology has the potential to completely change how banks operate while also boosting security, increasing efficiency, and possibly lowering costs. In this piece, we will explore the benefits of blockchain technology for the banking industry and look at a few possible use cases.

The technology behind blockchain is seen as a revolutionary development, especially in the financial industry, since it may drive the extension of the internet, promote financial innovation, and boost economic growth. The financial industry is in need of a digital transformation, and it must

Blockchain Technology: Unlocking the Value of 3Ts in the Business World.
Richa Chauhan & Vidhi Kaul (Eds.)

DOI: 10.1201/9781998511914-2

adopt effective strategies to ensure the successful integration of blockchain technology. This chapter explores the role that blockchain adoption plays as a mediator. It also discusses how blockchain technology has been used by the banking industry and how this has revolutionized banking technology.

2.1 INTRODUCTION: ADOPTION OF BLOCKCHAIN TECHNOLOGY

Blockchain adoption can act as a bridge that can ultimately lead to improved financial performance (Khalil, Khawaja, and Sarfraz, 2021). Here is a breakdown of the key terms in the statement, which include:

1. **Blockchain Adoption:** The process of implementing and using blockchain technology in a business.
2. **Digital Business Strategy:** A business strategy that uses digital technologies to improve efficiency, reach new customers, and create new products and services.
3. **Business Process Innovation:** The use of new or improved processes to enhance efficiency, quality, or customer service.
4. **Financial Performance:** The financial results of a business, such as revenue, profit, and return on investment.
5. **Information Technology Alignment:** The alignment of information technology (IT) with the overall business strategy.

Blockchain technology holds the promise of transforming numerous sectors, yet it encounters several obstacles that must be overcome for widespread adoption to occur. These challenges include technical limitations, social acceptance, economic viability, and regulatory uncertainty.

Blockchain technology is an emerging technology that is gaining traction from energy suppliers, start-ups, and technology developers. It offers several benefits that can be used to innovate new procedures and create secure and transparent business solutions (Ramchandra et al., 2022).

The author explores the idea and function of blockchain within the realm of the banking industry while also acknowledging the persistent critiques aimed at blockchain as a disruptive force in banking. The author contends that blockchain has the potential to be a crucial factor in reshaping the banking sector and its fundamental services (Kajla, Sarin, and Raj, 2022).

Statistical techniques, financial calculus, economic analysis, as well as information and communications technologies (ICTs) such as blockchain and artificial intelligence, are all used in financial engineering (FE; Chakravaram et al., 2021).

Because blockchain technology introduces more efficient processes than the ones that are already in place, it has the potential to improve the global financial infrastructure and support sustainable growth. In fact, many financial institutions are currently placing a strong emphasis on blockchain technology (Cocco, Pinna, and Marchesi, 2017).

Numerous research studies have delved into the potential advantages of blockchain technology, with some estimating substantial cost reductions in the financial system through blockchain-based systems (Lees and King, 2017; McLean, 2016). Yet, to date, there hasn't been any investigation specifically focusing on the effectiveness of the Bitcoin system or the obstacles impeding widespread adoption of blockchain technology (Accenture, 2017). This research presents an empirical analysis of the economic and energy implications associated with the Bitcoin system.

Blockchain technology offers the ability to create more efficient methods than those already in use, streamlining the global financial system and promoting economic development. As shown in "Bitcoin," the authors contend that resolving blockchain's present shortcomings is crucial to the technology's effective incorporation into financial procedures. These restrictions are mostly related to the expensive equipment and excessive energy usage (Guo and Liang, 2016).

Cutting-edge information technologies, often referred to as revolutionary or pioneering technologies, serve as the driving forces behind the expansion of the digital economy. Their impacts can be truly transformative, causing certain aspects of human endeavors to undergo profound and almost unrecognizable changes (Bansal and Suresh, 2017).

For instance, the utilization of the latest advancements in the realms of big data and data mining has given rise to a novel field known as bioinformatics. This discipline has opened up entirely new frontiers in the field of medicine. Decoding the genome of humans has begun with the use of big data analytic technologies, resulting in the development of powerful tools to fight genetic and inherited illnesses. In addition, there has been a notable increase in the development of novel medications (Garg et al., 2023).

The Internet of Things (IoT) ecosystem is expanding rapidly, with smart applications connecting over 80 billion objects by 2026. Sensors in IoT networks supply data, and apps need to exchange money to compensate data producers. This emphasizes how crucial IoT markets and payments are to enabling microtransactions across billions of linked devices in the diverse, decentralized, and heterogeneous IoT ecosystem. Nevertheless, implementing such an ecosystem will not be easy. These challenges include:

1. **Poor Interoperability:** The fact that various IoT platforms and devices frequently employ disparate protocols for communication and data formats makes it challenging for them to exchange information and make payments.
2. **Resource Constraints:** IoT devices are often battery-powered and have limited computing power, which makes it difficult to run complex payment and market applications on them.
3. **Security and Privacy Vulnerabilities:** IoT devices are often vulnerable to security and privacy attacks, which could be exploited to steal data or payments.

The research conducted offers an extensive examination of IoT payments and markets driven by blockchain technology. The investigation commences by introducing the fundamental concepts of IoT payments and IoT markets. It proceeds to delve into the technical hurdles involved in implementing IoT payments and markets, with a particular emphasis on blockchain-driven solutions. Moreover, the study introduces smart applications that are based on blockchain and make use of IoT marketplace and IoT payment principles. Each of these applications underscores the significance of blockchain. Lastly, the research delves into the challenges of integrating these solutions and explores potential remedies (Haque and Shoaib, 2023).

Individuals generate a lot of valuable data online, which can be used as proof of their creditworthiness. However, this data is currently controlled by large internet companies, which prevents individuals from owning or using it. Privacy concerns make it difficult for data to flow between companies, resulting in isolated data silos.

Blockchain technology has the capability to encrypt data, enabling individuals to have authority over their extensive data, confirm ownership, and maintain its authenticity. Additionally, it has the potential to decrease the expenses associated with data gathering by credit agencies. Utilizing blockchain technology, extensive data can be converted into credit resources with a transparent system of individual ownership, potentially serving as the foundation for forthcoming credit systems (Trivedi, Mehta, and Sharma, 2021).

DLT functions as a sequence of interconnected blocks, with each block containing a reference to the previous one, along with time-stamped digital information that acts as a distinctive identifier.

The technology behind blockchain has been hailed as having the potential to revolutionize the financial sector by increasing transparency, efficiency, and security. Some of the specific ways in which blockchain could be used in banking include:

1. **Simplifying International Payments:** Blockchain has the potential to simplify international payments by removing the necessity for intermediaries like banks and clearinghouses. This could lead to cost reductions and enhance the speed and efficiency of transactions.

2. **Improving Fraud Detection:** Blockchain's tamper-proof nature could be used to improve fraud detection by tracking the movement of money and assets. This could help to prevent fraud and money laundering.

3. **Enhancing Customer Due Diligence:** Blockchain could be used to enhance customer due diligence by storing and sharing customer data securely and efficiently. This could help to prevent financial crimes such as terrorism financing.

4. **Innovating Financial Offerings:** Blockchain has the capacity to innovate financial products and services, including smart contracts and DeFi. These innovations could introduce novel methods for handling finances and assets (Han, Yuan, and Wang, 2019).

2.2 BLOCKCHAIN AND BANKING

Blockchain technology holds the potential to revolutionize conventional business models in a variety of ways. Indian banks are actively constructing cooperative blockchain ecosystems to establish a pioneering business model that challenges the established norm and garners additional competitive benefits.

Historically, the banking sector has been a trailblazer in embracing new technologies to enhance customer satisfaction and operational efficiency. Currently, the banking sector is looking at the potential that blockchain technology offers. This technology has the ability to revolutionize the financial sector.

Blockchain technology holds the promise of establishing a secure and cooperative infrastructure that can tackle issues like operational risk and administrative expenses. Furthermore, it has the capacity to enhance transparency and ensure the permanence of processes. In addition to boosting operational effectiveness, blockchain technology offers the potential to generate fresh revenue prospects and strategic expansion for the banking industry.

Blockchain technology was first created as the technological basis for virtual currencies. Since then, it has matured into a decentralized record-keeping system that enables safe and open payments between two parties without any requirement for a centralized governing body. It is currently hailed as the next big thing in the banking industry since it has the ability to completely change how banks operate (Ma, 2000). These features are:

1. **Cryptographic Hash Functions:** These functions are used to create unique identifiers for each block of data on the blockchain.
2. **Distributed Ledger:** This is a database that is shared across multiple computers.
3. **Consensus Algorithm:** This feature makes sure that everyone using the blockchain is in agreement on the ledger's current state.

   The technology behind blockchain has a number of advantages over traditional systems, including:

1. **Security:** Blockchain is a very secure system, as it is difficult to tamper with the data on the ledger.
2. **Transparency:** All transactions on the blockchain are transparent, as they are visible to all participants.
3. **Efficiency:** Blockchain can streamline transactions and reduce costs.
4. **Immutability:** Once data is added to the blockchain, it cannot be changed.

Blockchain technology offers banks the means to overhaul their organizational structures and service offerings to cater to the demands of the contemporary, digitally literate clientele. Additionally, it can be employed for tracking assets and transactions without the necessity of a central authority, resulting in a secure, tamper-proof, transparent, dependable, and unalterable record of ownership. This has the potential to address critical challenges confronting the banking sector, including cross-border payments, deposit and loan insurance, and KYC procedures.

According to an Accenture analysis, by 2025, blockchain technology may save the biggest investment banks in the world's infrastructure costs by somewhere in the range of 8 billion to 12 billion dollars yearly. This suggests that banks might save money and improve their operational effectiveness by implementing blockchain technology.

All things considered, the blockchain system has the power to completely transform the financial sector. It may assist banks in enhancing their security, efficiency, and transparency as well as in resolving some of their current problems (IBEF, 2021).

The Indian banking system is strong, with 12 government-owned banks, 22 privately owned banks, 44 foreign banks, 43 rural banks in the region, 96,000 rural cooperative banks, 1484 urban cooperative banking institutions, as well as cooperative credit institutions making up the country, according to data provided by the Indian Banks' Association. With its 7.7% GDP contribution, the banking industry in India is a key player in the major economy that is expanding at the fastest rate in the world (Singh and Malik, 2018).

The 15 Indian banks that make up Indian Banks' Blockchain Infrastructure Co Pvt. Ltd. (IBBIC) are divided into 10 private sector banks, four public sector banks, and one international bank. This consortium was formed to handle significant issues that the traditional banking system was facing, such as trade finance, e-way bills, GST invoicing, and Letters of Credit (LCs). Although the blockchain system remains in its early phases of development in India, IBBIC is one of the leading attempts to use blockchain in the banking industry. The group is working hard to develop a blockchain-driven system that will make trade finance, LCs, GST invoices, and e-way bills easier to handle. It is anticipated that this project would save costs, improve operational effectiveness, and increase transparency in the banking industry.

The creation of IBBIC represents a major advancement in India's acceptance of blockchain technology. The consortium, which consists of some of the top institutions in the nation, is dedicated to applying blockchain technology to address some of the most urgent issues confronting the banking industry.

A bank secures a competitive edge by executing its operations with greater cost efficiency compared to its rivals. This cost efficiency can be realized through the utilization of innovative technology frameworks that provide swiftness, security, inalterability, traceability, and transparency. Blockchain technology represents one of these platforms that can empower banks in attaining a competitive advantage.

Blockchain technology can be used to address industry-wide problems and bottlenecks, such as the slow processing of transactions and the high cost of cross-border payments. By using blockchain, banks can reduce costs, improve efficiency, and increase transparency. The question of how blockchain can contribute to competitive advantage in the banking sector is a complex one. The competitive advantage could be diminished if every bank has a connection to a similar blockchain platform. However, it is also possible that the blockchain platform will create new opportunities for differentiation and innovation. Upon the universal adoption of blockchain platforms within the industry, there will be a significant transformation in cost and pricing mechanisms. This transformation has the potential to bring about a more streamlined balance within the banking sector.

Once blockchain platforms are widely adopted, the competitive advantage will not come from simply using the platform. Rather, the capacity to provide a better and quicker service experience will set you apart.

In a competitive market environment, organizations that neglect to leverage blockchain technology may find themselves at a competitive disadvantage over time. Therefore, it is important for organizations to constantly innovate and find new ways to use blockchain technology to improve their products and services. Indian banks will probably employ blockchain technology to boost organizational performance and change their business models in order to obtain a competitive edge. This could include using blockchain to:

1. **Streamline Processes:** Blockchain can be used to streamline processes and reduce costs.

2. **Improve Customer Service:** Blockchain can be used to improve customer service by making it faster and easier to interact with banks.
3. **Provide New Services and Goods:** Blockchain technology enables the provision of new services and products that are not feasible with conventional financial systems.

Many success stories have been published by academics and blockchain vendors that highlight the capabilities of blockchain to enhance competitive advantage and organizational performance. However, the relationship between blockchain capabilities, competitive advantage, and organizational performance has not been empirically examined (Crawley, 2021; IBEF, 2021; Oracle Netsuite, 2019; Springer, 2019; Vega, 2021).

2.3 BLOCKCHAIN CAN REVOLUTIONIZE BANKING TECHNOLOGY

The banking sector is facing a number of challenges, both internal and external. These challenges include:

1. **Rising Competition from Non-Bank Financial Institutions (NBFIs):** NBFIs are challenging the traditional banking model.
2. **Evolving Regulations:** The banking industry is constantly influenced by intricate and dynamic regulatory conditions, making it challenging for banks to introduce new ideas and remain competitive.
3. **Security Risks in the Digital Realm:** The banking sector is a prime target for cyber threats, which can disrupt operations, pilfer sensitive information, and harm the reputation of financial institutions (Garg et al., 2021).

The primary transformation that internet finance has introduced to the conventional banking sector lies in the expansion of application possibilities. Internet finance enterprises have harnessed the power of big data analysis to offer more tailored and user-friendly services, thus attracting a larger customer base (Singh and Malik, 2018). Nevertheless, a considerable portion of these application scenarios has already been established, rendering the process of introducing novel innovations in this field more challenging. Consequently, substantial technical advancements are necessary to pave the way for fresh avenues of growth.

Interbank payments often rely on intermediary clearing firms, which can lead to complex processes and significant time and cost involved. Cross-border payments, for example, can take nearly three days due to varying clearing procedures in different countries (Vega, 2021).

Blockchain systems like Ripple have been successfully tested by financial organizations including Standard Chartered and National Australia Bank. These platforms have the potential to drastically cut the time and expense of cross-border payments. For example, Ripple has been able to complete settlements in as little as 10 seconds, compared to the current banking system's two days.

2.4 FUTURE OF BLOCKCHAIN IN DIGITAL ECONOMY

The blockchain contains blocks. Each block has an identifier linked to the block before it and a timestamp. It basically functions as an electronic log of transactions that is replicated and dispersed over the blockchain's whole network of computer systems. Every block on the blockchain contains the following information:

- A timestamp;
- A link to the previous block;
- Transactional data, including the sender, the recipient, and amount;
- A cryptographic hash of the previous block.

Every block has a unique identification, which is the cryptographic hash. It is made with a function of mathematics that accepts the block's contents and outputs a fixed-length character string. The blockchain's integrity is preserved by the usage of the cryptographic hash.

Blockchain technology can be used in:

1. **Financial Sector:** Blockchain can be used to record financial transactions, such as investments, exchanges, and banking operations. This can help to reduce fraud and errors, and improve transparency.

2. **Land Registry and Real Estate Transactions:** Blockchain can be used to record land ownership and track the movement of real estate. This can help to reduce fraud and improve transparency.

3. **Identification Systems:** These that are trustworthy and safe may be built with blockchain. This can help a number of businesses, like travel and healthcare, become more efficient and secure.
4. **Education and Healthcare:** Blockchain technology has the potential to store and monitor medical records and academic qualifications, which can enhance efficiency and security.
5. **Worldwide Registries:** Blockchain technology may be used to create global registries for data management, processing, and storage. This can enhance efficiency and transparency in a number of industries, including supply chain operations and the protection of intellectual property.

As the decentralized ledger, blockchain technology can potentially solve the following financial problems in the first case, which include:

- Currently, the control over the economy is predominantly in the hands of governments and financial institutions.
- International transactions are often burdened by substantial expenses and slow processing times.
- Transparency is lacking across various facets of the financial system's operation.
- Extensive bureaucracy and corruption are prevalent.
- There exists a widespread lack of confidence among the public towards both governments (Figure 2.1).

2.5 APPLICATION

Without the need for a middleman, untrustworthy parties may agree on a database's state thanks to blockchain technology. Because of its self-regulating ledger, a blockchain's intrinsic architecture may be able to offer some financial services, like payment or securitization, without the need for a financial institution. Furthermore, blockchain makes it easier to use technologies like "smart agreements," which are self-executing blockchain-based contracts that may automate anything from processing claims and compliance to sharing information from a will. In cases where a high degree of decentralization isn't required but improved coordination is beneficial, distributed ledger technology, akin to blockchain, has the potential to assist businesses in creating more robust governance and regulations for the purpose of sharing data and fostering collaboration (Figure 2.2).

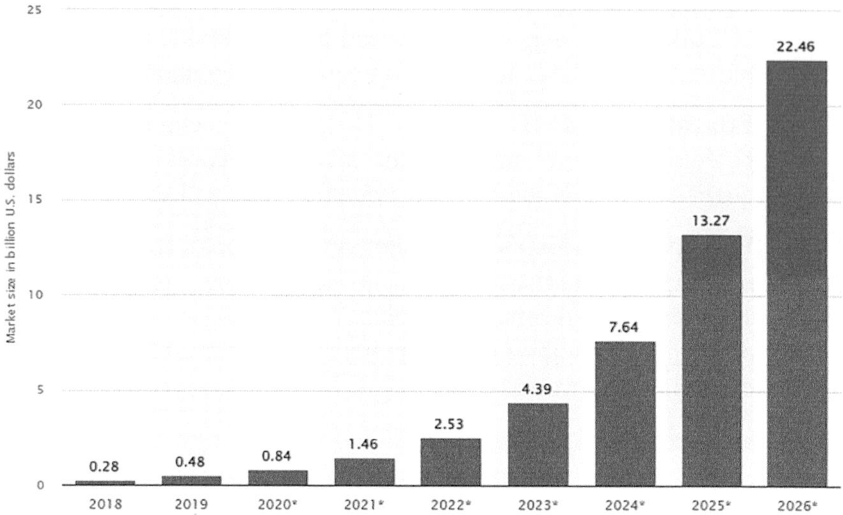

FIGURE 2.1 Market stats of blockchain in the financial sector.

Source: Devtechnosys https://devtechnosys.com/insights/future-of-blockchain-in-banking-industry/.

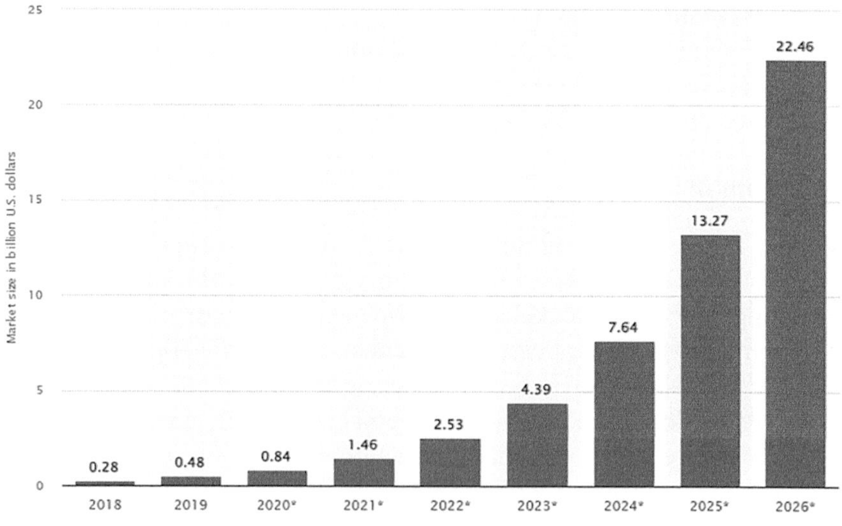

FIGURE 2.2 Cases of blockchain technology in banking security.

Source: Devtechnosys. https://devtechnosys.com/insights/future-of-blockchain-in-banking-industry/.

The services in the financial industry include:

1. **Payments:** Compared to traditional banking, it offers the potential to allow quicker and more affordable payments through the use of a decentralized payment ledger like Bitcoin.
2. **Settlement and Clearing Mechanisms:** Distributed ledgers hold the potential to lower operational expenses and advance us towards realizing instantaneous transactions between financial institutions.
3. **Capital Procurement:** Pioneering funding models, such as Initial Coin Offerings (ICOs), are exploring a novel method that disconnects fundraising from the traditional financial services provided by established companies.
4. **Financial Instruments:** By transforming traditional investments like equities, bonds, and non-traditional assets into tokens on public blockchains, blockchain technology can establish more streamlined and interconnected capital markets.
5. **Regenerate Loans and Credit:** Through the removal of intermediaries in the finance and banking sector, blockchain technology can enhance the safety of borrowing money while also providing reduced interest rates.
6. **Trade Finance:** Through the simplification of the laborious and documentation-heavy bills of lading procedures within the trade finance sector, blockchain technology can promote increased openness, privacy, and trust among participants in international trade.
7. **Customer Authentication and Deterrence of Unlawful Activities:** By retaining customer information on decentralized networks, blockchain technology can expedite and safeguard the transfer of data between financial institutions (Figure 2.3).

2.6 BLOCKCHAIN-BASED CURRENCY

Blockchain technology is swiftly revolutionizing the methods of transferring, storing, and managing assets. Both public and private cryptocurrencies based on blockchain are being created and adopted by individuals,

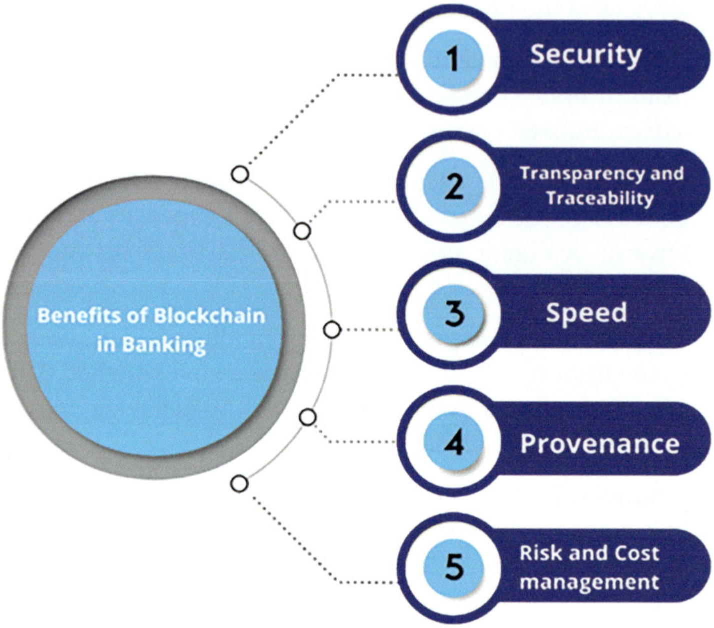

FIGURE 2.3 Benefits of blockchain in the banking industry.

Source: Devtechnosys. https://devtechnosys.com/insights/future-of-blockchain-in-banking-industry/.

businesses, and even entire nations (Rajnak and Puschmann, 2020). These advancements mark the initial phases of a completely new financial sector, brimming with entrepreneurial possibilities. With the increasing market capitalization and backing from institutional players, banks that put in place the essential framework for banking services related to cryptocurrencies, such as secure storage, payment handling, and lending options, will be well-prepared to serve customers participating in this groundbreaking category of assets (Sharma, Sharma, and Bansal, 2023).

As an increasing number of individuals and entities engage in the crypto asset market, the importance of offering services related to crypto assets will continue to expand. The inherent complexity and novelty associated with holding crypto assets are likely to generate greater demand for trusted intermediaries to provide custody services. Beyond merely securing these assets, as various types of crypto assets gain prominence, banks and regulators should be prepared for the introduction of additional financial services centered around crypto assets.

In recent years, cryptocurrency and blockchain technologies have advanced in tandem, bringing about significant innovations in currency creation, payment mechanisms, and currency storage. Nevertheless, it's important to note that existing cryptocurrencies cannot fully substitute traditional fiat currencies (Saputhanthri, De Alwis, and Liyanage, 2022).

2.6.1 DECENTRALIZED CRYPTO ASSETS

Due to its distinctive financial characteristics, including inherent durability, limited availability, interchangeability, easy transferability, and divisibility, Bitcoin is rapidly solidifying its position as an innovative store of value (BlockFi, n.d.). Notably, corporations have recently begun including Bitcoin in their asset holdings, with some asserting that it can serve as a safeguard for shareholder value against the impact of expansive global monetary policies (Crawley, 2021).

2.6.2 CLEARANCE AND SETTLEMENT SYSTEMS

Blockchain networks are now making it possible for transactions, including intricate cross-border dealings involving multiple parties, to settle within seconds, around the clock, and with significantly decreased fees. These advancements hold the potential to profoundly transform the way individuals transfer money and assets.

The expenses linked to settlement are particularly notable in the context of international transactions. For cross-border payments, the average transaction fees exceed 10%, and the typical processing time spans two to three days (Klebanov, 2021). Conversely, blockchain networks, equipped with their improved settlement capabilities, are expected to enable users to conduct transactions instantly, with ultimate settlement at nearly negligible costs. These attributes have the potential to turn previously unimaginable applications, like real-time cross-border payment streaming and immediate settlement of financial instruments, into everyday practices in the near future (Poon and Dryja, 2016).

2.6.2.1 PRIVATE BLOCKCHAINS

Unlike open, public blockchains, entities like financial institutions and governments have the capability to establish blockchain-based networks with the objective of facilitating global asset transfers and the exchange of information (JPMorgan, n.d.).

2.7 CHALLENGES AND OPPORTUNITIES OF IMPLEMENTATION OF BLOCKCHAIN IN BANKING SECTOR

The scarcity of customer information poses challenges in the credit creation process. This issue constrains growth opportunities for commercial banks and creates obstacles. Additionally, it contributes to a notable increase in non-performing loans, as banks face difficulties in evaluating and managing loan quality (Chauhan and Kaul, 2021). On the other hand, internet finance firms like Alipay enjoy an edge, as they can access comprehensive data regarding customer spending and financial habits. The insufficiency of customer information stands as a significant drawback within the banking sector, impeding the process of generating credit.

Blockchain technology has endeavored to revolutionize various industries since its inception, achieving varying levels of success. The financial sector finds blockchain especially attractive due to its core features of decentralization and immutability (PYMNTS, 2021).

In theory, blockchain seems like the perfect solution for most of the current challenges facing the financial sector. It is no wonder that, as per an analysis by Greenwich Associates, the financial services industry invests approximately $1.7 billion each year in blockchain solutions.

2.7.1 GOVERNANCE-RELATED LIMITATIONS

While blockchain's decentralization is often touted as its primary advantage, it does come with certain drawbacks for the financial sector. When there is no central authority, individuals engaged in a financial transaction on a blockchain may possess conflicting interests, resulting in significant delays.

This is precisely what the DAO Hack, which took place in 2016, involved. For additional context, DAO is an acronym for decentralized autonomous organization. Its purpose was to establish an automated decision-making system driven by pre-established rules, aiming to remove the necessity for conventional governance.

A DAO operates by having a team of experts create smart contracts to autonomously manage the organization. To fund the project, an ICO is conducted, and token owners subsequently become part of the organization. Afterward, participants can propose how the DAO should utilize its funds, and project backers can approve or reject these proposals through voting. Due to a programming flaw in the DAO, developed by the German company Slock.it, the system was breached, resulting in the theft of approximately $50 million worth of bitcoins. The majority of the community supported a "hard fork," which is a protocol modification that renders previously valid blocks invalid, to reimburse the DAO investors, disregarding the rest of the community.

This incident raised several warning signs. The primary objective of the system was to establish efficient and fair governance without human intervention. However, it soon became apparent that such a system could still be vulnerable to lobbying and traditional majority-based voting procedures, a notion that most financial institutions find unacceptable.

Several red flags were apparent in this situation. The fundamental objective of the system was to institute effective and unbiased governance that was free from human intervention. Nevertheless, it became apparent that even such a system could be susceptible to influence and conventional majority-based voting practices. This awareness renders the idea unsuitable for most financial institutions.

2.7.2 SCALABILITY ISSUES

The challenge of scalability intensifies as blockchain continues to grow in popularity. While various strategies have been proposed to address this issue, each has its limitations. One well-known example is "sharding."

With the aim of significantly boosting transaction throughputs, this strategy seeks to alleviate the burden on each node, removing the need to process the entire blockchain.

However, there are two fundamental challenges associated with sharding: security and communication. Each shard created when a blockchain is divided functions as an independent blockchain network. This complexity makes development more challenging due to the unique communication protocols required. Security presents another concern, as splitting the hash power of blockchain nodes diminishes overall security.

This makes it easier for cybercriminals to potentially compromise a single shard, leading to irreversible data loss.

While the brightest minds in the field are working to find solutions to these issues, only time will reveal whether blockchain can be scaled to levels comparable to Visa's transaction processing capacity.

2.7.3 POLICIES AND REGULATIONS

Furthermore, financial institutions grapple with the lack of well-defined regulations. Presently, there are no specific limitations governing transactions carried out through cryptocurrency and smart contracts. The integration of blockchain into financial institutions will not be feasible without the establishment of an appropriate regulatory framework.

For instance, the European Union reintroduced data privacy safeguards for its citizens last year through the enactment of the GDPR. Blockchain advocates immediately took notice of this new EU regulation for the following two reasons:

- The GDPR, which requires a well-defined entity to be responsible for ensuring EU citizens can exercise their data protection rights, poses a challenge in blockchain due to the absence of centralized governance structures. This intrinsic trait poses challenges for blockchain technology in aligning with EU regulations.
- A fundamental stipulation of the GDPR is that personal data should be erasable and changeable, which directly conflicts with the unchangeable nature of blockchain data.

2.8 EMERGING ISSUES AND CHALLENGES IN EMBRACING BLOCKCHAIN TECHNOLOGY

Security issues – blockchain systems face security vulnerabilities such as 51% attacks, where malicious actors gain control of over 51% of the processing power and can reverse transactions or modify blocks. Flash loan attacks are another concern, particularly in DeFi ecosystems. Coding loopholes and centralization of information also pose security risks.

One of the prevailing issues in blockchain technology is the existence of distinct architectural designs, which determine their level of security.

This has raised concerns, especially among cryptocurrency enthusiasts who prefer the security of decentralized chains. Notably, malicious actors behind these attacks often aim to engage in double-spending, diverting funds without compromising embedded crypto wallets. Several blockchain networks, including Bitcoin Cash ABC (BCHA), BCH, and Ethereum Classic, have fallen victim to these attacks.

2.8.1 IS IMPLEMENTING BLOCKCHAIN TECHNOLOGY A COMPLEX TASK?

At the present moment, the adoption of blockchain technology poses considerable challenges across major industries, such as banking, healthcare, and accounting. These challenges stem from the difficulties associated with finding and hiring professionals well-versed in blockchain technology, making the successful implementation of blockchain a formidable task.

The demand for blockchain specialists is high, and this scarcity of expertise has become a bottleneck for the widespread adoption of blockchain in these sectors. Notably, implementing blockchain technology necessitates a workforce of highly trained individuals with the requisite knowledge and skills, further compounding the challenge.

Despite these hurdles, it is noteworthy that the blockchain industry is experiencing a significant surge in investment. In the year 2021 alone, the reported development expenditure in this domain exceeded a substantial $16 billion, indicating a growing commitment to advancing blockchain technologies. As innovative blockchain solutions continue to emerge and gain prominence, it is expected that capital inflows will rise in tandem with the industry's expansion.

2.9 BLOCKCHAIN INTEGRATION IN SOCIAL MEDIA

The 21st century has witnessed a remarkable transformation in communication services, primarily attributed to the proliferation of social media platforms. These platforms have redefined contemporary moments and facts that dominate our world. However, it's crucial to acknowledge that traditional social media services operate within a centralized framework. They exert control over all information, dictate user navigation, and oversee all activities on their networks. This centralized approach has far-reaching implications for privacy, limitations, and regulatory concerns, sparking calls for a reevaluation of how these networks function and provide their services.

Addressing these concerns is an ongoing process, and any substantial changes are likely to materialize in the distant future. Researchers are currently exploring the fusion of blockchain technology with social media to address privacy and security issues within these platforms.

2.9.1 BLOCKCHAIN AND SOCIAL MEDIA

The convergence of blockchain technology and social media has given rise to a rapidly expanding sector. This integration brings decentralized and permissionless blockchain protocols to platforms designed for connecting people and facilitating content sharing. Below, we provide a brief overview of some of these blockchain-integrated social media platforms that are already in existence, highlighting their distinctions from conventional platforms and showcasing their unique features. It's important to note that this list is not exhaustive.

2.9.2 IS DECENTRALIZED SOCIAL MEDIA THE FUTURE?

While current decentralized social media platforms may not yet boast the same level of network effects as their ubiquitous centralized counterparts, they offer novel functionalities and modes of interaction. Enhanced privacy, resistance to censorship, and integrated monetization features are pivotal factors contributing to the adoption of crypto social media networks. The borderless and permissionless nature of blockchain protocols may usher in a more interconnected, open, and collaborative global social media ecosystem.

2.9.3 SOCIAL AND MANAGERIAL APPLICATIONS OF BLOCKCHAIN TECHNOLOGY

Blockchain technology finds application in a wide range of social and managerial contexts, as exemplified by the following scenarios:

1. **Supply Chain Management:** Blockchain enhances supply chain transparency by recording transactions and details at each supply

chain stage. This reduces fraud, assures product authenticity, and enhances product provenance.

2. **Voting Systems:** Blockchain creates secure and transparent voting systems by recording votes on an immutable ledger, ensuring a tamper-proof and auditable record of the voting process.
3. **Identity Verification:** Blockchain streamlines identity verification processes through self-sovereign identity systems. Individuals manage their digital identities, sharing only necessary information while upholding privacy and security.
4. **Smart Contracts:** Blockchain enables self-executing smart contracts, which automatically enforce predefined conditions. They automate processes, improve efficiency, and reduce costs in various managerial and social applications.
5. **Decentralized Finance (DeFi):** Blockchain empowers the development of decentralized financial applications that provide services like lending, borrowing, and trading without intermediaries, increasing accessibility and transparency.

It's essential to recognize that while blockchain technology offers numerous advantages, there are also challenges and considerations, including scalability, energy consumption, and regulatory frameworks, that require careful attention.

2.10 BLOCKCHAIN FOR MANAGEMENT

The period since 2020 has convincingly demonstrated the importance of digitalization, even to the most skeptical critics of the concept. Our lives have increasingly migrated to the online realm, and businesses that have not adapted to this shift are failing to adequately address their customers' needs.

This somewhat abrupt change in our environment has pushed companies to embrace a variety of technologies, some of which they may not have been familiar with previously. This evolution has notably shed light on the potential applications of blockchain technology in the business world. While it may have taken some time, the past few years have offered a unique opportunity for this technology to solidify its presence in multiple industries.

2.11 ENTERPRISE BLOCKCHAIN: REVOLUTIONIZING BUSINESS OPERATIONS

When considering the potential applications of blockchain technology in the corporate world, it might initially appear challenging, given the technology's original intent. Indeed, blockchain was initially conceived with a public model in mind, and numerous projects have utilized it for such purposes. However, a fortunate evolution has occurred, tailoring blockchain to meet the specific needs of companies.

As a logical departure from the open-access nature of public blockchains, a private variant of blockchain emerged, catering to the requirements of organizations. This specialized iteration is often referred to as "enterprise blockchain," a term that succinctly conveys its target audience.

In contrast to public blockchains, which are open to all internet users, enterprise blockchains are designed exclusively for organizational use and are accessible solely to their authorized personnel. These users are typically granted various levels of access, ensuring robust security measures. Notably, there is a significant departure from public blockchains in this context; a central authority figure assumes a more prominent role.

This central authority figure, whether an appointed administrator or a notary, holds the pivotal responsibility of overseeing the legitimacy of all operations within the enterprise blockchain. They have the authority to approve or disapprove the execution of transactions.

While this centralization may appear to deviate from the decentralized essence of blockchain technology, its fundamental advantages, including encryption, stability, and traceability, remain intact. Storing a company's entire dataset, whether using blockchain or any other technology, in a fully public manner is simply impractical. Thus, it is entirely reasonable for companies looking to harness the potential of blockchain to adapt and leverage the applications tailored to their specific business needs.

2.12 EXPLORING THE FASCINATING UTILIZATIONS OF BLOCKCHAIN IN BUSINESS

The multifaceted potential of blockchain applications extends far beyond cryptocurrency projects. With businesses of all sizes embracing enterprise blockchain, the scope of its use cases has expanded significantly, which include:

1. **Smart Contracts:** These are ingenious programs or protocols engineered to automate processes in accordance with the stipulations of a contract or agreement. These can encompass event monitoring, action documentation, or even full-fledged execution of prescribed actions. Their inclusion in discussions about blockchain applications for business is justified by their ability to be conceived and executed within blockchain networks.

   Businesses stand to gain from smart contracts due to their capacity to reduce reliance on intermediaries, mitigate the risks of malicious or accidental errors, and lower arbitration costs. A testament to their potential is the burgeoning global smart contracts market, which is expected to soar from $397.8 million in 2022 to a projected $1460.3 million in the ensuing seven years.

2. **Government:** The inclusion of blockchain within governmental contexts might seem unconventional to those unfamiliar with the expansive landscape of blockchain applications for businesses. However, the technology serves as a significantly superior alternative to conventional data storage methods, including chapter-based or other digital formats. Integration of enterprise blockchain offers governments the opportunity to expedite administrative processes, reduce bureaucratic red tape, and curtail financial expenditures.

   Governments can leverage blockchain to securely store documents, citizen information, and develop accessible systems for healthcare or social security. At a more advanced level, blockchain applications are actively employed within the European Union, which is spearheading the creation of the European Blockchain Partnership (EBP) through a public blockchain infrastructure. The EU, recognizing the transformative potential of this technology since 2018, has embarked on modernizing various facets of European citizens' lives in collaboration with established blockchain leaders.

3. **Voting:** Expanding upon the array of blockchain applications for businesses, blockchain can be a pivotal solution for enhancing the electoral process. This can be applicable to community, organizational, or governmental elections, with governments worldwide

actively exploring the digitalization of elections through various technologies. Yet, a unified global approach remains elusive.

Blockchain presents a compelling answer, offering transparency and the prevention of duplicate voting. The network can filter and permit voting solely for eligible participants, making manipulation of results more challenging compared to chapter-based voting. Comprehensive data in blocks, from timestamps to user information, further streamlines the process. Furthermore, voters can participate from the convenience of their electronic devices, eliminating the need for physical presence.

4. **Internet of Things (IoT):** The concept of the IoT is not novel, but it merits inclusion in the roster of blockchain applications for businesses. Decentralized technology can expedite the adoption of IoT, which encompasses physical objects embedded with sensors or software connecting to the internet and other devices.

   While instructing a smart assistant to perform tasks is convenient, it can be equally convenient for hackers to compromise devices and steal data. Blockchain offers a robust solution, enhancing the security of IoT devices through encryption and transparency. It maintains accessibility while providing enhanced security compared to conventional IoT practices.

5. **Personal Data Storage:** This blockchain application is valuable both for internal enterprise needs and safeguarding customer data. Despite technological advancements, many companies continue to centralize their data storage, encompassing employee records and client information. This practice has exposed organizations to vulnerabilities, with many falling victim to malicious data breaches.

   Enterprise blockchain offers an unequivocal solution for safeguarding information. It secures records through encryption and renders them nearly immutable. Its transparency enables employees to trace data processes seamlessly. Moreover, it is readily accessible, eliminating the need for cross-referencing multiple sources or the risk of losing chapter documentation.

6. **Money Transfer:** As originally intended, blockchain-based financial transactions stand out as a conspicuous business application.

This pertains to more than just cryptocurrency firms, as enterprise blockchains can challenge traditional banking with their swifter and often more cost-effective money transfer capabilities.

Employing blockchain in financial transactions not only reduces costs for both customers and organizations but also simplifies cross-border transactions. The technology operates on a global scale, devoid of reliance on third-party intermediaries, resulting in streamlined, borderless transactions.

7. **Supply Chain Management:** Among the numerous blockchain applications for businesses, one deserving mention is supply chain management. Firms in this sector can enhance service quality and streamline internal processes by implementing a reliable enterprise blockchain.

Blockchain serves as an unerring record of every stage in the supply chain, facilitating the early detection of potential issues and reducing the risk of delays. The transparency of this network simplifies supply management operations and automates critical processes.

2.13 CONCLUSION

The findings indicate a positive correlation between a digital business strategy, the enhancement of business processes, and a firm's financial outcomes. The incorporation of blockchain technology plays a mediating role in connecting a digital business strategy, process improvement, and financial performance. Furthermore, the alignment of information technology serves as a moderating element in the connection between adopting blockchain and improving processes.

Given the substantial possibilities that blockchain presents for the financial services industry, it is crucial for both the sector and government authorities to work together to promote the widespread and secure adoption of blockchain technology, ensuring strong safeguards for consumers. While a cautious approach in the past was justifiable due to the novelty of the technology, blockchain has now matured to a point where immediate action is required.

Crucially, the advantages of integrating blockchain technology into the financial services sector will be experienced by all parties involved, encompassing both financial institutions and the wider public. By potentially reducing the expenses tied to financial services, including cost savings achieved through cooperation among financial institutions and the implementation of standardized procedures across the sector, blockchain has the potential to extend more inclusive and uniform access to individuals who might otherwise be excluded from financial services.

The banking sector has a unique chance to undergo significant modernization by integrating both public and private blockchains into its array of services. By forging partnerships between the public and private sectors and implementing suitable governmental regulations, the current legal ambiguities within this domain can be resolved. This, in turn, would enable the U.S. banking industry to further harness blockchain technology, delivering more efficient and secure offerings to both its existing and potential customer base. The present moment presents an opportune time to seize these possibilities in the United States.

KEYWORDS

- **banking sector**
- **blockchain technology**
- **digital business**
- **financial outcomes**
- **financial services**
- **internet of things**
- **money transfer**
- **smart contracts**
- **supply chain management**

REFERENCES

Accenture. (2017). *Banking on Blockchain: A Value Analysis for Investment Banks* (Technical report). Accenture Consulting.

Bansal, D., & Suresh, E. (2017). Transforming India: A digitally empowered society. *Adarsh Journal of Management Research*, 1–6. Doi: 10.21095/ajmr/2017/v0/i0/121725.

BlockFi. (n.d.). Crypto loans. Retrieved from: https://blockfi.com/crypto-loans/.

Chakravaram, V., Ratnakaram, S., Agasha, E., & Vihari, N. S. (2021). The role of blockchain technology in financial engineering. In *ICCCE 2020: Proceedings of the 3rd International Conference on Communications and Cyber Physical Engineering* (pp. 755–765). Springer Singapore.

Chauhan, R., & Kaul, V. (2021). Blockchain technology: A panacea for medical distribution ailments. In *Blockchain for Healthcare Systems* (pp. 43–60). CRC Press.

Chauhan, R., & Kaul, V. (2022). Blockchain technology: Unlocking the business model maze for evolving businesses and startups. In *Transformations Through Blockchain Technology: The New Digital Revolution* (pp. 291–309). Springer International Publishing.

Cocco, L., Pinna, A., & Marchesi, M. (2017). Banking on blockchain: Cost savings thanks to blockchain technology. *Future Internet, 9*(3), 25.

Crawley, J. (2021). Tether passes a $50B market cap. *Coin Desk*. Retrieved from: https://www.coindesk.com/markets/2021/04/26/tether-passes-50b-market-cap/ (accessed on 14 December 2024).

Garg, P., Gupta, B., Chauhan, A. K., Sivarajah, U., Gupta, S., & Modgil, S. (2021). Measuring the perceived benefits of implementing blockchain technology in the banking sector. *Technological Forecasting and Social Change, 163*, 120407.

Garg, P., Gupta, B., Kapil, K. N., Sivarajah, U., & Gupta, S. (2023). Examining the relationship between blockchain capabilities and organizational performance in the Indian banking sector. *Annals of Operations Research*, 1–34.

Guo, Y., & Liang, C. (2016). Blockchain application and outlook in the banking industry. *Financial Innovation, 2*, Article No. 24.

Han, X., Yuan, Y., & Wang, F. Y. (2019). A blockchain-based framework for central bank digital currency. In *2019 IEEE International Conference on Service Operations and Logistics, and Informatics (SOLI)* (pp. 263–268). IEEE.

Haque, M. A., & Shoaib, M. (2023). e₹—The digital currency in India: Challenges and future prospects. *BenchCouncil Transactions on Benchmarks, Standards and Evaluations*, 3(1), 100107.

IBEF. (2021). Indian banking industry analysis. Retrieved from: https://www.ibef.org/industry/banking-presentation (accessed on 14 December 2024).

JPMorgan. (n.d.). Liink network. Retrieved from: https://www.jpmorgan.com/onyx/liink.htm (accessed on 14 December 2024).

Kajla, T., Sarin, V., & Raj, S. (2022). Blockchain in the banking sector: Revolution or digital disruption? In *Applications, Challenges, and Opportunities of Blockchain Technology in Banking and Insurance* (pp. 151–159). IGI Global.

Khalil, M., Khawaja, K. F., & Sarfraz, M. (2021). The adoption of blockchain technology in the financial sector during the era of the fourth industrial revolution: A moderated mediated model. *Quality & Quantity*, 1–18.

Klebanov, S. (2021). A look at blockchain in cross-border payments. *Payments Journal*. Retrieved from: https://www.paymentsjournal.com/a-look-at-blockchain-in-cross-border-payments/ (accessed on 14 December 2024).

Lees, A., & King, M. (2017). *World Payments* (Technical report). Capgemini and The Royal Bank of Scotland, Dublin, Ireland.

Ma, H. (2000). Competitive advantage and firm performance. *Competitiveness Review: An International Business Journal, 10*(2), 15–32.

McLean, J. (2016). Banking on blockchain: Charting the progress of distributed ledger technology in financial services. Technical report, Finextra Research Ltd, London, UK.

Oracle Netsuite. (2019). *What Blockchain Technology Can Do for Your Business, According to Experts*. Retrieved from: https://www.netsuite.com/portal/resource/articles/business-strategy/blockchain-technology.shtml (accessed on 14 December 2024).

Poon, J., & Dryja, T. (2016). *The Bitcoin Lightning Network: Scalable Off-Chain Instant Payments* (Kindle Edition, 112 pages).

PYMNTS. (2021). JPMorgan uses the Liink network to help FIs innovate payment economics. *PYMNTS*. Retrieved from: https://www.pymnts.com/blockchain/2021/jpmorgan-uses-liink-network-to-help-fis-innovate-payments-economics/ (accessed on 14 December 2024).

Rajnak, V., & Puschmann, T. (2020). The impact of blockchain on business models in banking. *Information Systems and e-Business Management, 19*, 1–53.

Ramchandra, M. V., Kumar, K., Sarkar, A., Mukherjee, S. K., & Agarwal, K. (2022). Assessment of the impact of blockchain technology in the banking industry. *Materials Today: Proceedings, 56*, 2221–2226.

Saputhanthri, A., De Alwis, C., & Liyanage, M. (2022). Survey on blockchain-based IoT payment and marketplaces. *IEEE Access, 10*, 103411–103437.

Sharma, A., Sharma, D., & Bansal, R. (2023). Emerging role of blockchain in banking operations: An overview. *Contemporary Studies of Risks in Emerging Technology, Part A*, 1–12.

Singh, D., & Malik, G. (2018). Technical efficiency and its determinants: A panel data analysis of Indian public and private sector banks. *Asian Journal of Accounting Perspectives, 11*(1), 48–71.

Springer. (2019). How can blockchain help your business gain a competitive advantage?

Trivedi, S., Mehta, K., & Sharma, R. (2021). Systematic literature review on the application of blockchain technology in e-finance and financial services. *Journal of Technology Management & Innovation, 16*(3), 89–102.

Vega, M. (2021). How blockchain is transforming the banking industry. Retrieved from: https://www.hexacta.com/blockchain-in-banking-transforming-the-industry/#:~:text=Using%20Blockchain%20in%20the%20banking,a%20wide%20array%20of%20clients (accessed on 14 December 2024).

CHAPTER 3

Design and Development of Blockchain Smart Contracts for Multi-Level Marketing

RAGHU VAMSI POTUKUCHI*

Department of Computer Science and Engineering, Jaypee Institute of Information Technology, Noida, Uttar Pradesh, India

*Corresponding author

ABSTRACT

Trust has become a significant concern in today's business world. Recent research highlights a decline in trust towards brands globally. Traditionally, trusted third parties played a role in ensuring trust in business transactions. However, with the advent of digitization, peer-to-peer networks, and cryptographic technologies, blockchain technology emerged as a trust layer embedded in the Internet's protocol. It serves as the missing trust layer in the evolution of the Internet. Blockchain's implications on marketing are profound. Customers are no longer just consumers; they can be value creators and brand ambassadors. Marketers must adapt to this evolving landscape, which includes rewarding participants through mechanisms like gamification. On the other hand, blockchain technology presents new challenges, with customer data becoming customer-owned. To succeed, brands must appeal to their audience with compelling stories and causes. Blockchain facilitates these concepts on a technological level. The blockchain has become a viable platform for creating decentralized applications that are ready for the market. Its features, such as consensus, provenance, immutability, and finality, make it ideal for creating apps for decentralized

Blockchain Technology: Unlocking the Value of 3Ts in the Business World.
Richa Chauhan & Vidhi Kaul (Eds.)

DOI: 10.1201/9781998511914-3

markets, primary markets, and trade forums for product sales, income sharing, salary distribution, and record-keeping. This chapter focuses on building blockchain smart contracts (SCs) for multi-level marketing (MLM), a revenue-generating model where non-salaried employees sell a company's products and earn money through commissions. The design and development of MLM SCs have not been fully explored, but this chapter covers it by first discussing blockchain technology and SCs. Then, the MLM model is defined as a problem statement, followed by the methodical design and development of blockchain SCs to automate core MLM processes such as franchise, membership, and commission management. The SC is written in Solidity and works with the Ethereum blockchain.

3.1 INTRODUCTION

Marketing models are classified into four categories: direct selling, multi-level marketing (MLM), network marketing, and party planning. Direct selling is when a product is sold directly from the manufacturer, producer, or agent to the customer without involving wholesalers or retailers (Deviacita, 2022; Grade, 2019; Gregor and Wadlewski, 2013). In direct marketing, the distributor earns profit from direct sales. MLM is a business model where a dealer earns commission from direct sales and a multi-level commission structure from their recruited sellers. MLM and direct marketing differ in the sales force structure. Network marketing, like MLM, focuses on attracting customers but uses different techniques for commission. Party planning is a combination of direct sales and MLM commission structure where parties are hosted to sell products and distributors act as hosts to earn commission. Nowadays, sellers and their recruits host parties in MLM and party plan methods. Customers not only buy products at these parties but also become part of the sales force because of the attractive commission structure. All of these models aim to generate revenue for companies by selling their products, either directly or indirectly.

3.1.1 STRUCTURE AND COMMISSION SHARING IN MLM

MLM, although legal, has faced controversy as a marketing model. Companies like Amway, Avon, Tupperware, Herbalife, and Nutrition, among others, solely depend on and benefit from MLM. However, this model comes with a pyramid-like commission system, where new recruits at the bottom pay those at the top, rather than being rewarded for their sales. The main issue with this structure is whether the company sells products to customers or to recruits for recruitment purposes. If the latter, the MLM is considered illegal and not a legitimate pyramid (Keep and Vander Nat, 2014; Purcaru et al., 2022).

MLM commission sharing can be divided into two categories:

- Binary (uni-level) commission system; and
- Hierarchical (multi-level/matrix) commission system (Bäckman and Hanspal, 2022; Legara et al., 2008).

The uni-level system allows unlimited network growth vertically while limiting each member to two positions. This means that adding more members requires them to be registered under one of the previous members' first available positions. An example of this category is the company happy.co. The structure of the binary commission system resembles a binary tree and is depicted in Figure 3.1. The matrix or multi-level system is designed to keep fewer active participants motivated and is often combined with the uni-level system. This combination makes it an effective system for network building while ensuring fair commission distribution. It has a horizontal limit set by the company, usually three free positions for each partner (Droney, 2016; Jyoti, 2022).

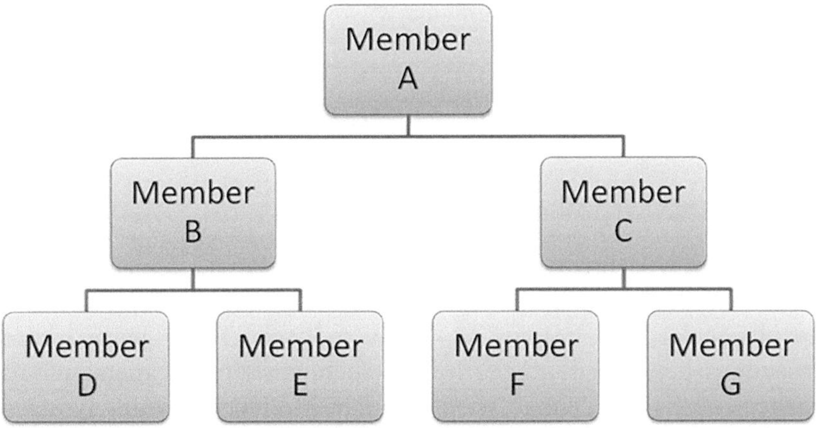

FIGURE 3.1 Hierarchical structure in MLM.

Figure 3.2 illustrates the commission sharing pyramid structure in MLM, starting with the company and branching out to recruited independent distributors who can then recruit a sales force and so on. Figure 3.3 shows the level structure of the MLM, where a company can have multiple franchises and each franchise can have multiple members, who may recruit even more members. However, if the focus is only on recruiting without product marketing, the MLM will not generate revenue and will result in loss. Limiting the size of the sales force under each franchise or member is necessary for success. A study (Muncy, 2004) suggests limiting the size and maintaining an 80–20 commission share ratio for profitability. It's important to keep track of the franchise, sales force, and their recruits, and to have secure records of the chain from the company to the last sales force. Trust in commission sharing and record keeping is crucial for new recruits and upward in the pyramid (Andriati, 2015; Grade, 2019).

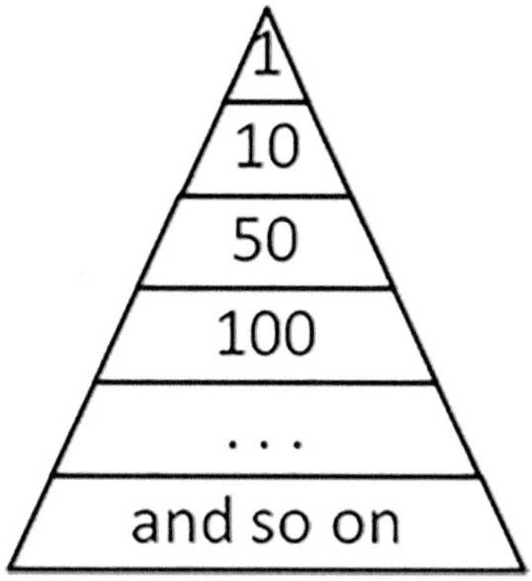

FIGURE 3.2 Pyramid structure of members in MLM.

3.1.2 BLOCKCHAIN AND ITS FEATURES

Blockchain is a secure, decentralized ledger used to track business activities. Its aim is to automate processes in a distributed environment and map physical record-keeping to digital ledgers. Digital ledgers record transactions between multiple entities, such as suppliers, manufacturers, exporters, consumers, and distributors in the product marketing field, with distributors playing a crucial role in establishing product standards. Logistics management requires tracking physical items and providing clear

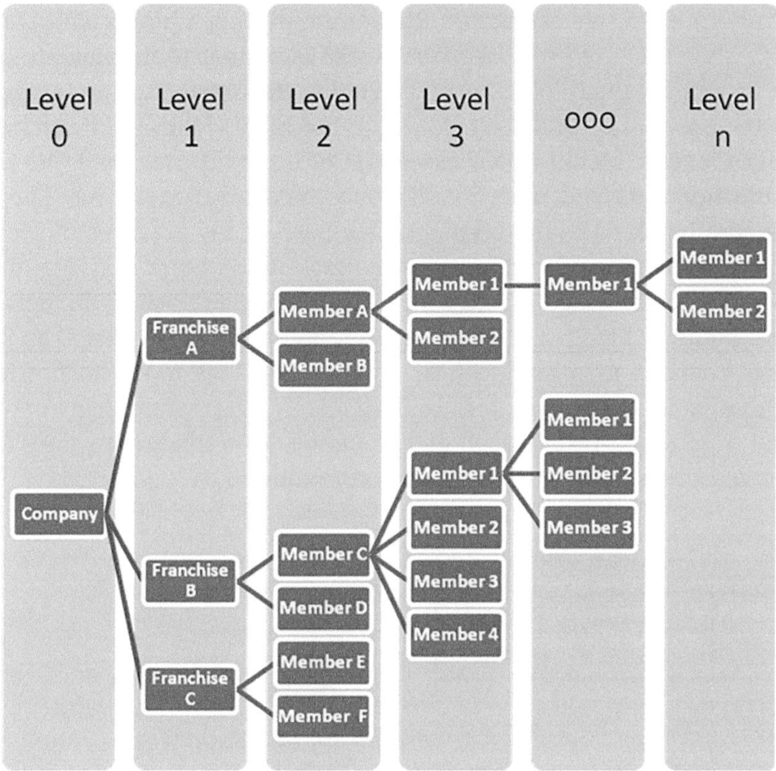

FIGURE 3.3 Level-wise franchise and members in MLM.

information, which a secure system facilitates for better optimization in company and trade transactions (Garg and Khadse, 2020; Pop et al., 2020).

Comparing traditional transaction recording using databases (depicted on the left side of Figure 3.4) with decentralized transaction recording using distributed ledger technology (depicted on the right side of Figure 3.5), it is clear that in traditional web applications, users view data based on the access features set by the application. Data is processed by business logic and then stored in a central database. In contrast, in a blockchain network, participating entities share a copy of the entire data consisting of transaction records. If a user initiates a transaction, all users in the network are updated with that transaction. Miners, who are special users, check the validity of the newly created transaction and reach a consensus to decide whether the transaction should be included in a block in the blockchain or not (Antal et al., 2021).

In Figure 3.5, we can see an illustration of how a blockchain is organized. The structure of blockchain blocks is similar to the linked lists in data structures. Every block is connected to the one that came just before it in the chain. The initial block within the blockchain, often referred to as the Genesis block, holds essential information including the chain ID, creation date and time, initial nonce, and other metadata. The next generated block takes the cryptographic hash of the previous block hash value as a reference for finalizing the block. Each block consists of four important fields, which identify the contents of the block in the network: magic number, block size, block header, and transaction data. The block header contains metadata such as version, previous block hash, Merkle tree root hash, timestamp, bits, and nonce. The transaction data contains actual transaction details such as the number of transactions, transaction list, transaction lock time, and script information.

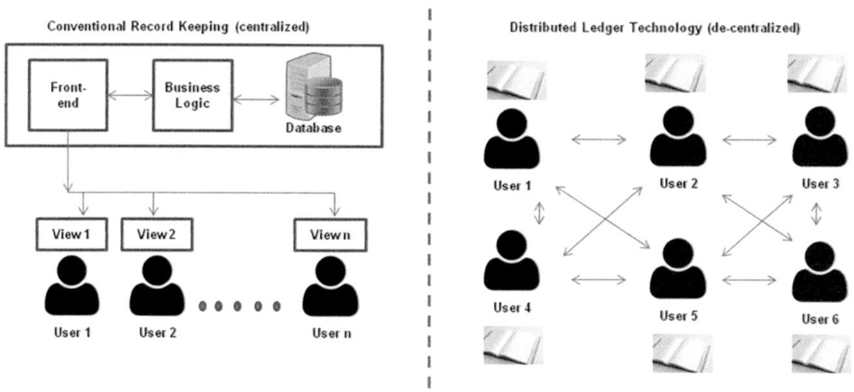

FIGURE 3.4 Centralized record keeping (left) versus Distributed ledger technology (right).

Blockchain technology provides a secure, shared, and transparent way of recording transactions in a deployed application. The technology is based on four key features that ensure reliability and integrity for users (Gorkhali et al., 2021).

1. **Consensus:** To ensure the validity of transactions, participating nodes must reach an agreement. The process of reaching agreement is known as consensus. For example, the Bitcoin network uses proof-of-work (PoW) consensus to determine if the transaction is valid and that all nodes agree on it.

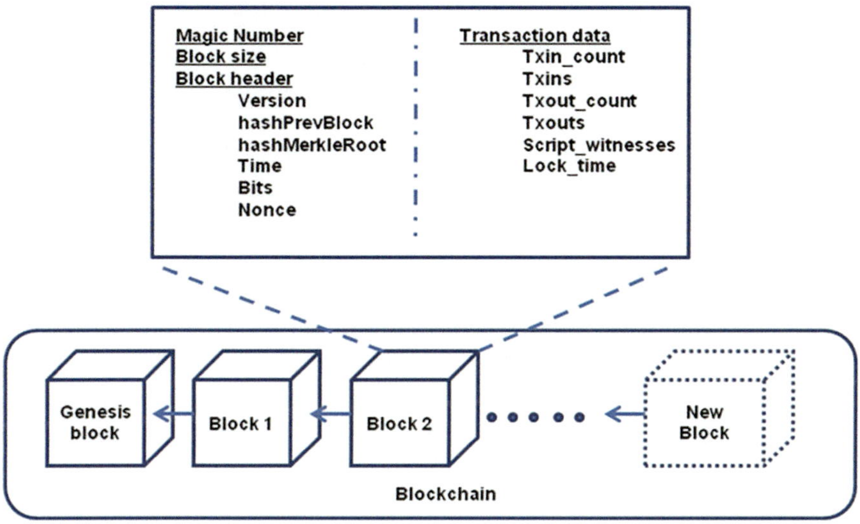

FIGURE 3.5 Block structure in blockchain.

2. **Provenance:** This feature helps preserve ownership history. All nodes in the blockchain network verify each block, and ownership information about the block is available to the network before it is added.
3. **Immutability:** Transactions stored in the blocks in the network are permanent and cannot be deleted, amended, or misrepresented. Only new transactions can be added to the network once the previous information has been provided.
4. **Finality:** All copies of the shared ledger contain the same accurate record of transaction information. The blockchain structure is updated anytime there is a change in the network.

3.1.3 SMART CONTRACTS (SCS)

A smart contract (SC) is a computer program that resides on the blockchain and automatically executes predefined actions when certain conditions are met (Gorkhali et al., 2020). SCs are key components of blockchain technology and have gained significant attention due to their potential to revolutionize various industries, including marketing. Advantages of SCs are as follows (Zheng et al., 2020; Zou et al., 2019):

- SCs provide increased transparency, allowing all parties involved to have visibility into the terms and conditions of an agreement. This transparency helps build trust among stakeholders and reduces the need for intermediaries.
- SCs enable automation of transactions, thereby eliminating the need for manual processes and reducing human error. This automation can streamline marketing operations and improve efficiency.
- SCs have the ability to enforce rules and ensure compliance. Once the conditions specified in the contract are met, the associated actions are automatically executed, leaving no room for misinterpretation or manipulation. This feature can help marketing professionals ensure fair and consistent execution of agreements and campaigns.
- SCs offer enhanced security by leveraging the cryptographic features of blockchain technology. The decentralized nature of blockchain ensures that data and transactions recorded in SCs are tamper-proof and resistant to hacking or unauthorized modifications. This level of security can protect marketing professionals from fraudulent activities and unauthorized access to sensitive information.

3.1.4 ADVANTAGES OF USING SMART CONTRACTS IN MLM

Blockchain technology has gained widespread recognition and adoption since its introduction by Satoshi Nakamoto in 2008. Not limited to cryptocurrencies, blockchain has emerged as a novel solution for various industries, such as finance, supply chains, healthcare, and social services (Jain et al., 2021; Moghrabi et al., 2023; Peres et al., 2022). One of the fundamental goals of blockchain technology is to reduce or eliminate the need for intermediaries or middlemen. One of the primary needs for blockchain in MLM is to enhance transparency. Blockchain allows for the creation of a distributed ledger that records all transactions and interactions within the MLM network. This transparent and immutable record ensures that all participants have access to accurate and up-to-date information, such as sales volumes, commissions, and bonuses. This transparency builds trust among MLM participants, as it eliminates the potential for manipulation or hidden information. Also, SCs play a crucial role in MLM by automating and enforcing the business rules and compensation plans. With SCs, MLM companies can program predefined conditions and actions into the contracts, which are executed automatically once the conditions are met. This automation ensures fair and consistent distribution of commissions, bonuses, and rewards, eliminating the potential for human error or bias. The decentralized nature of blockchain and SCs also addresses the issue of accountability in MLM. Since the data and transactions are recorded on a distributed ledger, no single entity or central authority can control or manipulate the information. This decentralization ensures that MLM participants can trust the accuracy and integrity of the system, reducing the risk of fraudulent activities or unethical practices (Epixel MLM Software, 2022; How can Blockchain Transform the MLM Industry, 2023; Javaid et al., 2022; Liu et al., 2021; MLM Platform with Smart Contract, 2022).

3.1.5 OBJECTIVE AND SCOPE OF THE WORK

The objective of this work is to explore the design and development of blockchain SCs for MLM and to automate core MLM processes such as franchise, membership, and commission management. The scope of the work includes discussing blockchain technology and SCs, defining the MLM model as a problem statement, and implementing methodical design and development processes for creating MLM SCs using Solidity and the Ethereum blockchain. The focus is on leveraging the features of blockchain, including consensus, provenance, immutability, and finality, to create decentralized applications for revenue generation, product sales, income sharing, salary distribution, and record-keeping within the MLM industry. The aim is to provide a comprehensive exploration of MLM smart contract development within the context of blockchain technology.

3.1.6 CONTRIBUTIONS OF THE WORK

The contributions of the current work can be summarized as follows:

1. **Exploration of MLM on Blockchain:** This chapter covers the design and development of MLM SCs on the blockchain, which has not been fully explored before.

2. **Automation of MLM Processes:** The SCs developed in this work will automate core MLM processes such as franchise, membership, and commission management, making the system more efficient and transparent.
3. **Use of Ethereum Blockchain and Solidity:** The SCs are written in Solidity and work with the Ethereum blockchain, making the system secure and reliable.
4. **Accessible Source Code:** The source code of the work presented in this chapter can be accessed from (Potukuchi, 2023) for further improvements and investigation by the researchers.

After discussing related work in Section 3.2, Section 3.3 presents the blockchain smart contract (SC) development methodology for multi-level marketing (MLM), including an analysis of the business problem, the selection of the appropriate blockchain category, requirements analysis, and smart contract design. Section 3.4 details the coding and execution of the designed contract, followed by testing in Section 3.5. Finally, Section 3.6 addresses the deployment and maintenance of the smart contract, and the chapter concludes with a summary of findings and future research directions in Section 3.7.

3.2 LITERATURE SURVEY

In the fourth industrial revolution, Industry 4.0 is reshaping how businesses operate across value chains (Chauhan et al., 2022). Industries are becoming smarter through data exchange, predictive analytics, and IoT technologies. Optimization and automation of business processes offer cost reduction, increased productivity, and error reduction. Digital technologies in a business-to-customer (B2C) setting offer opportunities to innovate business models and create new value and revenue prospects. Blockchain is a technology that came to address such issues in the recent COVID-19 pandemic. This has exposed limitations in producing and distributing critical medical goods and the challenges of relying on foreign supply. Until that time, comprehensive tracking of products and services was a distant possibility. However, the pandemic has highlighted the need for better quality pharmaceutical and healthcare supply chain management and tracing systems. Today's medical supply chains face significant logistical complexities, including a lack of real-time visibility, unpredictable weather, quality control, and intricate customs and payments. Emerging technologies offer solutions that can revolutionize medical supply chains, ensuring faster, transparent, and secure access to critical medicines. Blockchain, with its readiness and cost-effectiveness, plays a pivotal role in enabling companies to build solutions that accurately, transparently, and effectively track and trace every stock-keeping unit (SKU) (Chauhan and Kaul, Blockchain Technology: A Panacea for Medical Distribution Ailments, 2021).

With the rise of distributed energy generation through solar PV applications, new energy markets have emerged. Consumers and producers are no longer separate entities, giving rise to energy prosumers. A study presented by Boumaiza and Sanfilippo (2022) introduced a blockchain-based energy trading simulation environment integrated with a geographic information system (GIS) for Qatar's Education City Community Housing (ECCH). The data collected allows the recreation of spatiotemporal characteristics of trading in a small marketplace, providing stakeholders with insights into real trading dynamics. The model can be adjusted to tailor transaction characteristics, and large-scale simulations can be performed for more realistic outcomes. The technology offers a scalable environment for analyzing energy blockchain from the perspectives of Qatari society, finance, and technology.

The Federal Trade Commission (FTC) provides guidance on MLM businesses in the US, highlighting the potential for confusion between MLM and pyramid schemes (Business Guidance Concerning Multi-Level Marketing, 2022; Commission). Multiple studies have analyzed MLM from a marketing perspective, including a study by Lee and Loi (2016) that examines MLM from a marketing perspective, specifically focusing on distributor satisfaction. It investigated factors that contribute to the satisfaction of distributors within MLM companies. A study by Jyoti (2022) examines the influence of demographics on key drivers to join an MLM company. The authors investigated how factors such as age, gender, income, or educational background influence people's decision to participate in MLM. Other works explore the earning potential in MLM enterprises (Legara et al., 2008) and the use of electronic word-of-mouth in promoting MLM businesses (Deviacita, 2022). Far et al. (2022) presented a privacy-preserving framework for anonymous and confidential rewarding, exploring the application of blockchain technology in MLM in the chapter. The article by Peres et al. (2022) explored the opportunities and threats of blockchain in marketing. The impact of MLM on consumers is also investigated, including a case study by Chetty (2019) that looks at World Ventures under the Consumer Protection Act (2008), and a study by Bäckman and Hanspal (2022) that analyzes participation and losses in MLM. Additionally, the article by Andriati (2015) investigates the manipulative cooperation politics of MLM companies in Surabaya City. All these works are focused on studying various dimensions of MLM and exploring the advantages and limitations. However, these works do not have any technical implementation of the process flow of MLM.

Articles published online, such as “How can Blockchain Transform the MLM Industry” (2023) and “MLM Platform with Smart Contract” (2022), have focused on the use of blockchain in the MLM process and highlighted the advantages of using smart contracts in MLM. Epixel has developed software for MLM flow, but the technicalities of developing the contract were not discussed (Epixel MLM Software, 2022). The digital transformation certainly impacts the traditional flow of existing systems; however, it is necessary to adapt to the new technical challenges and automate the existing flow. To this end, Moghrabi et al. (2023) explored the impact of digital transformation on businesses, specifically in the manufacturing sector. It highlights how various aspects, including Cloud Computing, integration, data communication, and the internet, have led to sustainable manufacturing by decentralizing supply chains. Additionally, digital transformation innovations like predictive analysis and big data analytics have optimized manufacturing processes by predicting customer demands and reducing overproduction or underproduction. Thontirawong and Chinchanachokchai (2021) demonstrated how marketing educators can introduce AI and ML concepts in their classes by utilizing a cloud-based platform called AzureML Studio. The authors also showcased the implementation of ML models for customer churn prediction, resulting in improved student learning and positive outcomes in terms of career preparation, traditional learning goals, time utilization, and overall satisfaction. Shrestha et al. (2020) introduced a novel user modeling platform utilizing blockchain technology, specifically designed for maintaining data privacy, user control, and incentivizing data sharing. The platform addresses three key challenges: preserving privacy and user autonomy, ensuring verifiable sharing, and creating incentives for collaboration. The work presented by Ertemel (2018) explored the implications of blockchain technology on the field of marketing from various angles. It identifies key strategies that marketers need to adopt to thrive in the rapidly evolving business landscape. Finally, the chapter draws conclusions about the future direction of marketing influenced by these technological advances.

It is observed that over time, sales teams emerged, leading to the birth of MLM business models. Today, MLM is a multi-billion-dollar industry, with crypto assets (CAs), especially in crypto trading (CT), fully embracing it. Despite the MLM model facing challenges over the years, it has the potential to restore its reputation and prove its effectiveness in African CT. Network marketing (NM) is at the heart of MLM and has been used by many large businesses to increase awareness and sales. In African CT, the MLM model enhances investment, employment, and entrepreneurship opportunities in the crypto space. To this end, a study presented by Masama (2023) discussed the sanitization of MLM and its effectiveness in African countries.

In a recent work by Jain et al. (2023), the authors presented the development of a blockchain smart contract for collaborative knowledge sharing in project teams. Kushwaha et al. (2021) presented the development of a blockchain smart contract for the human resource recruitment process. In this, the procedure of the HR recruitment process is automated using a smart contract. The work reported by Ravindran et al. (2021) presented the development of smart contracts for financial applications. The authors focused on automating the process of financial loan processing using smart contracts. In the work by Ravindran et al. (2021), the authors provided a model for automating the process of the coffee retail business using smart contracts. The existing studies indicate that researchers are actively engaged in automating business processes to align with emerging industry technologies like blockchain. However, it is evident that there is still limited exploration into the design and development of MLM smart contracts. Consequently, the objective of this research is to investigate the design and development of blockchain-based Smart Contracts for MLM and to automate essential MLM processes, including franchise, membership, and commission management. This study encompasses an in-depth discussion on blockchain technology and Smart Contracts, articulates the MLM model as a problem statement, and employs a systematic design and development approach to create MLM Smart Contracts utilizing Solidity and the Ethereum blockchain. The subsequent section presents the construction of the blockchain Smart Contract for the MLM process using Solidity programming.

3.3 BLOCKCHAIN SMART CONTRACT FOR MLM

In this section, the utilization of the software development life cycle (SDLC) methodology for the creation of smart contracts for multi-level marketing is depicted in Figure 3.6. The SDLC methodology offers a systematic and repeatable approach for developing software systems and is fitting for the development of a smart contract in the MLM field. Adhering to this methodology, the project team can guarantee that the end result is of excellent quality, meets the needs of stakeholders, and is delivered promptly and within the set budget (Almeida et al., 2019; Faruk et al., 2022; Marchesi et al., 2018; Yutia et al., 2022). The SDLC methodology consists of several phases, each with its own distinct tasks and objectives, such as:

- Understanding the MLM business problem, including the presentation of the problem and definition of the problem statement.
- Selecting the appropriate blockchain category for SC development.
- Analysis and design, where requirements are analyzed based on the problem statement and the contract design is outlined.
- Coding and testing, where the contract design is implemented through Solidity code and test cases are devised for testing the code.
- Deployment, where the tested SC is deployed to assess its functionality.

3.3.1 UNDERSTANDING BUSINESS PROBLEM

Understanding the current systems and identifying areas for improvement is crucial in driving innovation and progress. The current manual system has been found to lack trust and ethical practices, leading to various issues and limitations. To address these concerns, a proposed system has been designed using blockchain technology. The use of blockchain ensures that trust and ethical practices are achieved by providing a secure and transparent environment for transactions and interactions. This new system has the potential to revolutionize the way things are done, bringing about a more trustworthy and ethical future (Figure 3.6).

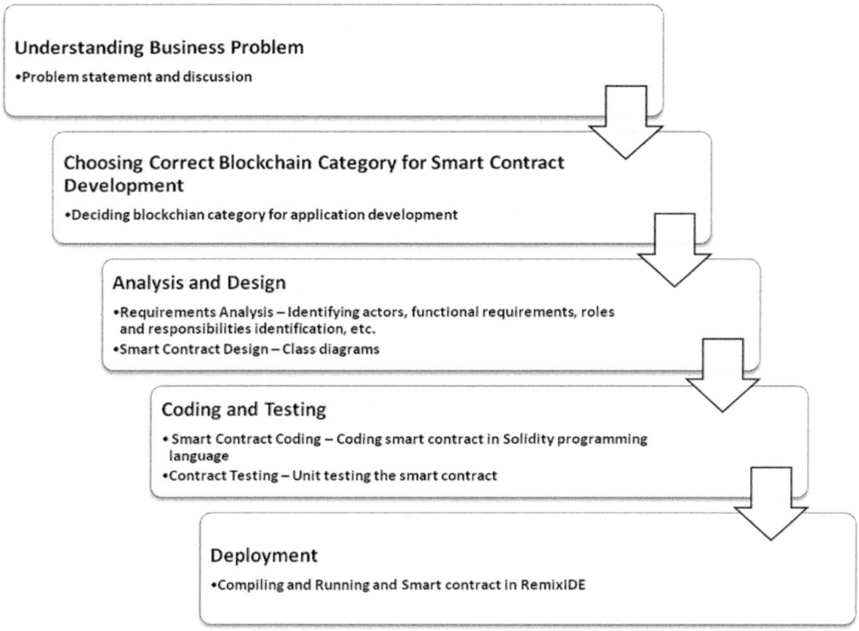

FIGURE 3.6 MLM smart contract development methodology.

This work aims to address specific business problems and challenges in the promotion and marketing of a product manufacturing company. The current traditional method of public advertisements and demonstrations, while effective in some aspects, may not always be the most effective way to reach the common person. To better connect with the target customers and effectively educate them about the company's products, a new marketing model is needed. The project will focus on implementing the MLM model, which utilizes a non-salaried workforce to promote and sell the products. This model is unique in its use of a pyramid-shaped commission system to provide incentives for the sales force. By addressing these business challenges, the first project aims to provide a more effective and efficient marketing solution for the product manufacturing company.

The traditional method of handling MLM operations is prone to errors and lacks transparency. Distribution of profits among members is manual and prone to errors. To address these challenges, the company is looking to develop a SC for MLM operations. The SC should automate the approval process for franchisees and members' fees, and sharing of profits among members. The company, which is facing difficulties in managing the process efficiently, keeping track of the members' contributions and payments, and ensuring a fair payment of franchise and membership fees, should distribute the profits among members as per the agreed-upon ratios. The contract should enforce the maximum size of the group and ensure that the franchisee's share of the membership fee and profits is accurately credited to their account. The contract should also provide transparency and efficiency in managing the MLM operations.

The above statement is formulated in a concrete way with numerical values as follows: "*Anyone interested in MLM should seek approval from the company's marketing manager to start a franchise. The company has set the franchise registration fee at 500 Wei. The franchisee can build small groups for daily marketing purposes. To this end, franchisees can join members in the group. In addition, the approved group members can invite anyone to join them. A new member must pay a membership fee of 100 Wei for this. If the new member joined the group through a franchisee, the franchisee receives a full refund of the membership fees. Otherwise, 30% of the membership fee will be credited to the franchisee and the remaining 70% to the group member who recruited the new member. The same share is applicable to the profits of the products sold by members. Further, to have control over the group, the maximum group size is fixed at 25.*"

Once the problem statement is defined, the next step is to choose the blockchain category for implementation. Broadly, there are three categories available such as:

1. **Public Blockchain:** These types of networks are open to anyone and allow anyone to participate as a node in the network. Bitcoin and Ethereum are examples of this category. They offer a decentralized and transparent system where transactions are validated through consensus mechanisms like PoW or proof of stake (PoS). Public blockchains are often used for cryptocurrencies and applications that require a high level of decentralization and security.

2. **Private Blockchain:** These are restricted to a specific group of participants. They are typically operated by a single organization or a consortium of organizations. Private Blockchains provide higher privacy and control over the network compared to public blockchains. They are commonly used in enterprise settings where data privacy and limited access are important considerations.

3. **Consortium Blockchain:** These networks are said to be a hybrid approach that combines features of both public and private blockchains. In a consortium blockchain, multiple organizations form a network where participation is restricted to a pre-selected group of participants. These blockchains offer greater scalability, efficiency, and control compared to public blockchains while maintaining a certain level of decentralization.

3.3.2 CHOOSING BLOCKCHAIN CATEGORY

Selecting the appropriate blockchain for a specific use case is crucial for its successful implementation. Figure 3.7 illustrates the sequential process involved in selecting an appropriate category of blockchain, considering factors such as reliance on trusted third parties, data storage requirements, multiple writers, and the need for access control and consensus (Wüst and Gervais, 2018). The sequence of steps involves:

1. **Determining the Need for Blockchain:** The first step in the blockchain selection process involves evaluating whether there is a need to rely on a trusted third party. If the answer is no, then it implies that trustless transactions are desired and blockchain technology becomes a viable solution.
2. **Data Storage and Multiple Writers:** The next considerations involve the necessity of storing data and the existence of multiple writers. If data storage or a distributed database is required and there are multiple writers contributing to the system, then the blockchain can address these requirements effectively.
3. **Untrusted and Conflicting Writers:** Furthermore, if the writers in the system are unknown, untrusted, or likely to have conflicts of interest, blockchain provides a transparent and immutable ledger that can prevent fraudulent activities and ensure the integrity of the data.
4. **Sequence of Requirements Checking:** After confirming the need for blockchain based on the above considerations, the following requirements should be evaluated in a sequential manner.
5. **Access Control Requirement:** The first requirement to assess is the need for access control. If access control is necessary to

restrict the participants who can engage in transactions, then further evaluation is needed.

6. **Public Transactions:** If access control is required, the next consideration is whether transactions need to be public. If the transactions must be publicly visible, then it becomes crucial to examine the trade-off between immutability and efficiency.
7. **Immutability versus Efficiency Trade-Off:** If immutability is a critical requirement, indicating that the data must be tamper-proof and irreversible, a public blockchain should be chosen. However, it is essential to consider the potential impact on efficiency, as public blockchains often involve longer confirmation times and higher transaction costs.
8. **Multiparty Consensus Requirement:** If access control is not necessary or if public transactions are not required, the next requirement to evaluate is the need for multiparty consensus. If consensus among multiple participants is essential for decision-making and transaction validation, a consortium blockchain becomes a suitable option.
9. **Private Blockchain Selection:** In the absence of a requirement for multiparty consensus, a private blockchain is recommended. Private Blockchains offer increased privacy and control, as they are restricted to a select group of participants or a single organization.

In this way, the blockchain selection process involves a series of considerations and steps to determine the most appropriate category of blockchain for a given use case. From the current case, it can be understood that the marketing process is happening in a closed environment. This means that this marketing group does not need to be monitored by a third party. Each network participant must be held accountable for the amount paid; the company and franchisee must both keep track of the fee. These transactions must be recorded for future reference, for which a database is needed. The marketing manager, franchisee owner, and members are the participants who write on the ledger. In general, the company issues a franchise, and the franchise in turn maintains the groups. Here, group members are in no way related to the company, and it is the choice of the member to move from one franchise to another or

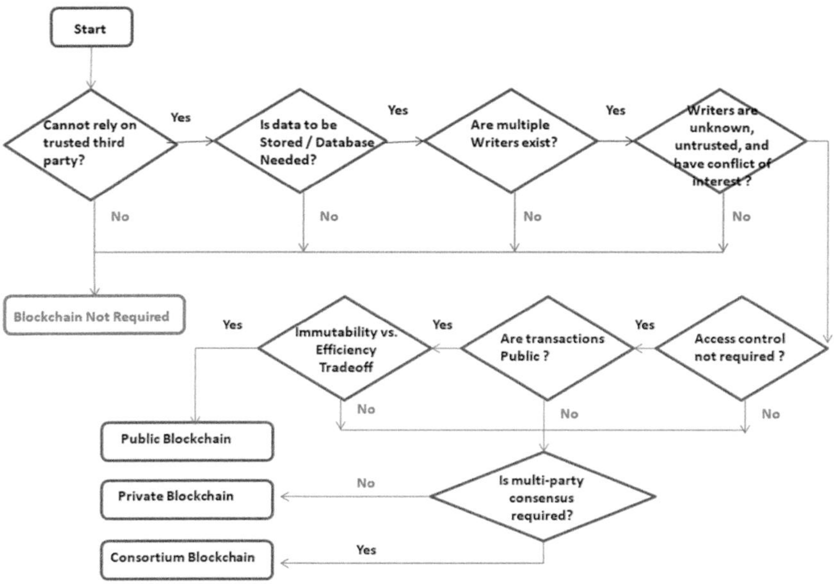

FIGURE 3.7 Flow graph showing selection of blockchain.

to start his/her own franchise. To this end, the participants in the network have conflicts of interest and are completely unknown. Since the entire activity is happening in a closed environment and to maintain trust in the system, only registered participants should have access to the transactions. It is possible when the transactions are private. Further, a shared ledger is needed to keep track of the updated transaction. A transaction made by the member will be approved by the franchise, and a request for a franchise will be approved by the marketing manager. In both approvals, only one authority is responsible for action. To this end, the most suitable blockchain for application development is a private blockchain.

3.3.3 REQUIREMENTS ANALYSIS AND DESIGN

During the requirements gathering, the project team will gather requirements from stakeholders, including the marketing manager, franchisees, and group members. This information will be used to create a comprehensive understanding of the needs and objectives of the project. In the analysis stage, the project team will analyze the gathered requirements to determine the feasibility and scope of the project. They will also create a detailed design of the SC, including the rules for franchisee registration, group creation, membership fee payment, profit sharing, and group size control (Al-Ashmori et al., 2021).

According to the problem statement, there are four roles:

- Company;
- Supervisor;
- Franchisee; and
- Member.

Except for the manager, everyone has their own account to hold fees and profit shares. The following is a discussion of each role:

1. **Company:** It has an account into which the franchise registration fee will be deposited. It maintains a depositors' list and deposited amount for auditing purposes. At any time, a company can view the registration account balance and can also transfer the registration account balance (i.e., from the SC account) to its own address (i.e., an externally owned account). An additional provision has been made while paying the fee that any excess amount other than the franchise registration fee will be credited back to the applicant's account. The company's address is public, and anyone can view it. The companies recruit area managers according to pin code to approve the franchise applications. The company can also change the managers of the corresponding pin code. Also, the company maintains a list of approved franchisee addresses, pin code to manager mapping, franchisee address to approved manager mapping, and others. To support company tasks, various bean (set, get, and is) functions are defined. The entire MLM contract can be destroyed (or killed) only by the company.

2. **Manager:** Each manager has an address and an associated pin code. Applications for a franchise in a particular pin code will be processed by the corresponding manager. Once a franchise is approved by the manager, a new entry will be made in the franchise list with franchise information such as address, account creation, franchise size, member list, and so on. At the same time, franchise application details are removed from the depositors' list.

3. **Franchisee:** Each franchisee has an address and associated data, such as a member list, franchise size (or group size), a franchisee

account to hold membership shares, and others. A franchise can destroy itself. While doing so, the left-over membership share will be transferred to the franchisee account.

4. **Member:** Each member has an address and associated data, such as a franchisee address, the address of who joined him/her in the group, an account to hold their membership share, and others. A member can destroy himself. While doing so, the remaining membership share will be transferred to the account.

Figure 3.8 depicts the use case diagram of the MLM system using the above-mentioned four roles. Separate mapping and data structures are maintained for accessing the details of franchisees and members. Once the destructor is called, the related entries are removed from the data structures. Anyone who is interested in joining a franchise can directly apply for membership by mentioning the franchisee's address or a known member's address and providing the membership fee. If an applicant applies for membership using a franchisee address, the entire membership fee will be transferred to the franchisee account. In the event that a member joins another member, 70% of the membership fee will be transferred to his or her account. And the remaining 30% share will be transferred to the corresponding franchisee account.

3.3.4 SMART CONTRACT DESIGN

In the design stage, the project team will create a detailed design of the SC, including the technical specifications, architecture, and data structures. They will also identify the necessary components, such as the blockchain platform, database, and user interface. Figure 3.9 depicts the basic class diagram of the SC design.

The main contract in an MLM system encompasses the Company and Manager, which are declared as structure members. Both the Company and Manager are integral parts of the MLM contract, contributing to the overall functioning of the MLM network. The Company is identified by its name and address, serving as a unique identifier within the system. The MLM contract incorporates various multiplicities and relationships to establish the hierarchical structure and facilitate interactions within the network. The Company can have any number of Managers, denoted

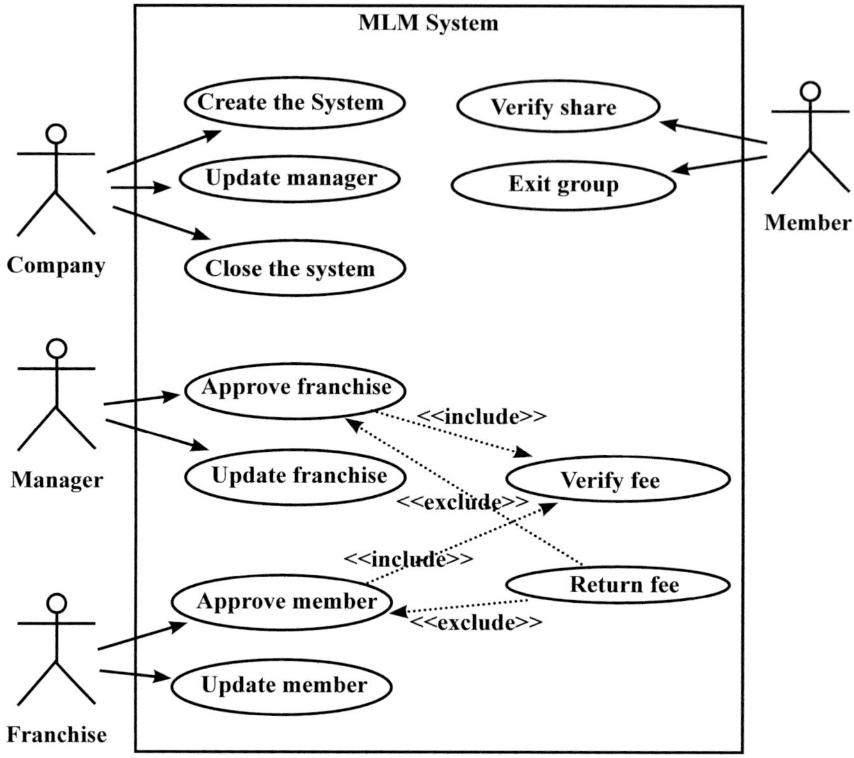

FIGURE 3.8 Use case diagram of the MLM system.

by one or more multiplicities, indicating the flexibility to accommodate multiple managerial positions. This allows for effective management and oversight of the MLM operations.

To handle transactions and maintain accurate records, the Company, Franchisee, and Member possess individual accounts. These accounts enable functions such as address validity verification and balance validation, ensuring the integrity of transactions within the MLM system. By implementing robust account management features, the MLM network can enhance trust and reliability. A Manager holds the authority to approve multiple Franchisees, as indicated by the one-to-many multiplicity. This relationship empowers Managers to effectively monitor and support the activities of Franchisees, promoting consistency and compliance within the MLM structure.

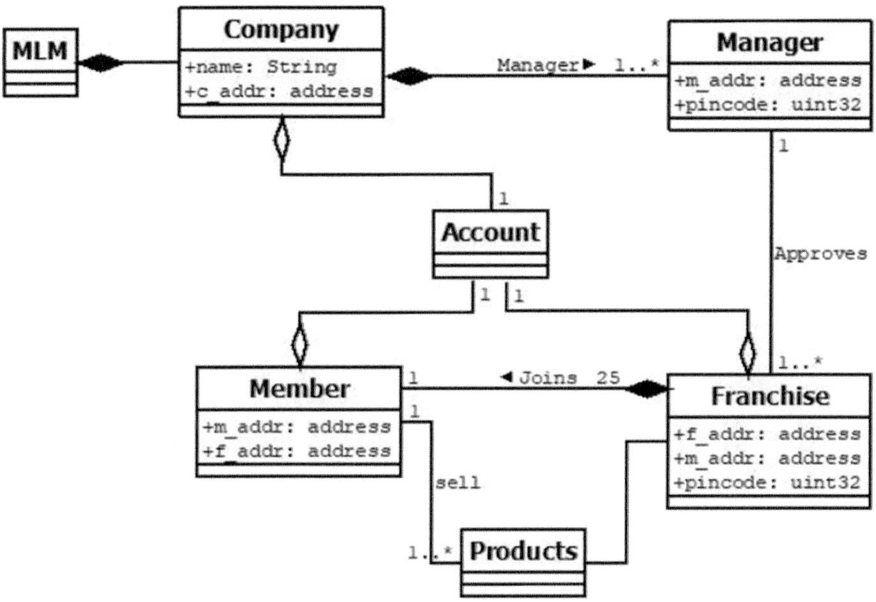

FIGURE 3.9 Basic class diagram of MLM SC.

Each Franchisee is capable of accommodating a maximum of 25 members, denoted by the 1-to-25 multiplicity. This hierarchical structure allows for effective management of the Member base within each Franchisee, ensuring scalability and control. Members within a Franchisee can actively engage in product sales and earn commissions based on their performance.

The structure of MLM products is defined using product IDs and names, enabling clear identification and categorization of products within the MLM network. By establishing a well-defined product structure, MLM systems can effectively manage inventory, pricing, and commission calculations.

Figure 3.10 depicts the complete class diagram of MLM SC in which the structure and relationships within the MLM contract are labeled with specific stereotypes, providing additional information and context. The detailed meanings of each stereotype are elaborated in Table 3.1, enabling interpretation and understanding of the purpose and behavior associated with each structure or association.

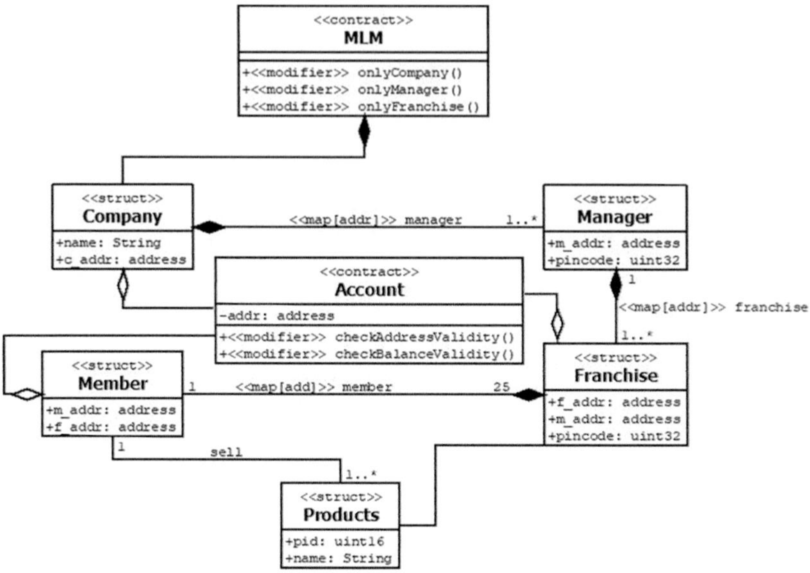

FIGURE 3.10 Enhanced class diagram of the MLM smart contract.

TABLE 3.1 Description of Stereotypes

Sl. No.	Stereotype	Description
1.	<<contract>>	It represents the contract structure. It is displayed in the first section of the class symbol.
2.	<<struct>>	It holds the data structure of the roles defined in the case. It consists of functions and is displayed in the first section of the class symbol.
3.	<<interface>>	This is a contract with function declarations.
4.	<<enum>>	This is enumerated data containing values to be used in the program. It is displayed in the first section of the class symbol.
5.	<<library contract>>	It represents the contract used from predefined or user-defined libraries. It is displayed in the first section of the class symbol.
6.	<<modifier>>	It is used along with functions. It is displayed in the last section of the class symbol.
7.	<<array>>	It represents the mapping of one-to-many relations. It is represented on the association.
8.	<<map>>	It represents the mapping of one-to-many relations. It is represented on the association.
9.	<<map[uint]>>	It represents the mapping of one-to-many relations with integer-to-value mapping. It is represented on the association.
10.	<<map[addr]>>	It represents the mapping of one-to-many relations with address-to-value mapping. It is represented on the association.

The following are the three scenarios for which the sequence diagrams are provided in Figures 3.11:

- **Scenario 1:** Franchisee registration.
- **Scenario 2:** Franchisee adds a new member to the group.
- **Scenario 3:** Members share the profits from product sales.

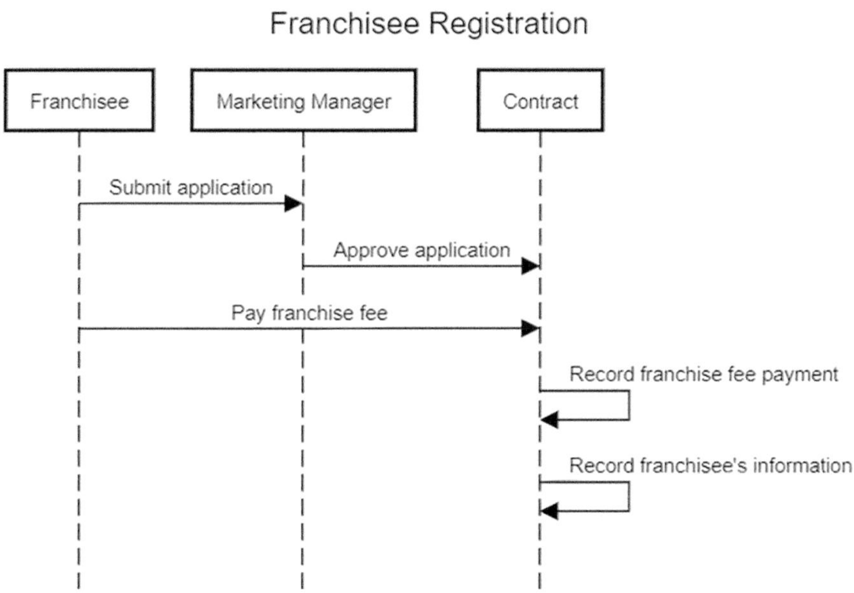

FIGURE 3.11 Sequence diagram for franchisee registration.

In Scenario 1 (Figure 3.11), users initiate the process by submitting their franchisee applications along with the required fee. This step marks the beginning of the franchisee application process, which aims to evaluate and approve qualified applicants. Upon submission, the relevant details of the user and application are recorded in the contract. This recording ensures the immutability and transparency of the application data, preventing unauthorized modifications and facilitating auditing and verification processes. The manager is responsible for overseeing the franchisee application process and periodically checks for newly submitted applications. This check ensures that the applications are promptly reviewed and processed, contributing to the efficiency of the overall system. If the user details and paid fee are deemed sufficient, then the franchisee application undergoes an evaluation process. This evaluation typically considers factors such as the applicant's qualifications, financial capacity, and adherence to predefined criteria. If the application meets the necessary requirements, then the franchisee application is approved.

In Scenario 2 (Figure 3.12), the franchisee takes the initiative to add a new member to their MLM group. The process begins with the franchisee sending a request to the SC, indicating their intention to add a new member. The SC, upon receiving the request, performs a check to ensure that the group size is within the maximum limit specified by the MLM system. Once the check is successful, the SC notifies the franchisee that they can proceed with adding the new member. The franchisee then directs the new member to pay the required membership fee. The SC verifies the payment and records the membership fee payment along with the relevant information of the new member. In this way, the SC ensures that the franchisee receives their rightful share of the membership fee. It credits the franchisee's account with the corresponding amount, reflecting their share of the fee for successfully adding the new member to the MLM group.

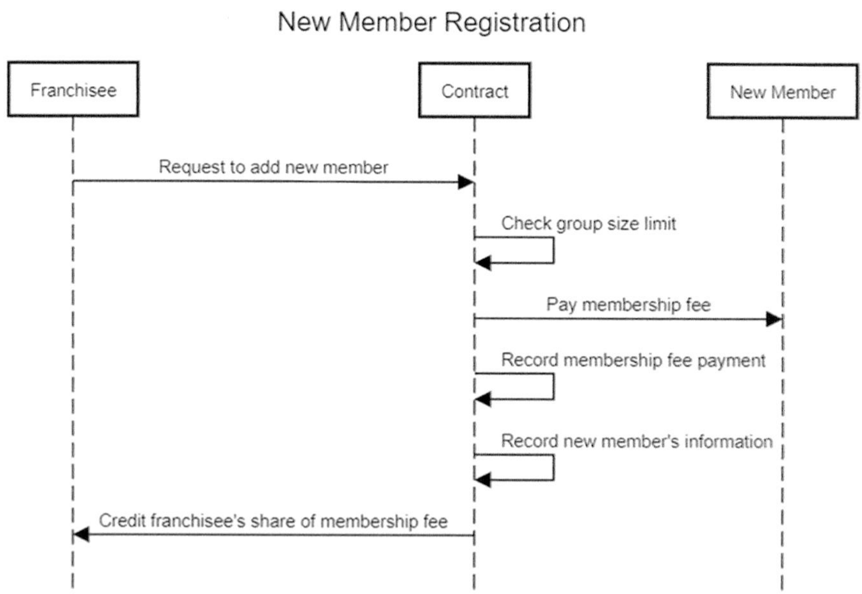

FIGURE 3.12 Sequence diagram for new member registration.

In Scenario 3 (Figure 3.13), the franchisee takes the initiative to distribute the profits among the members based on the sales of products within the MLM system. The smart contract comes into action by calculating the profit ratios for each member, ensuring a fair and transparent distribution. Once the calculations are complete, the contract executes the distribution process, allocating the profits according to the predefined ratios. Throughout the process, the contract diligently records the details of the profit distribution, maintaining a reliable and auditable record of the transactions. By leveraging the power of smart contracts, MLM systems can ensure accurate and efficient profit sharing among the members, promoting trust and accountability within the network.

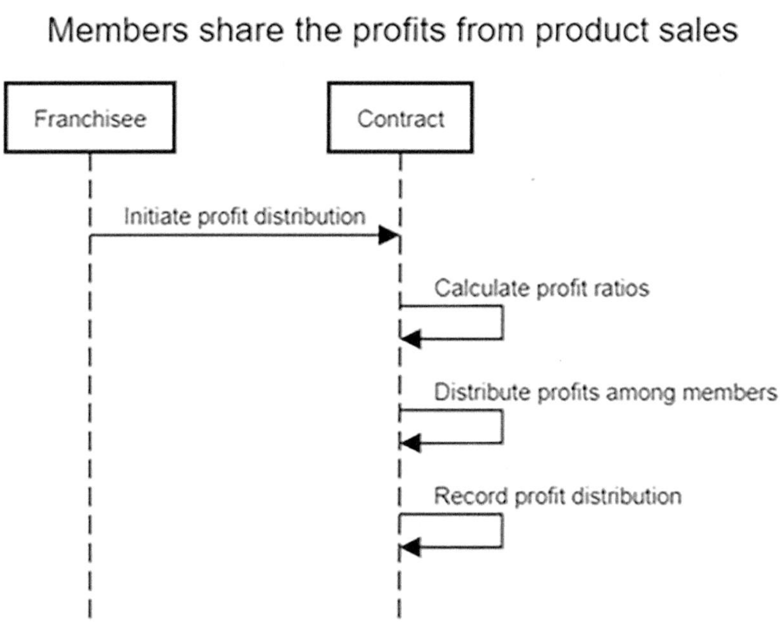

FIGURE 3.13 Sequence diagram for members sharing the profit from product sales.

3.4 CODING AND EXECUTION

During implementation, the project team will write the code for the smart contract (SC) and implement it on the blockchain platform. They will also create the database and user interface, and integrate all of the components into a single system. The following code shows the implementation of the class diagram shown in Figure 3.10. Account contract: Company, Franchisee, and Members have their own accounts. To this end, the following SC (Account) defines the required functions. The code is developed in Solidity version 0.5.0. Readers may refer to books such as (Chittoda, 2019) for Solidity programming constructs. Due to space constraints, the source code of the implementation Solidity program for implementing Multi-Level Marketing is provided in the GitHub repository (Potukuchi, 2023). Readers may go through the MLM.sol file for the Solidity code for testing the MLM smart contract.

Remix IDE is used to execute and run the code (Remix, 2023). The Remix IDE is a powerful and widely used integrated development environment for programming and testing SCs on the Ethereum blockchain. It provides a user-friendly interface and a comprehensive set of features to facilitate the development, debugging, and deployment of Solidity SCs. With Remix IDE, developers can write, compile, and deploy contracts directly from the user's web browser, making it accessible and convenient for both beginner and experienced blockchain developers. The IDE offers a range of tools such as code highlighting, syntax checking, and debugging capabilities, enabling developers to write secure and efficient SCs. Additionally, Remix IDE supports integration with popular Ethereum test networks, allowing developers to test and interact with their contracts in a simulated blockchain environment. Overall, Remix IDE is an invaluable tool for blockchain developers, providing a seamless and efficient development experience for building decentralized applications on the Ethereum platform. Figure 3.14 shows the selection of a suitable Solidity compiler (in this case, the 0.5.0 compiler) to compile the SC code.

Once the compilation is successful, navigate to the "Deploy and Run Transactions" tab in Remix IDE. Figure 3.15 shows the selection of the SC that is to be deployed from the dropdown menu. Choose the appropriate account from the available accounts and specify any constructor arguments required. Click on the "Deploy/Transact" button to initiate the deployment process. Since the MLM SC constructor is defined with two parameters, such as name and address, these values need to be supplied before clicking the Deploy/Transact button, as shown in the figure. Once the deployment is successful, a message will appear in the debug section showing the address of the contract instance created (*see* Figure 3.15 – bottom right corner). From now on, interaction with the contract will be done using the instance of the MLM contract created after deployment.

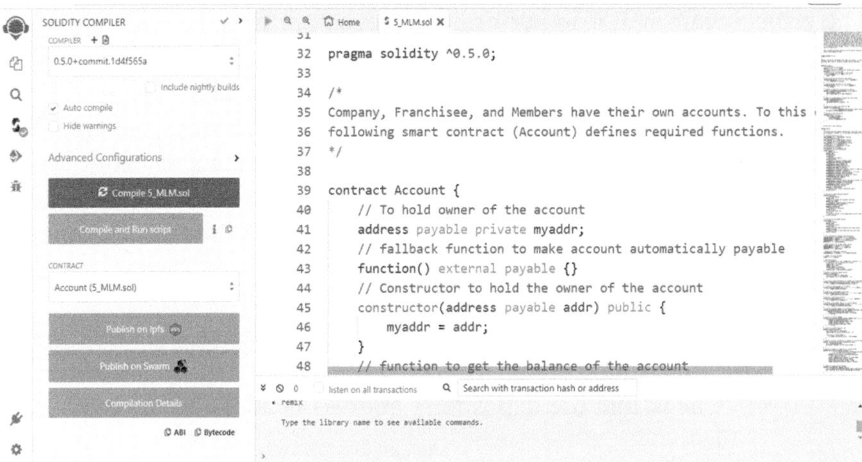

FIGURE 3.14 Screenshot of the MLM Solidity smart contract compilation in Remix IDE.

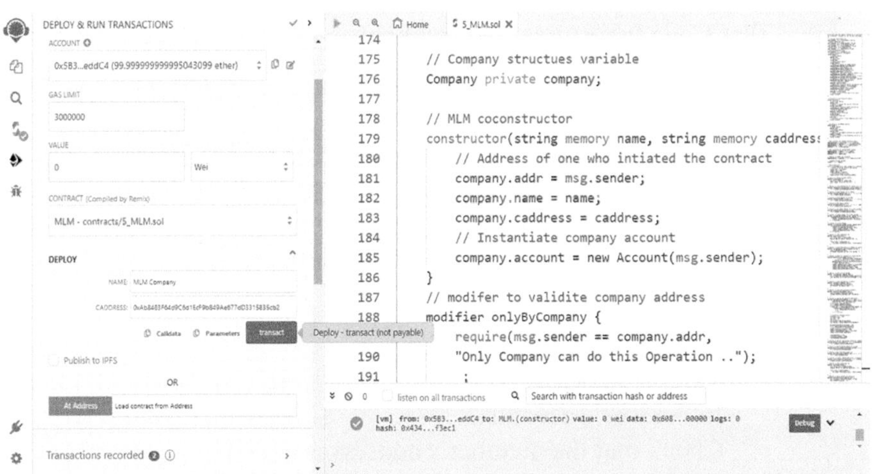

FIGURE 3.15 Screenshot of the deployment of the MLM smart contract in Remix IDE.

3.5 SMART CONTRACT TESTING

In the testing stage, the project team will perform various tests on the SC, including unit tests, integration tests, and end-to-end tests. They will ensure that the SC functions as intended and meets the requirements of the stakeholders. In the deployment stage, the SC will be deployed on the blockchain platform, and the system will be made available to users.

The project team will also perform any necessary training and support to ensure a smooth transition. Maintenance will be done after deployment; the project team will provide ongoing maintenance and support for the SC. This may include fixing bugs, adding new features, and updating the system as necessary. The following code represents a test suite for an MLM (multi-level marketing) SC, which is written in Solidity and utilizes the Truffle framework to validate the functionality of the MLM contract:

1. **Test Case for Applying Franchise:**
   - Verify that the franchise registration fee is correctly transferred to the company account.
   - Check that the depositor's address is added to the depositors list.
   - Ensure that the franchise applicant's address is added to the franchise list.

2. **Test Case for Approving Franchise:**
   - Verify that only the company manager can approve a franchise application.
   - Check that the franchise applicant's details are correctly removed from the depositors list.
   - Ensure that the franchisee's address is added to the franchisee list.
   - Verify that the franchisee's account balance is updated.

3. **Test Case for Applying Membership:**
   - Verify that the membership fee is correctly transferred to the franchisee's account.
   - Check that the member's address is added to the member list.
   - Ensure that the member's details are correctly added to the franchisee's member list.
   - Verify that the franchisee's group size is incremented.

4. **Test Case for Joining as a Member Under an Existing Member:**
   - Verify that the membership fee is correctly distributed between the franchisee and the member who joined.
   - Check that the new member's address is added to the member list.

- Ensure that the new member's details are correctly added to the franchisee's member list.
- Verify that the franchisee's group size is incremented.

5. **Test Case for Changing Franchise Manager:**
   - Verify that only the company manager can change the franchise manager.
   - Check that the new manager's address is correctly updated for the given pin code.
6. **Test Case for Getting Franchisee Details:**
   - Verify that the correct franchisee address is returned for a given pin code.
7. **Test Case for Getting Manager Details:**
   - Verify that the correct manager address is returned for a given pin code.
8. **Test Case for Killing the MLM Contract:**
   - Verify that only the company owner can kill the contract.
   - Check that the remaining balance is correctly transferred to the owner's address.

Due to space constraints, the source code of the implementation of test cases in Solidity programming is provided in the GitHub repository (Potukuchi, 2023). Readers may go through the MLM_test.sol file for the Solidity code for testing the MLM smart contract. These test cases require an already deployed MLM contract instance to test using Truffle or Hardhat and have access to the deployed contract address. It is important to understand that in the parameter DeployedAddresses.MLM passed to the MLM contract, the "Deployed Addresses" need to be replaced with the actual contract address if the testing is performed using a different testing framework or deployment method other than Remix IDE.

3.6 DEPLOYMENT AND MAINTENANCE

SC deployment and maintenance are critical aspects of ensuring the success and longevity of decentralized applications. Through meticulous planning, careful code development, and proactive maintenance, developers can maximize the security, efficiency, and reliability of SCs. Regular audits and code reviews are essential to identify and rectify any vulnerabilities or issues that may arise during the contract's lifecycle (Chakraborty et al., 2018). This helps in preventing potential exploits or attacks and maintaining the contract's integrity. By adhering to best practices, continuously monitoring performance, and adapting to evolving requirements, stakeholders can leverage the full potential of SCs to drive innovation and foster trust in decentralized ecosystems. SCs are instrumental in incorporating the key business logic and serving as the foundation for blockchain decentralized applications (DApps) (Infante, 2019). These applications rely on a combination of technologies to create a full-stack decentralized web application. Consequently, it becomes imperative to distinguish between the development requirements for SCs and those for external systems, which act as interfaces for users, such as web or mobile applications. SCs function in the background, executing predefined conditions and facilitating the recording of critical transactions on the blockchain. The invocation and execution of these contracts are triggered by inputs received from the external system, which is interconnected with storage mediums like databases and file systems. Therefore, it is essential that both SC development and external system development progress simultaneously to construct a robust full-stack decentralized application (Singh, 2023).

3.7 CONCLUSION AND FUTURE WORK

The concept of NM traces its roots back to the dawn of civilization when people relied on each other for information about available products and services. However, there was no reward for sharing such information. Over time, sales teams were organized, emphasizing the importance of a vibrant sales force. This led to the establishment of businesses relying solely on sales representatives, giving rise to the MLM model. Today, the MLM industry is a multi-billion-dollar industry, attracting numerous start-up businesses. One industry that has embraced the MLM model significantly is the CAs industry, particularly in CT. However, the MLM model's reputation has suffered over the years due to various reasons. To stay competitive and provide long-term value, businesses are navigating the need for continuous improvement. To achieve this, blockchain has become a viable platform for creating decentralized applications that are ready for the market. Blockchain technology is a key player in these innovations, enabling collaboration and the formation of consortia. Blockchain tokens and entrepreneurship are also being reshaped.

The choice between public and private blockchains presents distinct advantages. Customized blockchain innovations present a challenge, especially in analyzing, customizing, selling, or purchasing intangible products or services. To explore this, this chapter first presented an introduction to MLM and then discussed the related work on applying blockchain to the MLM scenario. From the literature, it is observed that the design and development of MLM SCs have not been fully explored. To this end, this research work focused on the development of an SC for the MLM use case by utilizing the SDLC and the Solidity programming language. The significance of blockchain technology in revolutionizing the MLM industry is highlighted, with SCs offering a wide range of advantages such as enhanced transparency, automated transactions, and elimination of intermediaries. The process of developing MLM SCs is thoroughly presented, encompassing key considerations such as compensation plan design, hierarchical structure implementation, and integration with external systems. The importance of implementing robust security measures and conducting thorough testing to ensure reliability is emphasized. By harnessing the potential of blockchain technology and SCs, MLM companies can overcome challenges related to fraud, scalability, and trust, ultimately establishing a more transparent and sustainable business model. Ongoing monitoring and maintenance of SCs are crucial to address potential vulnerabilities and accommodate necessary updates. The adoption of MLM SCs holds transformative potential for the industry, fostering fairness, efficiency, and trust among participants. As future work, it is proposed to extend the current research and integrate the developed SC with an external system to create a comprehensive decentralized application.

KEYWORDS

- **blockchain**
- **Ethereum**
- **multi-level marketing**
- **smart contract**
- **solidity**
- **sustainable business model**
- **web application**

REFERENCES

Al-Ashmori, A., Basri, S., Dominic, P. D., Muneer, A., Al-Tashi, Q., & Al-Ashmori, Y. (2021). Blockchain-oriented software development issues: A literature review. *Software Engineering Application in Informatics: Proceedings of 5th Computational Methods in Systems and Software 2021, Vol. 1*, 48–57.

Almeida, S., Albuquerque, A., & Silva, A. (2019). An approach to develop software that uses blockchain. *Software Engineering and Algorithms in Intelligent Systems: Proceedings of 7th Computer Science On-Line Conference 2018, Volume 1 7*, 346–355.

Andriati, R. (2015). Manipulative cooperation politics of MLM companies in Surabaya City. *HUMANIORA: Journal Budaya, Sastra dan Bahasa (Journal of Culture, Literature, and Linguistic)*, *27*, 243–251.

Antal, C., Cioara, T., Anghel, I., Antal, M., & Salomie, I. (2021). Distributed ledger technology review and decentralized applications development guidelines. *Future Internet, 13*, 62.

Bäckman, C., & Hanspal, T. (2022). Participation and losses in multi-level marketing: Evidence from a Federal Trade Commission settlement. *Financial Planning Review, 5*, e1137.

Belchior, R., Vasconcelos, A., Guerreiro, S., & Correia, M. (2021). A survey on blockchain interoperability: Past, present, and future trends. *ACM Computing Surveys (CSUR), 54*, 1–41.

Boumaiza, A., & Sanfilippo, A. (2022). AI for energy: A blockchain-based trading market. *IECON 2022–48th Annual Conference of the IEEE Industrial Electronics Society*, 1–6.

Chakraborty, P., Shahriyar, R., Iqbal, A., & Bosu, A. (2018). Understanding the software development practices of blockchain projects: A survey. *Proceedings of the 12th ACM/IEEE International Symposium on Empirical Software Engineering and Measurement*, 1–10. https://doi.org/10.1145/3239235.324029.

Chauhan, R., & Kaul, V. (2021). Blockchain technology: A panacea for medical distribution ailments. In *Blockchain for Healthcare Systems* (pp. 43–60). CRC Press.

Chauhan, R., & Kaul, V. (2022). Blockchain technology: Unlocking the business model maze for evolving businesses and startups. In *Transformations Through Blockchain Technology: The New Digital Revolution* (pp. 291–309). Springer.

Chetty, O. S. (2019). Network marketing case study under the Consumer Protection Act, 2008: WorldVentures. https://researchspace.ukzn.ac.za/handle/10413/18628 (accessed on 14 December 2024).

Chittoda, J. (2019). *Mastering Blockchain Programming with Solidity: Write Production-Ready Smart Contracts for Ethereum Blockchain with Solidity*. Packt Publishing Ltd.

Deviacita, Q. A. (2022). Use of electronic word of mouth in promoting Jafra multi-level marketing business. *International Journal of Multicultural and Multireligious Understanding, 9*, 436–446.

Droney, D. (2016). Networking health: Multi-level marketing of health products in Ghana. *Anthropology & Medicine, 23*, 1–13.

Epixel MLM Software. (2022). How blockchain will transform the MLM industry. Retrieved from: https://www.epixelmlmsoftware.com/blog/how-blockchain-will-transform-the-mlm-industry (accessed on 14 December 2024).

Ertemel, A. V. (2018). Implications of blockchain technology on marketing. *Journal of International Trade, Logistics and Law, 4*, 35–44.

Far, S. B., Rad, A. I., & Asaar, M. R. (2022). ACR-MLM: A privacy-preserving framework for anonymous and confidential rewarding in blockchain-based multi-level marketing. *Data Science and Management, 5(4), 219–231.*

Faruk, M. J., Subramanian, S., Shahriar, H., Valero, M., Li, X., & Tasnim, M. (2022). Software engineering process and methodology in blockchain-oriented software development: A systematic study. *2022 IEEE/ACIS 20th International Conference on Software Engineering Research, Management and Applications (SERA)*, 120–127.

Federal Trade Commission. (2022). Business guidance concerning multi-level marketing. Retrieved from: https://www.ftc.gov/business-guidance/resources/business-guidance-concerning-multi-level-marketing (accessed on 14 December 2024).

Garg, P., & Khadse, N. (2020). Solidity Essentials. In *Bitcoin and Blockchain* (pp. 115–136). CRC Press.

Gorkhali, A., Li, L., & Shrestha, A. (2020). Blockchain: A literature review. *Journal of Management Analytics, 7*, 321–343.

Grade, G. F. M. (2019). *Multi-Level Marketing and The Impact on Distributors' Loyalty of (Un)Success Factors: An Approach to Measure Loyalty* [Master's dissertation, Iscte–Instituto Universitário de Lisboa]. Repositório do Iscte. http://hdl.handle.net/10071/19671 (accessed on 14 December 2024).

Gregor, B., & Wadlewski, A. A. (2013). Multi-level marketing as a business model. *Marketing Instytucji Naukowych i Badawczych*, 2–19.

Infante, R. (2019). *Building Ethereum Dapps: Decentralized Applications on the Ethereum Blockchain*. Simon and Schuster.

Jain, D., Dash, M. K., Kumar, A., & Luthra, S. (2021). How is blockchain used in marketing: A review and research agenda. *International Journal of Information Management Data Insights, 1*, 100044.

Jain, S., Vamsi, P. R., Agarwal, Y., & Goel, J. (2023). CodeBlockS: Development of collaborative knowledge sharing application with blockchain smart contract. *International Journal of Information Engineering and Electronic Business, 14*, 1.

Javaid, M., Haleem, A., Singh, R. P., Suman, R., & Khan, S. (2022). A review of blockchain technology applications for financial services. *Bench Council Transactions on Benchmarks, Standards and Evaluations*, 2(3), 100073.

Jyoti, J. (2022). Influence of demographics on key drivers to join multi-level marketing company. *International Journal of Economic Perspectives, 16*, 171–189.

Keep, W. W., & Vander Nat, P. J. (2014). Multilevel marketing and pyramid schemes in the United States: A historical analysis. *Journal of Historical Research in Marketing, 6(2), 188–210.*

Kushwaha, R., Patel, A. R., & Vamsi, P. R. (2021). Transparent recruitment model with blockchain technology. *International Journal of Information Security and Software Engineering, 7*, 11–23.

Lee, K. F., & Loi, K. Y. (2016). Towards satisfying distributors in multilevel marketing companies. *International Journal of Management and Applied Research, 3*, 48–64.

Legara, E. F., Monterola, C., Juanico, D. E., Litong-Palima, M., & Saloma, C. (2008). Earning potential in multilevel marketing enterprises. *Physica A: Statistical Mechanics and its Applications, 387*, 4889–4895.

Liu, J., Xu, Z., Li, R., Zhao, H., Jiang, H., Yao, J., & Chen, S. (2021). Applying blockchain for primary financial market: A survey. *IET Blockchain, 1*, 65–81.

Marchesi, M., Marchesi, L., & Tonelli, R. (2018). An agile software engineering method to design blockchain applications. *Proceedings of the 14th Central and Eastern European Software Engineering Conference Russia*, 1–8.

Masama, B. (2023). The sanitization of multi-level marketing and its effectiveness in African crypto trading. Available at: https://ssrn.com/abstract=4407913 (accessed on 14 December 2024).

Maticz. (2022). *Smart Contract MLM Software | MLM Platform with Smart Contract.* Retrieved from: https://maticz.com/smart-contract-mlm-software (accessed on 14 December 2024).

Moghrabi, I. A., Bhat, S. A., Szczuko, P., AlKhaled, R. A., & Dar, M. A. (2023). Digital transformation and its influence on sustainable manufacturing and business practices. *Sustainability, 15*, 3010.

Muncy, J. A. (2004). Ethical issues in multi-level marketing: Is it a legitimate business or just another pyramid scheme? *Marketing Education Review, 14*, 47–53.

Nakamoto, S. (2008). Bitcoin: A peer-to-peer electronic cash system. *Decentralized Business Review*, *21260*, 1–11.

Peres, R., Schreier, M., Schweidel, D. A., & Sorescu, A. (2022). *Blockchain meets marketing: Opportunities, threats, and avenues for future research. International Journal of Research in Marketing, 40*(1), 1–11.

Pop, C., Cioara, T., Anghel, I., Antal, M., & Salomie, I. (2020). Blockchain-based decentralized applications: Technology review and development guidelines. *IEEE Open Journal of the Computer Society, (99), 1–12.*

Potukuchi, R. V. (2023). GitHub–prvonline/mlm: Solidity smart contract for multi-level marketing. Retrieved from: https://github.com/prvonline/mlm (accessed on 14 December 2024).

Purcaru, I. M., Urdea, A. M., Constantin, C. P., & Brătucu, G. (2022). Building long-term business sustainability: The influence of experiential marketing on sales representatives' loyalty to multi-level marketing systems. *Sustainability, 14*, 9507.

Ravindran, N. (2023). How can blockchain transform the MLM industry. Retrieved from: https://www.antiersolutions.com/how-can-blockchain-transform-the-mlm-industry/ (accessed on 14 December 2024).

Ravindran, U., & Vamsi, P. R. (2021). A secure blockchain-based finance application. *2021 Thirteenth International Conference on Contemporary Computing (IC3–2021)*, 99–109.

Ravindran, U., Bhardwaj, P., & Vamsi, P. R. (2021). Blockchain design for securing supply chain management in coffee retailer network. *International Journal of Scientific Research in Science and Technology.* Doi: 10.32628/CSEIT2174119.

Remix. (2023). Remix Online IDE. *Remix Online IDE.* Retrieved from: https://remix.ethereum.org/ (accessed on 14 December 2024).

Shrestha, A. K., Vassileva, J., & Deters, R. (2020). A blockchain platform for user data sharing ensuring user control and incentives. *Frontiers in Blockchain, 3*, 497985.

Singh, H. (2023). DApps: Decentralized applications for blockchains. In *Distributed Computing to Blockchain* (pp. 87–104). Elsevier.

Thontirawong, P., & Chinchanachokchai, S. (2021). Teaching artificial intelligence and machine learning in marketing. *Marketing Education Review, 31*, 58–63.

Wüst, K., & Gervais, A. (2018). Do you need a blockchain? *Crypto Valley Conference on Blockchain Technology (CVCBT)*. IEEE.

Yutia, S. N., Fathiana, R. Z., & Fajriyah, S. Z. (2022). Blockchain-based smart contract for decentralized marketplace. *JOIV: International Journal on Informatics Visualization, 6*, 475–480.

Zheng, Z., Xie, S., Dai, H. N., Chen, W., Chen, X., Weng, J., & Imran, M. (2020). An overview on smart contracts: Challenges, advances, and platforms. *Future Generation Computer Systems, 105*, 475–491.

Zou, W., Lo, D., Kochhar, P. S., Le, X. B. D., Xia, X., Feng, Y., & Xu, B. (2019). Smart contract development: Challenges and opportunities. *IEEE Transactions on Software Engineering, 47*, 2084–2106.

CHAPTER 4

The Future of Fashion in the Age of the Metaverse

DEEP SAGAR VERMA,[1,*] SATYA BANERJEE,[2] and PRATEEK JANGID[3]

[1]*Department of Fashion Management Studies, NIFT Panchkula, Haryana, India*

[2]*London College of Fashion, University of the Arts London, United Kingdom*

[3]*Zudio Tata-Trent, Mumbai, Maharashtra, India*

*Corresponding author

ABSTRACT

The chapter explores the transformative potential of the Metaverse in reshaping the fashion industry, highlighting its evolution since gaining global attention with Facebook's rebranding as Meta. The Metaverse integrates advanced technologies like VR, AR, Blockchain, and NFTs, enabling immersive virtual environments where users can engage through customizable avatars. With major industries and fashion brands entering the Metaverse ecosystem, it has opened avenues for digital fashion, interactive events, and personalized experiences. Notable developments, such as the first-ever Metaverse Fashion Week and the rise of digital fashion platforms like DressX, underscore its impact. As the industry recovers from pandemic-induced disruptions, the Metaverse promises to redefine the future of fashion, offering a blend of creativity, interactivity, and economic potential.

Blockchain Technology: Unlocking the Value of 3Ts in the Business World.
Richa Chauhan & Vidhi Kaul (Eds.)

DOI: 10.1201/9781998511914-4

4.1 INTRODUCTION

In October 2021, the global spotlight turned towards the evolving Metaverse, capturing widespread interest as the tech giant Facebook underwent a significant rebranding, adopting the name Meta (2021). This transformation resonated with audiences worldwide, highlighting the growing significance of the concept. Nevertheless, the term "Metaverse" was initially introduced in 1992 by Neal Stephenson in his science fiction novel "Snow Crash." In this literary work, individuals have the ability to forge digital avatars and engage with one another within an expansive virtual reality realm (Joshua, 2017).

It is visualized to have a never-ending immersive virtual world, where anyone can be a part of it with customized and stylized avatars navigating freely from one space to another with ease (Ball, 2022). While the Metaverse is said to be realized through existing hardware like mobile phones and computers, its true impact can be fully experienced when used with virtual reality glasses, providing a complete 360° view of the virtual world and enhancing the sense of immersion (Dwivedi et al., 2022). Users have been immersed in virtual worlds for an extended period through the utilization of virtual reality headsets like the Oculus series (Meta Quest), HTC Vive, and PlayStation VR. The recent unveiling of Apple Vision Pro further strengthens the concept of the Metaverse with a promise to bring the digital and real world closer and interoperable (Benjamin, 2023).

The expansive concept of the Metaverse unfolds as a computer-generated simulated realm, seamlessly supplanting the physical world through the sophisticated fusion of virtual reality (VR), augmented reality (AR), and mixed reality (MR) technologies. This immersive digital space offers a multifaceted experience, transcending conventional boundaries and reshaping the landscape of interactive virtual environments (Maan et al., 2018), under the umbrella term of Web 3.0 (Murray et al., 2023). Associated with the core technologies that define the landscape of Web 3.0 such as blockchain, cryptocurrency (Huynh-The et al., 2023), non-fungible tokens (NFTs; Far et al., 2022), artificial intelligence (Huynh-The et al., 2023), spatial computing (Greenwold, 2003), haptics (Wiederhold, 2023), and interoperability (World Economic Forum, 2023) are powering the Metaverse.

By the year 2030, the Metaverse is anticipated to generate a business value of $5 trillion, with both consumers and brands actively participating in its ecosystem, as reported by McKinsey and Company in 2022. Various industries, including high tech, media and entertainment, tourism, healthcare, public services, financial services, telecommunications, construction, energy and resources, transport and logistics, automotive machinery, professional services, and apparel, footwear, and luxury (AF&L), as well as consumer and retail, have already embraced the Metaverse, allocating a portion of their digital budgets for future investments (McKinsey and Company, 2022). Additionally, an industry report by Masoni and Chatterjee (2021) projects that by 2030, the digital market for fashion and luxury brands is forecasted to experience substantial growth, potentially reaching a valuation of $50 billion.

In March 2022, the first-ever Metaverse fashion week was conducted in Decentraland, a blockchain-based gaming platform grabbing major attention from the fashion industry (McDowell, 2022). It opened avenues for further exploration of digital fashion and its use cases in real life, as digital avatars could wear and style themselves with ease. Digital fashion store DressX, which recently received funding of US $15 million for developing digital fashion for the Metaverse, has more than 3500 digital clothing and accessories items at present, which customers can wear through digital avatars in virtual worlds like Decentraland and Ready Player Me (Takahashi, 2023).

A digital avatar is a form of self-representation in computer-mediated communication, particularly in collaborative virtual environments (CVEs) (Bailenson and Beall, 2006) that revolves around three key concepts: representation, presence, and immersion (Mitchell et al., 2009). The representation of an avatar includes its appearance, the surrounding environment, and the interactions between avatars and the environment (Mitchell et al., 2009). The presence of an avatar pertains to the interactive immersive visualization in a three-dimensional virtual environment (Steuer, 1992). The immersion refers to the state where an individual perceives themselves as actively interacting within the virtual environment, surpassing awareness of the environment around them (Guadagno et al., 2007). Avatars, being one of the major pillars of the Metaverse, are designed to be highly customized, attractive, and stylized with creative freedom and expressions following fashion trends and norms (Ante et al., 2023). As people care about their presentation in real life, virtual avatars also play a crucial part in influencing the online persona through virtual fashion and desired body type and size (Ling et al., 2011).

Fashion, being one of the largest segments of businesses, has already entered into the ecosystem of the Metaverse in late 2021 and 2022. Fashion designers, retailers, and brands are anticipating huge growth in this latest technology, resulting in their fast entry and development in the space. Recovering from the huge losses because of the global pandemic, COVID-19, and subsequent changes in the buying behavior of customers, the Metaverse may prove to be a sigh of relief for the fashion industry in the future. It is conceptualized to give an immersive and real-time experience to customers with a high level of interactivity inside the Metaverse. In the changing scenario of the supply-demand gap, the Metaverse is poised to reshape the historical, current, and forthcoming dynamics of the fashion industry.

4.2 PRELIMINARIES OF THE METAVERSE TECHNOLOGY

4.2.1 WEB 3.0

The introduction of the internet not only made communication and information easily available but also ushered in a paradigm shift in our way of life. The development and maturation of the internet can be generally delineated into three distinct phases: Web 1.0, Web 2.0, and Web 3.0 (Murray et al., 2023). Termed the semantic web, Web 3.0 signifies the third stage in this transformative journey, indicating a deeper level of meaning and interconnectivity in the online experience. It's not ready yet, but it's under development, and when it's ready, it's expected to make big changes in many industries, including fashion. It's not going to replace Web 2.0 entirely; it's more like building on top of it. The start of Web 3.0 is connected with the introduction of blockchain technology (Huynh-The et al., 2023) and the creation of the first peer-to-peer decentralized cryptocurrency "Bitcoin" (Nakamoto, 2009). Web 3.0 is envisioned to function in a decentralized ecosystem without any centralized controlling authority (Murray et al., 2023). Decentralization in Web 3.0 further supports Decentralized Autonomous Organizations (DAOs; Santana and Albareda, 2022), Decentralized Finance (DeFi; Chen and Bellavitis, 2020), the creator economy (Petit, 2022), and Privacy and Security of Data (Chen et al., 2022).

4.2.2 BLOCKCHAIN AND CRYPTOCURRENCY

Blockchain acts like a super-secure digital ledger system. It is a digital notebook where every transaction gets written down in a block. Each block has its own special code, making it unique. Once a block is added, its code is stored in the next one, creating a connected chain of blocks (Huynh-The et al., 2023). This special coding ensures that no one can mess with the transactions, making the records super secure and unchangeable.

To make it work, blockchain relies on a bunch of computers scattered around the world. They all have to agree on the transactions happening, making the data record permanent and unchangeable. Some key parts of blockchain include a shared digital record book, records that can't be messed with, smart contracts (which are like self-executing agreements), and a way for all these computers to agree on what's happening (Shrimali and Patel, 2022).

Now, let's talk cryptocurrency. This is like digital money, but it's not controlled by any one person or government. Instead, it operates on a decentralized system, often using blockchain technology (Jalan and Barbesino, 2023). Cryptocurrencies are split into two types: coins and tokens. Coins are like the main currency of a blockchain, while tokens can serve different purposes. In the exciting world of the Metaverse, projects like Apecoin, the Sandbox, Axie Infinity, Decentraland, Enjin Coin, and Mask Network are making waves, all built on the Ethereum blockchain (CoinMarketCap, 2023).

4.2.3 *NON-FUNGIBLE TOKENS (NFTS)*

An NFT, or Non-Fungible Token, operates akin to a digital certificate, confirming one's ownership of a distinctive digital asset, such as unique art or collectibles, whose ownership validation is secured through blockchain technology (Collins, 2022). NFTs serve as digital certificates of authenticity, offering unequivocal details about ownership, previous ownership, and authorship of digital items, aiding in the identification of originals among duplicates. These tokens can be associated with various digital items like photos, videos, and audio files, and are utilized to transform these digital objects into valuable assets across diverse domains (Nadini et al., 2021). A Non-Fungible Token product can play a pivotal role in boosting brand recognition and fostering community engagement, especially when it involves existing brands leveraging their unique and identical digital assets (Colicev, 2023).

4.2.4 GENERATIVE ARTIFICIAL INTELLIGENCE (AI)

The advent of ChatGPT by OpenAI has rapidly increased the prominence of generative AI, prompting the emergence of numerous AI tools in its wake. Generative AI, using advanced deep learning models, crafts content that closely resembles human creations, such as images and text. It adeptly reacts to an extensive array of intricate prompts, spanning diverse languages, instructions, and inquiries with versatility and precision (Lim et al., 2023). In the evolving landscape of generative AI within the metaverse, the future directions will primarily revolve around three key areas. Primary efforts are directed towards advancing technical research to enhance the customization capabilities of generative AI, ultimately resulting in the generation of higher-quality content. A secondary focus involves exploring the applications of generated content across diverse fields, with a particular emphasis on its role in metaverse content production. Lastly, there is a deliberate and collaborative initiative to ensure the smooth integration of generative AI with other technologies, thereby magnifying its overall influence and expanding its potential applications (Lv, 2023).

4.2.5 EXTENDED REALITY (XR) AND DEVICES ASSOCIATED

The term X may be replaced with V (Virtual), A (Augmented), and M (Mixed) Realities. Meta (previously Facebook) has developed Oculus (Meta Quest) to provide an experience in the virtual universe. Companies like Sony, Apple, Ray-Ban, Microsoft, Samsung, and others are developing VR glasses to capture the market.

According to a Bloomberg report, Apple's VR headset "Reality Pro" will be launched in spring 2023 and shipped in fall 2023 (Goswami, 2023). Devices in current times vary in terms of hardware, resolution, and field of view, which determine the visibility and immersive experience for a user. Since headsets are limited in numbers in the current markets, prices are on the higher side; however, the adaptation rate is increasing (Figure 4.1).

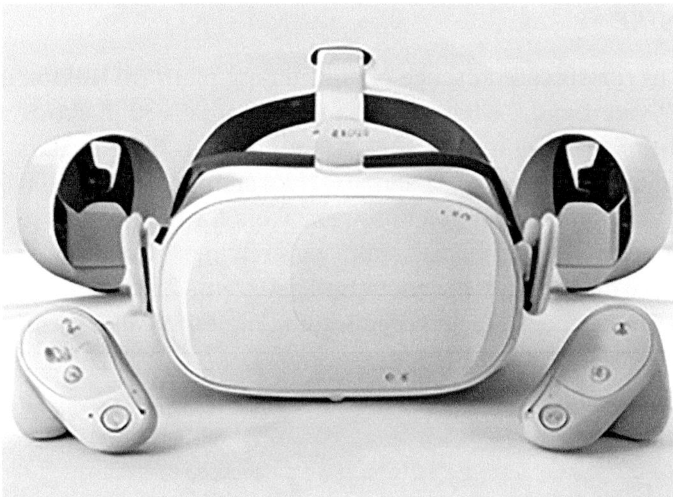

FIGURE 4.1 Created using generative AI.

Source: https://midjourney.com on 28/02/2023 Command/imagine prompt 'individual oculus picture with controllers HD, white.'

4.3 FASHION BRANDS ENTERING THE METAVERSE ECOSYSTEM

A rising array of fashion brands is expanding their presence by entering the world of gaming and the Metaverse. These brands are leveraging the popularity and reach of gaming platforms and virtual reality environments to connect with new audiences and offer unique experiences. Fashion and lifestyle, being one of the biggest industries, is witnessing tremendous growth and development through the Metaverse. The changing face of fashion is going to create a major influence on procurement, supply, usage, and disposal of products in the coming times. Metaverse platforms are actively collaborating with fashion brands. The fusion of fashion and technology is adding a new dimension to the overall brand and shopping experience. Post-COVID-19, there has been a drastic shift in consumer behavior, with individuals believing in digital fashion with a high level of customization and self-identity creation. Global brands immediately identified this as an opportunity and started co-creating with major gaming and metaverse companies. It is a perfect example of community-based economies advocating decentralization and co-creation. Some of the major brand partnerships are detailed in the subsections.

4.3.1 NIKE

Nike has a widespread presence on the gaming platform Roblox under the name of "Nike Land," where players can get involved in sports activities with real-time interaction. In this, users dress up their avatars in Nike-branded sneakers and apparel and use Nike sports gear. Nike is also planning to let customers design and sell sneakers for the Metaverse. In 2019, Nike ventured into the virtual realm by creating "Downtown Drop," an immersive world within Fortnite (Ball, 2022, p. 264). To strengthen its Metaverse presence, Nike acquired leading digital sneaker creator RTFKT Studios in December 2021 (Figure 4.2).

FIGURE 4.2 Nike Land on the gaming platform, "Roblox" (accessed and screenshot on 26/02/2023).

Source: https://www.roblox.com.

4.3.2 ADIDAS ORIGINALS

In December 2021, Adidas Originals launched the "Into the Metaverse" initiative with 30,000 Non-Fungible Tokens, which were minted within hours, and the brand sold more than $22 million worth of NFTs (The Verge, 2021).

Adidas has acquired 144 plots of virtual land inside The Sandbox worth 400 ETH (approx. $1.7 million) (Takahashi, 2021).

In November 2022, Adidas launched its first-ever digital apparel collection, Adidas Virtual Gear. The owners of Adidas Virtual Gear can style their NFT profile pictures (PFPs) (Adidas, 2022; Figures 4.3 and 4.4).

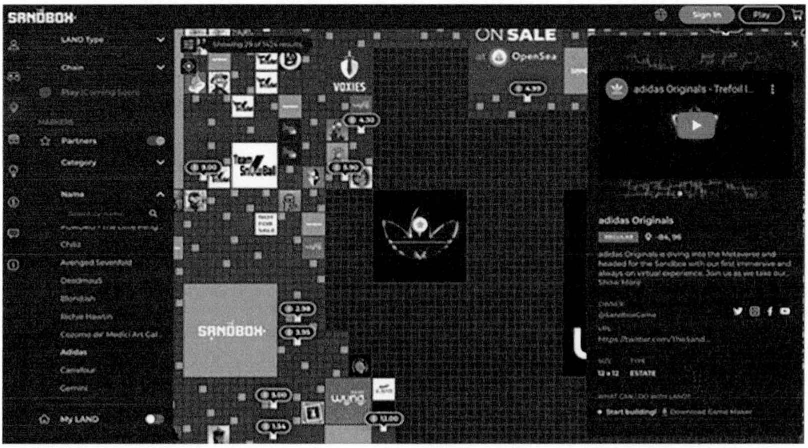

FIGURE 4.3 Adidas Originals on the gaming platform, "The Sandbox" (accessed and screenshot on 25/02/2023).

Source: https://www.sandbox.game/en/map/?lat=102&zoom=4&lng=78&partner=13&y=96&x=84.

FIGURE 4.4 Virtual gears for new realities (accessed and screenshot taken on 27/02/2023).

Source: https://news.adidas.com/originals/virtual-gear-for-new-realities-adidas-originals-launches-inaugural-nft-wearables-collection/s/b9c13f73-b7b4-412f-9e19-f8a4d9781621.

4.3.3 PUMA

PUMA, the renowned sports company, has unveiled its groundbreaking metaverse website experience, Black Station, marking a significant moment during New York Fashion Week as part of the "FUTROGRADE" show. This immersive platform not only showcases exclusive NFTs but also presents limited edition redeemable physical sneakers, providing a unique fusion of digital and tangible experiences for enthusiasts (Puma, 2022; Figure 4.5).

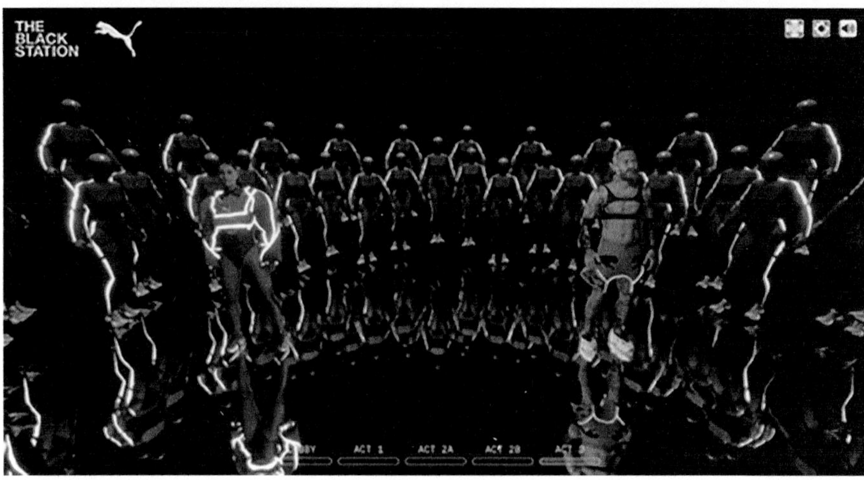

FIGURE 4.5 Puma on The Black Station platform (accessed and screenshot taken on 27/02/2023).

Source: https://blackstation.puma.com/.

4.3.4 BURBERRY

At the forefront of luxury fashion's presence in the Metaverse, Burberry, the esteemed British luxury brand, made a significant impact with the launch of its initial game, B Bounce, in October 2019, followed by the extension Ratberry in January 2020, both meticulously crafted in-house by Burberry's digital teams. The collaboration between Burberry and Blankos Block Party, a multiplayer game by Mythical Games, achieved notable success with their recent NFT drop. The collective release of 2250 NFTs sold out in an astonishing 30 seconds, highlighting their widespread popularity (Yordanova, 2021). In an expansive move, Burberry has disclosed a partnership with Mythical Games to unveil a fresh NFT collection in Blankos Block Party. Going beyond NFTs, Burberry is creating a distinctive virtual social space within the game, providing a unique gathering experience for its community members (Burberry, 2022; Figure 4.6).

FIGURE 4.6 Burberry Blanko NFTs (accessed and screenshot taken on 27/02/2023).
Source: https://www.burberryplc.com.

4.3.5 GUCCI

Gucci, the esteemed Italian fashion and leather goods brand, is actively constructing a digital realm to offer a captivating metaverse experience. Pioneering in the realm of virtual domains, Gucci initiated the creation of Gucci Garden on Roblox in 2020. Embracing innovation, Gucci has ventured into an experimental realm within The Sandbox, carving out a unique space for NFTs, vintage items, and emerging creators. This strategic move positions Gucci as a trailblazer, being the first major luxury brand to establish its distinctive presence within the metaverse platform (Schulz, 2022). In a creative collaboration, Gucci has also partnered with the multimedia company Superplastic to introduce unique NFTs known as Supergucci (Figure 4.7).

FIGURE 4.7 Gucci vault land (accessed and screenshot taken on 27/02/2023).

Source: https://register.sandbox.game/gucci-vault.

4.4 FASHION EVENTS IN THE METAVERSE

An integral aspect within the Metaverse is its ability to deliver a captivating and interactive live experience during events, fostering real-time inclusive participation. In November 2021, the celebrated pop artist Justin Bieber graced the virtual stage of the Metaverse through the WaveXR platform. A groundbreaking occurrence, the Metaverse Fashion Week (MVFW), unfolded on Decentraland from March 24–27, 2022. Representing an unprecedented four-day event, this widely discussed happening marked a pivotal moment for the retail sector, seamlessly merging the domains of fashion and technology in anticipation of the metaverse's future. Decentraland played a crucial role as the host for MVFW (EmperiaVR, 2022). Building on the success of the inaugural event, the Metaverse Fashion Week of 2023 took place from March 28–31, 2023.

This inclusive event permitted all participants to acquire NFTs of prominent fashion brands using Decentraland's native cryptocurrency, MANA. Notable brands partaking in the event included Tommy Hilfiger, Dolce and Gabbana, Bulova, Estée Lauder, and Selfridges. Through various social media channels, Decentraland actively encouraged attendees to identify Sophia, a famous robot, using the designated hashtag #findingsophiatherobot. Additionally, preparations are underway for the second iteration of MVFW slated for March 2023 in Decentraland (Figure 4.8).

FIGURE 4.8 Decentraland's first Metaverse Fashion Week (accessed and screenshot taken on 27/02/2023).

Source: https://www.vogue.com.

In November 2022, the Metaverse Fashion Summit was orchestrated by the metaverse developer firm Mext and a DAO known as the "Metaverse Fashion Council." This summit comprised key activities such as live sessions by industry experts, an exhibition floor, and a dedicated press room. The surge in popularity has led the majority of Metaverse companies to create dedicated spaces for hosting fashion shows, events, festivals, and exhibitions. The accessibility of attending global fashion events for free or at nominal rates is poised to attract a significant clientele. Furthermore, fashion is anticipated to collaborate more extensively with Metaverse entertainment, encompassing concerts, theme parks, and virtual movies (Figure 4.9).

FIGURE 4.9 Virtual Fashion Show Space in Engage XR (accessed and screenshot taken on 25/02/2023).

Source: https://engagevr.io/.

4.5 CUSTOMIZATION THROUGH DIGITAL FASHION IN THE METAVERSE

Digital fashion is at the core of the Fashion Metaverse. It advocates a circular economy, sustainability, and creative freedom. With the help of digital fashion, consumers can customize their looks and experience them through AR/VR. Before the actual purchase, customers may have unlimited options to try and be satisfied with the products as per their requirements. It will lower the actual production and consumption pattern, and fashion brands would only produce according to the requirements of the customers, thus advocating sustainability. Some of the leading companies in digital fashion are detailed in the subsections.

4.5.1 DRESSX

DRESSX, a global multi-brand retailer, specializes in digital fashion, curating collections from contemporary brands and designers. The landscape of digital fashion unfolds through the creation of virtual clothing using advanced 3D software, producing lifelike garments that can be visualized and simulated to closely resemble tangible apparel (Ellen MacArthur Foundation, 2021). Users have the option to link their digital wallets for purchasing specific designer wear or branded digital garments, which can also be ordered to obtain the corresponding physical product (Figure 4.10).

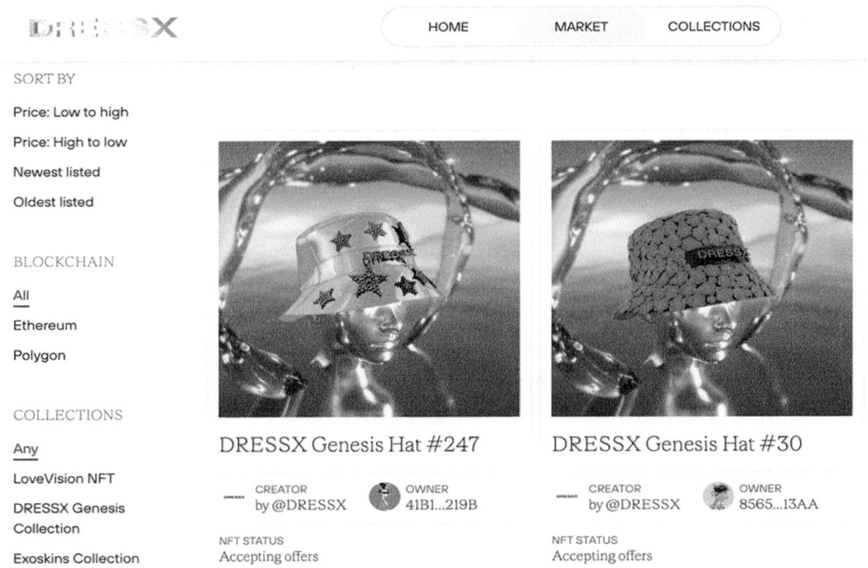

FIGURE 4.10 AR dressing in DRESSX (accessed and screenshot taken on 27/02/2023).
Source: https://nft.dressx.com/marketplace/nft/.

4.5.2 THE FABRICANT

It is a digital fashion studio exclusively designed for the Metaverse. Launched in November 2021, the platform aims to democratize virtual fashion design, making it accessible to individuals worldwide. It allows virtual outfits to be worn across diverse digital environments within the metaverse – a burgeoning digital realm offering users a parallel life experience (Crook, 2021). The Fabricant empowers users to collaboratively create, customize, and express their unique identities by decentralizing fashion through digital collections. NFT holders have the ability to craft, exchange, and don digital fashion collections, establishing a wardrobe within the Metaverse (Figure 4.11).

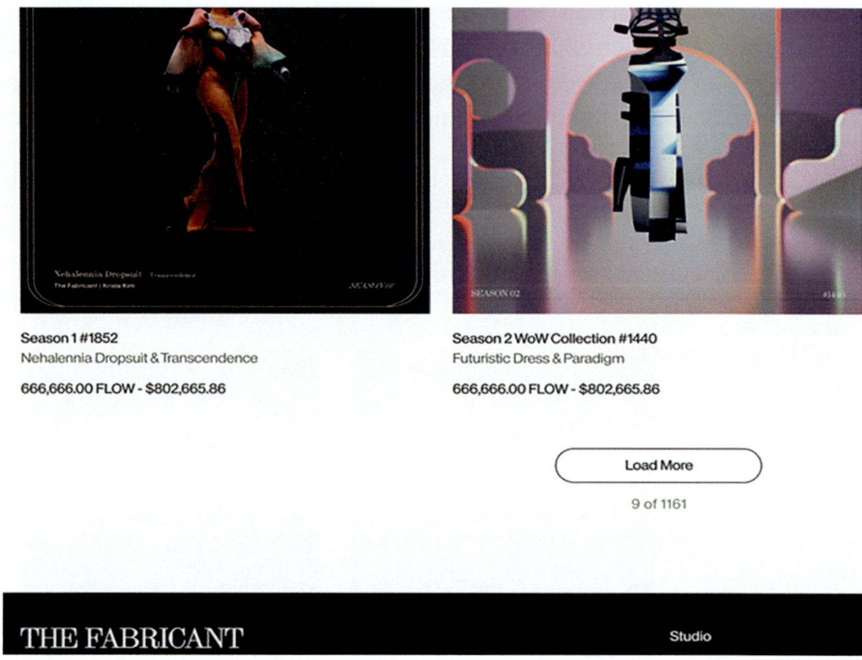

FIGURE 4.11 Marketplace of the Fabricant (accessed and screenshot taken on 27/02/2023).

Source: https://www.thefabricant.com/marketplace.

4.6 CONCLUSIONS

In conclusion, the early adopters of the Fashion Metaverse are positioned to unlock significant sales and profits within the industry, presenting an opportunity to redefine conventional business models. The Metaverse offers solutions to intricate challenges in the fashion sector, ranging from supply chain complexities to overproduction, wastage, and the costs associated with retail operations and discounts. Co-creation and interoperability in the Fashion Metaverse open up additional business channels, synergizing with existing brick-and-mortar and e-commerce models. The Metaverse, as a less intimidating setting, particularly resonates with aspirational consumers of premium and luxury products, especially the tech-savvy Gen Z demographic, creating a substantial opportunity for the fashion business. However, as the Metaverse heralds a new social order, potential challenges must be acknowledged. The decentralized nature of the Metaverse, while liberating, may introduce security and safety concerns. The anonymity afforded by avatars could potentially lead to an increase in crime rates, necessitating careful consideration of safety measures within these digital realms.

1. Positive Aspects:

- Creative opportunities for designers and creators in the Metaverse;
- Enhanced customer engagement through immersive virtual experiences;
- Potential for sustainable practices in the digital fashion realm; and
- Access to a global clientele and market expansion.

2. Key Considerations:

- Imbalance between real and virtual products;
- Potential job displacement due to Generative AI and NFT collections;
- Need for technical expertise for Metaverse implementation; and
- Accessibility challenges with VR headsets.

In embracing the Metaverse, it is crucial to recognize both the challenges and positive aspects. The transitional nature of the technology, marked by uncertainties about future use and acceptance, requires adaptability. The entry of luxury fashion brands into the Metaverse generates excitement among industry stakeholders, urging collaborative efforts to innovate and shape the future of fashion within this dynamic digital landscape.

4.7 THE FUTURE AHEAD

The convergence of fashion and the metaverse heralds an exhilarating future, according to research conducted by various consultancy firms, academicians, industry experts, and thought leaders. As seen in the dynamic landscape of Metaverse Fashion Week and the collaborative ventures of renowned brands, the authors envision a transformative journey where digital and physical realms seamlessly intertwine. Embracing the positive side, this fusion not only revolutionizes consumer experiences but also propels the fashion industry into an era of boundless creativity and innovation.

In the authors' optimistic perspective, the metaverse empowers individuals to transcend traditional boundaries, offering a playground for self-expression and immersive engagement. The positive industry-oriented outlook reveals a paradigm shift where NFTs, virtual fashion shows, and blockchain technology redefine the value chain. Brands foraying into the metaverse exemplify a forward-thinking approach that not only captivates consumers but also underscores the immense economic potential within this burgeoning space. As the metaverse continues to unfold, the authors envision a future where fashion transcends the physical, creating an inclusive and dynamic ecosystem that shapes the very essence of our digital existence.

KEYWORDS

- **augmented reality**
- **blockchain**
- **digital existence**
- **digital fashion realm**
- **economic potential**
- **fashion industry**
- **metaverse**
- **metaverse fashion week**
- **mixed reality**
- **virtual reality**

REFERENCES

Adidas. (2022). *Adidas PFP Styling Tool*. Available: https://atelier.adidas.com/ (accessed on 14 December 2024).

Altuzarra, J., Jonathan Simkhai, Kim Shui, & The Blonds. (2022). Afterpay drops designer NFT 'Keys' to New York Fashion Week. *Benzinga*. Available: https://www.benzinga.com/pressreleases/22/08/b28600802/afterpay-drops-designer-nft-keys-to-new-york-fashion-week (accessed on 14 December 2024).

Ante, L. (2023). *Avatars: Shaping Digital Identity in The Metaverse*. Blockchain Research Lab. https://www.blockchainresearchlab.org/wp-content/uploads/2020/05/Avatars-Shaping-Digital-Identity-in-the-Metaverse-Report-March-2023-Blockchain-Research-Lab.pdf (accessed on 14 December 2024).

Ante, L., Fiedler, I., & Steinmetz, F. (2023). *Avatars: Shaping Digital Identity in The Metaverse*. Blockchain Research Lab. Retrieved from: https://www.blockchainresearchlab.org/wp-content/uploads/2020/05/Avatars-Shaping-Digital-Identity-in-the-Metaverse-Report-March-2023-Blockchain-Research-Lab.pdf (accessed on 14 December 2024).

Ball, M. (2022). *The Metaverse: And How it Will Revolutionize Everything* (1st ed.). Liveright Publishing.

Barclays. (2021). *Decrypting Crypto*. Available: https://rise.barclays/news/reports/decrypting-crypto/ (accessed on 14 December 2024).

Benjamin, A. (2023). YouTube. Retrieved from: https://www.apple.com/apple-vision-pro/?&&cid=wwa-us-kwgo-VisionPro--slid-----Announce-&mtid=20925r2y69020&aosid=p238&mnid=44sHhjLc-dc_mtid_20925r2y69020_pcrid_%7Badid%7D_pgrid_151440461475_ (accessed on 14 December 2024).

Berkley. (2022). Beeple sold an NFT for $69 million. *The Verge*. Available: https://www.theverge.com/2021/3/11/22325054/beeple-christies-nft-sale-cost-everydays-69-million (accessed on 14 December 2024).

Breia, R. (2022). 15 fashion brands leveraging the metaverse. *Sensorium*. Retrieved from: https://sensoriumxr.com/articles/fashion-brands-in-the-metaverse (accessed on 14 December 2024).

Burberry. (2021). Burberry drops NFT collection in Mythical Games' Blankos Block Party. *Burberry*. Available: https://www.burberryplc.com/en/news/brand/2021/Blankos.html.

Burberry. (2022). Burberry X Blankos Block Party: New NFT collection and social space. *Burberry*. Available: https://www.burberryplc.com/en/news/brand/2022/burberry-x-blankos-block-party--new-nft-collection-and-social-sp.html (accessed on 14 December 2024).

Citi Global Insights. (2022). Life, the metaverse, and everything. Available: https://icg.citi.com/icghome/what-we-think/global-insights/insights/life-the-metaverse-and-everything (accessed on 14 December 2024).

Citigroup Inc. (2022). Metaverse and money. Available: https://www.citifirst.com.hk/home/upload/citi_research/AZRC7.pdf (accessed on 14 December 2024).

Clark, M. (2021). Beeple sold an NFT for $69 million. *The Verge*. Available: https://www.theverge.com/2021/3/11/22325054/beeple-christies-nft-sale-cost-everydays-69-million (accessed on 14 December 2024).

Collins, C. (2022). *Enter the Metaverse–The Beginner's Guide to Virtual Worlds: NFT Games, Play-to-Earn, GameFi, and Blockchain Entertainment Such as Axie Infinity, Decentraland, The Sandbox, Meta, Gala, Gods Unchained, Bloktopia, and More!* Publishing Forte.

Crook, L. (2021). The Fabricant Studio "allows anyone to become a digital fashion designer." *Dezeen*. Available: https://www.dezeen.com/2021/11/03/the-fabricant-studio-digital-fashion-dezeen-15/ (accessed on 14 December 2024).

Deloitte. (2016). Blockchain enigma. Paradox. Opportunity. Available: https://www2.deloitte.com/content/dam/Deloitte/uk/Documents/Innovation/deloitte-uk-blockchain-full-report.pdf (accessed on 14 December 2024).

Dwivedi, Y. K. K., Hughes, L., Baabdullah, A. M., Ribeiro-Navarrete, S., Giannakis, M., Al-Debei, M. M., Dennehy, D., Metri, B., Buhalis, D., Cheung, C. M. K., Conboy, K., Doyle, R., Dubey, R., Dutot, V., Felix, R., Goyal, D. P., Gustafsson, A., Hinsch,

C., Jebabli, I., & Wamba, S. F. (2022). Metaverse beyond the hype: Multidisciplinary perspectives on emerging challenges, opportunities, and agenda for research, practice, and policy. *International Journal of Information Management, 66*, 102542. https://doi.org/10.1016/j.ijinfomgt.2022.102542.

Ellen MacArthur Foundation. (2021). A digital-only, multi-brand fashion platform: DRESSX. Available: https://ellenmacarthurfoundation.org/circular-examples/dressx (accessed on 14 December 2024).

EmperiaVR. (2022). Highlights from the Metaverse Fashion Week 2022. Available: https://emperiavr.com/2022/03/30/highlights-from-the-metaverse-fashion-week-2022/ (accessed on 14 December 2024).

Far, S. B., Bamakan, S. M. H., Qu, Q., & Jiang, Q. (2022). A review of non-fungible tokens applications in the real world and metaverse. *Procedia Computer Science, 214*, 755–762. https://doi.org/10.1016/j.procs.2022.11.238.

Golden, J. (2021). Nike teams up with Roblox to create a virtual world called Nikeland. *CNBC*. Retrieved from: https://www.cnbc.com/2021/11/18/nike-teams-up-with-roblox-to-create-a-virtual-world-called-nikeland-.html (accessed on 14 December 2024).

Goswami, R. (2023). Apple reportedly set to announce the 'Reality Pro' VR headset this spring and will ship it in the fall. *CNBC*. Retrieved from: https://www.cnbc.com/2023/01/09/apple-vr-headset-to-launch-this-spring-and-ship-in-the-fall-report.html (accessed on 14 December 2024).

Griffin, C. (2022). *Metaverse: The Visionary Guide for Beginners to Discover and Invest in Virtual Lands, Blockchain Gaming, Digital Art of NFTs, and the Fascinating Technologies of VR, AR, and AI*. Top Notch International.

GSMA. (2022). Exploring the metaverse and the digital future. Available: https://www.gsma.com/asia-pacific/wp-content/uploads/2022/02/270222-Exploring-the-metaverse-and-the-digital-future.pdf (accessed on 14 December 2024).

Guadagno, R., Blascovich, J., Bailenson, J. N., & McCall, C. (2007). Virtual humans and persuasion: The effects of agency and behavioral realism. *Media Psychology, 10*, 1–22. https://doi.org/10.1080/15213260701300865.

Gucci Vault. (2022). Gucci Vault opens a metaverse world in The Sandbox with games and vintage fashion. *Vogue Business*. Available: https://www.voguebusiness.com/technology/gucci-vault-opens-metaverse-world-in-the-sandbox-with-games-and-vintage-fashion (accessed on 14 December 2024).

Hernandez, O. (2022). Tiffany & Co. turns CryptoPunks into a $12.5M luxury jewelry sale. *Blockworks*. Available: https://blockworks.co/news/tiffany-co-turns-cryptopunks-into-12-5m-luxury-jewelry-sale (accessed on 14 December 2024).

Huynh-The, T., Gadekallu, T. R., Wang, W., Yenduri, G., Ranaweera, P., Pham, Q. V., da Costa, D. B., & Liyanage, M. (2023). Blockchain for the metaverse: A review. *Future Generation Computer Systems, 143*, 401–419. https://doi.org/10.1016/j.future.2023.02.008.

Huynh-The, T., Pham, Q. V., Nguyen, X. Q., Han, Z., & Kim, D. S. (2023). Artificial intelligence for the metaverse: A survey. *Engineering Applications of Artificial Intelligence, 117*(Part A). https://doi.org/10.1016/j.engappai.2022.105581.

IBM. (n.d.). What is blockchain technology? Available: https://www.ibm.com/in-en/topics/what-is-blockchain (accessed on 14 December 2024).

Jalan, T., & Barbesino, P. (2023). Making the metaverse real. *Digital Business, 3*(2), 100059. https://doi.org/10.1016/j.digbus.2023.100059.

Joshua, J. (2017). Information bodies: Computational anxiety in Neal Stephenson's *Snow Crash. Interdisciplinary Literary Studies, 19*(1), 17–47. https://doi.org/10.5325/intelitestud.19.1.0017.

Lim, W. M., Gunasekara, A., Pallant, J. L., Pallant, J. I., & Pechenkina, E. (2023). Generative AI and the future of education: Ragnarök or reformation? A paradoxical perspective from management educators. *The International Journal of Management Education, 21*(2), 100790. https://doi.org/10.1016/j.ijme.2023.100790.

Ling, C., Beuys, J., & Bourriaud, N. (2011). Christine Liao dissertation. *Electronic Theses and Dissertations*. Retrieved from: https://etda.libraries.psu.edu/files/final_submissions/1035 (accessed on 14 December 2024).

Lv, Z. (2023). Generative artificial intelligence in the metaverse era. *Cognitive Robotics, 3*, 208–217. https://doi.org/10.1016/j.cogr.2023.06.001.

Maan, S., Furness, T., Yuan, Y., Iorio, J., & Wang, Z. (2018). All reality: Virtual, augmented, mixed (x), mediated (x, y), and multimediated reality. *arXiv preprint arXiv:1804.08386*.

Masoni, D., & Chatterjee, S. (2021). Metaverse: A $50 billion revenue opportunity for luxury. *The Business of Fashion*. Retrieved from: https://www.businessoffashion.com/news/technology/metaverse-a-50-billion-revenue-opportunity-for-luxury/ (accessed on 14 December 2024).

McDowell, M. (2022). Metaverse Fashion Week: The hits and misses. *Vogue Business*. Retrieved from: https://www.voguebusiness.com/technology/metaverse-fashion-week-the-hits-and-misses (accessed on 14 December 2024).

McKinsey & Company. (2022). Value creation in the metaverse. *McKinsey*. Retrieved from: https://www.mckinsey.com/capabilities/growth-marketing-and-sales/our-insights/value-creation-in-the-metaverse (accessed on 14 December 2024).

Meta. (2021). The Facebook Company is now Meta. *Meta*. Retrieved from: https://about.fb.com/news/2021/10/facebook-company-is-now-meta/ (accessed on 14 December 2024).

Mitchell, A., Murphy, J., Owens, C., & Whittaker, S. (2014). Beyond conversation: Theorizing contact in social VR. In *2014 ACM International Joint Conference on Pervasive and Ubiquitous Computing* (pp. 1183–1192).

Montfort, N., & Bogost, I. (2009). *Racing the Beam: The Atari Video Computer System*. MIT Press.

Naranovich, A. (2022). Gucci celebrates its centennial year in the metaverse with a virtual exhibition. *Robb Report*. Retrieved from: https://robbreport.com/style/fashion/gucci-celebrates-its-centennial-year-in-the-metaverse-1234642187/ (accessed on 14 December 2024).

Nasiri, N., Hooman, J., & Khazaei, S. (2023). A novel framework for sustainable artificial intelligence in the metaverse. *Electronics, 12*(7), 901. https://doi.org/10.3390/electronics12070901.

Nike. (2022). Nike files trademark for virtual sneakers, revealing continued interest in the metaverse. *CNBC*. Available: https://www.cnbc.com/2022/01/20/nike-files-trademark-for-virtual-sneakers-revealing-continued-interest-in-the-metaverse.html (accessed on 14 December 2024).

Okuda, T., Teranishi, Y., & Suzuki, R. (2023). A study on the application of blockchain and smart contracts for intellectual property protection in the metaverse. *Journal of Engineering Science and Technology Review, 16*(2), 18–32.

Puma. (2022). PUMA®–Black Station: PUMA reveals first-ever metaverse experience with exclusive NFTs at New York Fashion Week. *About PUMA*. Available: https://about.puma.com/en/newsroom/brand-and-product-news/2022/09-07-2022-puma-black-station (accessed on 14 December 2024).

Razer. (2021). Razer launches the world's first RGB-powered smart mask and gaming chair concept designs at CES 2021. *Razer*. Available: https://www.razer.com/press/press-releases/razer-concept-ces-2021 (accessed on 14 December 2024).

Reeve, E. (2022). Luxury fashion brands are moving into the metaverse. *CNN Business*. Retrieved from: https://www.cnn.com/style/article/metaverse-luxury-fashion-brands/index.html (accessed on 14 December 2024).

Schulz, M. (2022). Gucci Vault opens a metaverse world in The Sandbox with games and vintage fashion. *Vogue Business*. Retrieved from: https://www.voguebusiness.com/technology/gucci-vault-opens-metaverse-world-in-the-sandbox-with-games-and-vintage-fashion (accessed on 14 December 2024).

Shrimali, B., & Patel, H. B. (2022). Blockchain state-of-the-art: Architecture, use cases, consensus, challenges, and opportunities. *Journal of King Saud University–Computer and Information Sciences, 34*(9), 6793–6807. https://doi.org/10.1016/j.jksuci.2021.08.005.

Steuer, J. (1992). Defining virtual reality: Dimensions determining telepresence. *Journal of Communication, 42*, 73–93. http://dx.doi.org/10.1111/j.1460-2466.1992.tb00812.x.

Takahashi, D. (2023). DressX raises $15M for digital fashion. *VentureBeat*. Retrieved from: https://venturebeat.com/games/dressx-raises-15m-for-digital-fashion/ (accessed on 14 December 2024).

Tencent. (2022). Tencent's investment in the metaverse: What you need to know. *KrASIA*. Available: https://kr-asia.com/tencents-investment-in-the-metaverse-what-you-need-to-know (accessed on 14 December 2024).

The Business of Fashion. (2022). Fashion's metaverse fantasy faces a more practical reality. Retrieved from: https://www.businessoffashion.com/news/business-of-fashion/fashions-metaverse-fantasy-faces-a-more-practical-reality (accessed on 14 December 2024).

The Economist. (2023). NFTs are entering the metaverse and creating a new digital economy. Retrieved from: https://www.economist.com/special-report/2023/02/04/nfts-are-entering-the-metaverse-and-creating-a-new-digital-economy (accessed on 14 December 2024).

The Verge. (2021). Adidas sold more than $22 million in NFTs, but it hit a few snags along the way. Retrieved from: https://www.theverge.com/2021/12/17/22843104/adidas-nfts-metaverse-sold-bored-ape (accessed on 14 December 2024).

The Verge. (2021). Beeple sold an NFT for $69 million. Retrieved from: https://www.theverge.com/2021/3/11/22325054/beeple-christies-nft-sale-cost-everydays-69-million (accessed on 14 December 2024).

The World Bank Group. (2018). Cryptocurrencies and blockchain. *The World Bank Group*. [Online]. Available: https://documents1.worldbank.org/curated/pt/293821525702130886/pdf/Cryptocurrencies-and-blockchain.pdf (accessed on 14 December 2024).

Tiffany & Co. (2021). Tiffany & Co. turns CryptoPunks into a $12.5M luxury jewelry sale. *Blockworks*. Available: https://blockworks.co/news/tiffany-co-turns-cryptopunks-into-12-5m-luxury-jewelry-sale (accessed on 14 December 2024).

Vallance, C. (2022). ChatGPT: New AI chatbot has everyone talking to it. *BBC*. Retrieved from: https://www.bbc.com/news/technology-63861322 (accessed on 14 December 2024).

Wiederhold, B. (2023). Haptics: Making the Metaverse a touching experience. *Cyberpsychology, Behavior, and Social Networking*. https://doi.org/10.1089/cyber.2023.29278.editorial.

World Economic Forum. (2023). Interoperability in the metaverse. *World Economic Forum*. Retrieved from: https://www3.weforum.org/docs/WEF_Interoperability_in_the_Metaverse.pdf (accessed on 14 December 2024).

Yordanova, H. (2021). Burberry x Blankos NFTs sold out in 30 seconds. *DappRadar*. [Online]. Available: https://dappradar.com/blog/burberry-x-blankos-nfts-sold-out-in-30-seconds (accessed on 14 December 2024).

Zhang, Y., Liu, Y., & Huang, G. Q. (2022). Digital twin-enabled smart manufacturing: A comprehensive survey. *Journal of Manufacturing Systems, 64*, 119–139. https://doi.org/10.1016/j.jmsy.2022.10.006.

CHAPTER 5

Blockchain and Green Finance for Sustainable Development

SHASHI GUPTA[1,*] and VIPIN GUPTA[2]

[1]*UPID, Noida, Uttar Pradesh, India*

[2]*Wipro LTD, Edmonton, Canada*

*Corresponding author

ABSTRACT

Blockchain is seen as an unusual innovative technology. Regardless of the fact that numerous educators have recognized the significance of blockchain, blockchain exploration is still in its infancy stage. As a result, the study looks at the latest research on blockchain, in reference to business and the economy. By in-depth analysis of the literature, we find that blockchain is the most recent phenomenon. The emergence of a paradigm shift in not just the way we think about financial matters but also how we can impact our planet's future is being driven by the spread of green finance and blockchain technologies. Green finance is a financial system that supports sustainability by funding the power generation sector and encouraging the expansion of clean energy sources. Blockchain is a decentralized database that can be accessed by anyone.

5.1 INTRODUCTION

Blockchain is generally viewed as a technology that will radically change society. Although numerous academicians accept the significance of blockchain, this field of study is still in its infancy stage. Thus, the objective of

Blockchain Technology: Unlocking the Value of 3Ts in the Business World.
Richa Chauhan & Vidhi Kaul (Eds.)

DOI: 10.1201/9781998511914-5

this chapter is to develop an understanding for users regarding the use of blockchain technology and green finance for sustainable development. For this purpose, the chapter covers the latest work on blockchain with reference to business and the economy. The examination of publications that are most frequently used, the most productive nations, and the most popular keywords is conducted using a thorough analysis of studies gathered from various sources. The remaining study themes "profitable benefit," "blockchain technology," "initial coin offerings," "FinTech revolution," and "participating economy" are likewise determined using clustering. Currently, blockchain is classified into three versions, that is, 1.0, 2.0, and 3.0 based on the operations they support.

5.1.1 BACKGROUND

Engelhardt (2017), Hyvarinen et al. (2017), Kim and Laskowski (2018), O'Dair and Beaven (2017), Radanovic and Likic (2018), and Savelyev (2018) noted that nowadays blockchain technology is used for digital currency and finance. It is also being used to handle requests, manage supply chains, cover imprints, and provide smart energy.

Christidis and Devetsikiotis (2016), Cruz et al. (2018), and Kraft (2016) highlighted that numerous academic fields have investigated blockchain technology. For example, some researchers have explored the blockchain's foundational technologies, including distributed storage, peer-to-peer networking, cryptography, smart contracts, and consensus algorithms.

Kiviat (2015) and Paech (2017) explained that the rules and laws governing blockchain-related technology are of interest to legal researchers. As stated by earlier studies, academics from various fields "speak many different languages" and have diverse perspectives.

5.1.2 BASICS OF BLOCKCHAIN TECHNOLOGIES

The first and most fundamental blockchain technology was introduced in November 2008 as an electronic cash system, the distributed ledger behind Bitcoin transactions, by Satoshi Nakamoto (an unknown programmer or group of programmers). This system was viewed as a paradigm shift, given that it comprises four core factors: hash, digital signature, peer-to-peer network, and consensus mechanisms. The technology used in distributed digital ledger records transactions as an immutable list of blocks that are shared in a peer-to-peer network. It doesn't require a single server, thus eliminating a single point of failure. Each piece of information in the block is encrypted, and each block is linked to the previous block using a unique identifier created using the hashing algorithm. A new block of data can be created to validate its authenticity (Figure 5.1).

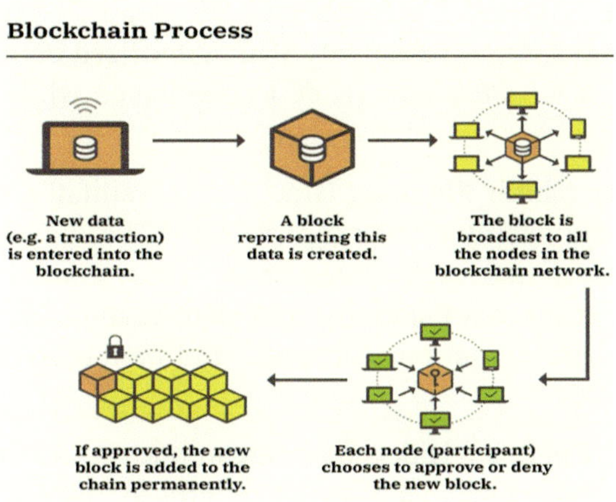

FIGURE 5.1 Blockchain process.

Source: Money. https://money.com/what-is-blockchain/

5.1.3 OPERATION OF BLOCKCHAIN TECHNOLOGY IN FINANCE

Blockchain technology has significantly impacted the finance industry, revolutionizing traditional processes and offering new opportunities for financial services. Some of the crucial operations of blockchain in finance include:

1. **Digital Payments and Cross-Border Transactions:** Blockchain enables fast, secure, and low-cost peer-to-peer digital payments. Cryptocurrencies like Bitcoin and Ethereum allow individuals to transfer funds globally without the need for intermediaries such as banks. Additionally, blockchain can streamline cross-border transactions by reducing processing times and foreign exchange costs.

2. **Smart Contracts:** These are self-executing agreements with the terms of the contract directly written into lines of code on the blockchain. They automatically execute when the predefined

conditions are met, eliminating the need for intermediaries and reducing transaction costs. Smart contracts find use in various financial operations, such as insurance claims, trade finance, and derivatives.

3. **Identity Verification and KYC:** Blockchain can enhance identity verification processes by creating decentralized and secure identity systems. Users can maintain control over their personal data and share it selectively with authorized entities, improving Know Your Customer (KYC) compliance while ensuring data privacy.
4. **Supply Chain Finance:** Blockchain facilitates transparent and traceable supply chains by recording the movement of goods and verifying each transaction. This transparency reduces fraud, ensures authenticity, and enables more accessible supply chain finance options for small and medium-sized enterprises (SMEs).
5. **Trade Finance:** Blockchain streamlines trade finance by digitizing trade documents (such as bills of lading and letters of credit) on an immutable ledger. This reduces the risk of fraud and increases the efficiency of trade agreements, benefiting exporters, importers, and banks involved in international trade.
6. **Tokenization of Assets:** Blockchain allows the fractional ownership and tokenization of assets, making illiquid assets (e.g., real estate, art, and collectibles) more accessible to a broader range of investors. Investors can buy and trade these tokens, promoting liquidity and diversification.
7. **Decentralized Finance (DeFi):** It is an ecosystem of financial operations built on blockchain networks, providing decentralized alternatives to traditional financial services. DeFi platforms offer lending, borrowing, yield farming, decentralized exchanges, and more, all accessible without intermediaries.
8. **Auditing and Compliance:** Blockchain's transparent and auditable nature simplifies regulatory compliance and auditing processes for financial institutions. Transactions recorded on the blockchain can be easily verified, enhancing trust and accountability.

9. **Crowdfunding and Initial Coin Offerings (ICOs):** Blockchain has enabled new crowdfunding models, such as ICOs, which allow companies and projects to raise capital by issuing digital tokens. This democratizes fundraising, enabling global participation in early-stage investment opportunities.
10. **Central Bank Digital Currencies (CBDCs):** Some central banks are exploring the use of blockchain technology to develop their digital currencies. CBDCs can enhance the efficiency and effectiveness of payment systems, enable direct monetary policy interventions, and improve financial inclusion.

The applications of blockchain technology in finance are continually evolving, and its disruptive potential in the financial sector continues to attract significant attention and investment. However, regulatory challenges, scalability concerns, and integration with existing systems remain critical areas for further development and research.

The entire verification process and addition of data to the blockchain is known as mining, which requires significant computational power. The consensus on the validity of a block is determined by a set of rules, such as proof of stake (PoS), proof of work (PoW), evidence of ceased time (minstrel), proof of elapsed time (PoET), delegated proof of stake (DPoS), and practical Byzantine fault tolerance (PBFT). Each block in the blockchain encrypts transactions using a cryptographic hashing function (Hash256), which enhances the security of blockchain technology. These factors contribute to the improved attributes of blockchain, including security, trust, privacy, and anonymity.

5.2 GREEN FINANCE: MEANING

It is known as sustainable finance or environmental finance, which means fiscal products, services, and investments that support environmentally friendly and sustainable systems. The primary goal of green finance is to promote a conversion to a low-carbon, climate-resilient, and environmentally sustainable economy. It involves directing financial resources towards systems and activities that have positive environmental impacts and contribute to the achievement of sustainable development goals (Figure 5.2).

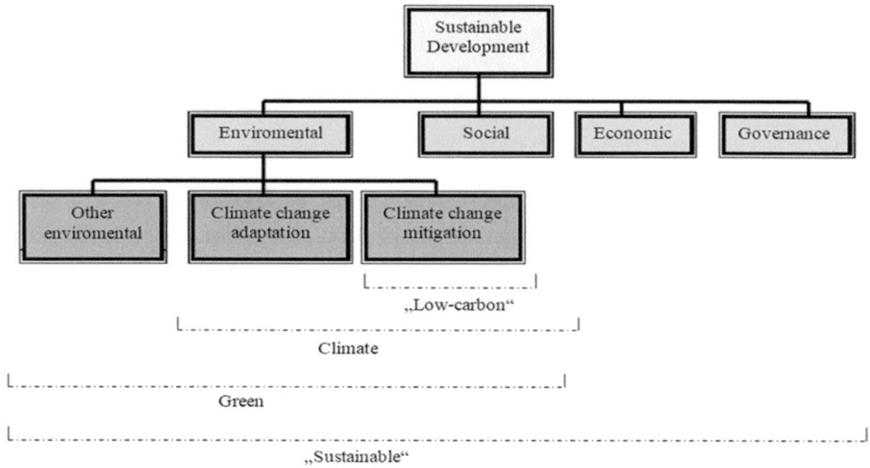

FIGURE 5.2 Blockchain and green finance.

5.2.1 CRUCIAL ASPECTS OF GREEN FINANCE

1. **Green Bonds:** These are debt instruments issued by governments, corporations, or financial institutions to raise funds for environmentally beneficial projects. These projects may include renewable energy infrastructure, energy efficiency initiatives, sustainable transportation, and green building projects. The proceeds from green bonds are allocated specifically for green projects.
2. **Climate Funds:** These are investment vehicles that pool financial resources to support climate-related projects and initiatives. These funds often target projects in developing countries that aim to address climate change impacts and promote sustainable development.
3. **Green Loans:** These are loans provided to finance environmentally friendly projects or activities. They are structured to offer better terms, such as lower interest rates or longer repayment periods, to incentivize borrowers to invest in green projects.
4. **Sustainable Investment Funds:** These are investment funds that incorporate environmental, social, and governance (ESG) criteria into their investment strategies. Sustainable investment funds seek

to invest in companies and projects that demonstrate strong environmental and social practices alongside financial performance.

5. **Carbon Markets:** These facilitate the trading of carbon credits or permits, allowing companies to buy and sell the right to emit greenhouse gases. These markets create financial incentives for companies to reduce their carbon emissions and invest in carbon offset projects.
6. **Green Insurance:** These products aim to cover and manage risks related to climate change and environmental impacts. For example, insurance products can cover losses due to extreme weather events or support climate-resilient agriculture.
7. **Green Microfinance:** This focuses on providing financial services to low-income individuals and communities for environmentally sustainable activities. It helps promote access to clean energy, sustainable agriculture, and eco-friendly practices among underserved populations.
8. **ESG Reporting and Disclosures:** Green finance also involves promoting transparency and accountability through ESG reporting. Companies and financial institutions disclose their environmental and social performance, allowing investors to make informed decisions based on sustainability criteria.

Green finance plays a vital role in combating climate change, promoting environmental protection, and achieving sustainable development goals. It aligns financial systems with sustainability objectives, channels capital to environmentally friendly projects, and encourages responsible business practices. As global awareness of climate and environmental challenges grows, green finance continues to gain traction as a crucial tool in the transition towards a greener and more sustainable future. Embracing green finance innovation is pivotal for fostering sustainable economic growth, addressing climate change, and achieving environmental objectives.

Green finance innovation involves developing and implementing new financial products, services, and technologies that support environmentally friendly initiatives and accelerate the transformation to a low-carbon and sustainable economy. There are several ways to embrace green finance innovation, such as:

1. **Encouraging Research and Development:** Governments, financial institutions, and research organizations should invest in research and development to explore new green finance solutions. This includes technologies like blockchain for transparent carbon markets, innovative financing instruments for sustainable projects, and advanced analytics for assessing climate-related risks.
2. **Supporting Green FinTech Startups:** FinTech startups focused on green finance need support and incentives. Governments can provide funding, regulatory sandboxes, and incentives for these startups to experiment with and scale innovative green finance solutions.
3. **Promoting Policy and Regulatory Frameworks:** Governments and regulators should create an enabling environment for green finance innovation. This involves establishing clear policies, standards, and incentives to encourage investment in green projects and facilitate the integration of sustainability factors into financial decision-making.
4. **Financial Institutions:** They should design and offer a variety of green financial products, such as green bonds, sustainable investment funds, and green loans, to cater to the growing demand for environmentally responsible investment options.
5. **Enhancing ESG Integration:** By integrating ESG factors into investment decision-making processes, which entails assessing the environmental impact and sustainability practices of companies and projects before making investment choices.
6. **Collaborating with Public and Private Sector Partners:** Collaborations between public and private sectors are essential for advancing green finance innovation. Governments can partner with financial institutions, corporations, and NGOs to co-create and finance large-scale green projects.

5.2.2 EMBRACING GREEN FINANCE INNOVATION

Embracing green finance innovation is pivotal for fostering sustainable economic growth, addressing climate change, and achieving environmental objectives. Green finance innovation involves developing and implementing new financial products, services, and technologies that support environmentally friendly initiatives and accelerate the transformation to a low-carbon and sustainable economy. There are several ways to embrace green finance innovation:

1. **Encouraging Research and Development:** Governments, financial institutions, and research organizations should invest in research and development to explore new green finance solutions. This includes technologies like blockchain for transparent carbon markets, innovative financing instruments for sustainable projects, and advanced analytics for assessing climate-related risks.
2. **Supporting Green FinTech Startups:** FinTech startups focused on green finance need support and incentives. Governments can provide funding, regulatory sandboxes, and incentives for these startups to experiment with and scale innovative green finance solutions.
3. **Promoting Policy and Regulatory Frameworks:** Governments and regulators should create an enabling environment for green finance innovation. This involves establishing clear policies, standards, and incentives to encourage investment in green projects and facilitate the integration of sustainability factors into financial decision-making.
4. **Developing Green Financial Products:** Financial institutions should design and offer a range of green financial products, such as green bonds, sustainable investment funds, and green loans, to cater to the growing demand for environmentally responsible investment options.
5. **Enhancing ESG Integration:** Investors can embrace green finance by integrating ESG factors into their investment decision-making processes. This entails assessing the environmental impact and sustainability practices of companies and projects before making investment choices.

Embracing green finance innovation isn't only essential for mitigating climate change and environmental challenges but also presents significant opportunities for financial institutions and investors. By aligning financial flows with sustainability goals, green finance innovation contributes to a more resilient, equitable, and environmentally responsible global economy.

5.3 COUNTRIES INVESTING IN GREEN TECHNOLOGY

As of September 2021, several countries were laboriously investing in green technology to address climate change, promote sustainable development, and transition towards a low-carbon economy. These investments were aimed at developing and adopting renewable energy sources, advancing energy efficiency technologies, and promoting environmentally friendly practices. Some of the leading countries investing in green technology include:

1. **China:** It has been one of the largest investors in green technology, with significant investments in renewable energy, electric vehicles (EVs), and green infrastructure. The country is a major producer and consumer of solar panels and wind turbines.
2. **Germany:** It has been a pioneer in green technology and renewable energy. The country has made substantial investments in solar, wind, and biomass energy sources. It also offers incentives and subsidies for energy-efficient technologies.
3. **United States:** The U.S. has been actively investing in green technology, particularly in the development of renewable energy sources like solar and wind power. Numerous states have their own renewable energy targets and incentives.
4. **Japan:** It has been a leader in green technology research and development. The country invests in clean energy technologies, energy-efficient infrastructure, and sustainable transportation.
5. **Sweden:** This country is committed to achieving carbon neutrality and has invested heavily in renewable energy, energy-efficient infrastructure, and sustainable transportation.
6. **Denmark:** It is known for its ambitious renewable energy targets and investments in wind energy. The country has made significant progress in transitioning to green technologies.
7. **India:** It has been increasing its investments in green technology, focusing on solar and wind power. The country is also working on promoting electric mobility and energy-efficient solutions.

8. **Norway:** It has invested in renewable energy and is a global leader in electric vehicle adoption. The country provides incentives for EVs and has a strong focus on sustainability.
9. **South Korea:** It has been actively promoting green technology, including renewable energy and energy-efficient infrastructure. The country has ambitious plans for reducing greenhouse gas emissions.
10. **United Kingdom:** The UK has made substantial investments in wind energy and has set ambitious targets for reducing greenhouse gas emissions.

These countries, among others, recognize the significance of investing in green technology to address environmental challenges and achieve sustainable development goals. As the world continues to focus on combating climate change, it's likely that more countries will increase their investments in green technology and adopt environmentally friendly practices. Please note that the information provided is based on data available up to September 2021, and there may have been further developments since that time.

5.4 SUSTAINABLE DEVELOPMENT THROUGH BLOCKCHAIN-DRIVEN GREEN FINANCE

Blockchain technology has the potential to revolutionize financial innovation by introducing transparency, efficiency, and security into traditional financial systems. Its decentralized and distributed ledger architecture provides several benefits that can transform various aspects of the financial industry.

There are several ways in which blockchain can drive financial innovation, such as:

1. **Improved Security and Data Integrity:** Blockchain's immutability ensures that once information is recorded on the ledger, it cannot be altered or deleted. This feature enhances security, reduces the risk of fraud, and protects sensitive financial data, creating a more robust and secure financial ecosystem.
2. **Blockchain Enables Faster and Cheaper Cross-Border Payments:** By eliminating intermediaries and enabling peer-to-peer transactions, blockchain technology can reduce settlement

times and transaction costs, enhancing the efficiency of global payment systems.

3. **Smart Contracts:** These are self-executing agreements with predefined conditions written into lines of code on the blockchain. They automate processes, such as loan approvals, trade agreements, and insurance claims, removing the need for intermediaries and streamlining financial transactions.
4. **Blockchain:** It can enable access to financial services for the unbanked and underbanked populations, especially in regions with limited banking infrastructure. Through blockchain-based platforms, individuals can access digital wallets, loans, and other financial services without relying on traditional banks.
5. **Tokenization of Assets:** Blockchain allows the tokenization of various assets, including real estate, art, and commodities. Tokenization enables fractional ownership, making traditionally illiquid assets more accessible to a broader range of investors and promoting liquidity in the market.
6. **Decentralized Finance (DeFi):** It is an ecosystem of financial applications built on blockchain networks that offer decentralized alternatives to traditional financial services. DeFi platforms provide lending, borrowing, yield farming, decentralized exchanges, and more, all accessible without intermediaries.
7. **Trade Finance and Supply Chain Management:** Blockchain streamlines trade finance by digitizing trade documents and verifying transactions on a tamper-resistant ledger. This reduces paperwork, improves transparency, and lowers the risk of fraud in international trade.
8. **Regulatory Compliance and Reporting:** Blockchain can facilitate real-time auditing and reporting by maintaining a transparent and immutable record of financial transactions. This feature can simplify regulatory compliance for financial institutions and improve accountability.
9. **Data Sharing and Interoperability:** Blockchain can enable secure and efficient data sharing across financial institutions, reducing duplication of efforts and enhancing collaboration. This

interoperability can lead to faster, more accurate decision-making processes.

10. **Green Finance:** Blockchain can be used to create tokenized carbon credits, enabling a decentralized and transparent carbon credit market. This can incentivize organizations to reduce their carbon footprint and contribute to sustainable finance initiatives.

While blockchain technology shows tremendous promise in transforming financial services, challenges related to scalability, regulatory compliance, and privacy remain. Overcoming these hurdles will be essential to fully harness the potential of blockchain and drive financial innovation towards a more efficient, inclusive, and sustainable future.

5.5 UNLOCKING MOBILIZATION OF GREEN FINANCE: PROSPECTS OF BLOCKCHAIN TECHNOLOGY

Blockchain technology has the potential to unleash the mobilization of green finance by addressing critical challenges and enhancing various aspects of sustainable finance. Its unique characteristics, such as transparency, immutability, and decentralization, can significantly improve the efficiency, accountability, and accessibility of green finance initiatives. Here are some ways blockchain can unleash the mobilization of green finance:

1. **Translucency and Responsibility:** Blockchain's transparent and inflexible nature enables traceability and responsibility in green finance systems. Investors and stakeholders can fluently track the inflow of finances, ensuring that money is used for its intended purpose and that systems are authentically environmentally beneficial.
2. **Smart Contracts for Automated Compliance:** Smart contracts can automate compliance with sustainability norms and regulations. For example, they can automatically trigger the release of finances when predefined environmental goals are achieved, enhancing the effectiveness and credibility of green finance enterprises.
3. **Tokenization of Green Assets:** Blockchain allows the fractional ownership and tokenization of green assets, such as renewable

energy systems or carbon credits. This makes it easier for investors of all sizes to participate in green finance opportunities, standardizing access to sustainable investments.

4. **Decentralized Green Finance Platforms:** Blockchain can power DeFi platforms that enable peer-to-peer green lending and investment. These platforms can connect borrowers and investors directly, bypassing intermediaries and reducing transaction costs.
5. **Global Availability:** Blockchain's borderless nature enables the seamless transfer of green finance across borders. This facilitates international cooperation and investment in global sustainability initiatives, serving both developed and developing economies.
6. **Increased Trust and Reduced Risk:** Blockchain's secure and tamper-resistant structure reduces counterparty risk and increases trust among stakeholders in green finance systems. This can attract more investors and financiers to participate in sustainable enterprises.
7. **Carbon Market Integration:** Blockchain can create decentralized carbon markets, allowing easy trading of carbon credits and promoting greater participation in carbon offset systems. This liquidity and availability can encourage more companies to offset their emissions and invest in sustainable practices.
8. **Enhanced Data Integrity:** Blockchain ensures the integrity of data related to environmental criteria and impact assessments. This reliability in data can strengthen the creditworthiness of green projects and attract more institutional investors.
9. **Green Supply Chain Finance:** Blockchain can facilitate green supply chain finance by providing a transparent and auditable record of sustainability efforts along the supply chain. This encourages companies to adopt environmentally friendly practices and opens up funding opportunities for sustainable suppliers.
10. **Carbon Accounting and Reporting:** Blockchain-based platforms can streamline carbon accounting and reporting processes for companies and governments, simplifying compliance with emission reduction commitments and easing emission trading.

While blockchain technology holds immense potential in unleashing the rallying of green finance, it's pivotal to address challenges such as scalability, interoperability, and regulatory frameworks. Cooperative efforts from governments, financial institutions, and technology providers are essential to fully realize the potential of blockchain in accelerating the transition to a sustainable and climate-resilient economy.

5.6 COMPONENTS OF BLOCKCHAIN TECHNOLOGY IN EMPOWERING GREEN FINANCE

Blockchain technology plays a vital role in empowering green finance by addressing challenges and creating opportunities for sustainable investment and development. Its unique features have the potential to revolutionize the financial industry and accelerate the transition to a more environmentally friendly and climate-resilient economy. There are some crucial aspects of blockchain technology in empowering green finance:

1. **Translucency and Traceability:** Blockchain's transparent and inflexible ledger ensures that all transactions related to green finance initiatives are recorded and visible to stakeholders. This translucency allows investors to track the use of finances and verify the impact of projects, promoting trust and accountability.
2. **Smart Contracts for Automated Compliance:** Smart contracts can automate compliance with environmental norms and regulations. They ensure that finances are released only when predefined green objectives are met, reducing the risk of greenwashing and ensuring the actual implementation of sustainable projects.
3. **Tokenization of Green Assets:** Blockchain allows the tokenization of green assets, such as renewable energy systems or carbon credits. This fractional ownership of assets makes it easier for investors of all sizes to participate in green finance opportunities, standardizing access to sustainable investments.
4. **DeFi for Green Projects:** Blockchain can power DeFi platforms that enable peer-to-peer green lending and investment. These platforms can connect borrowers and investors directly, bypassing intermediaries and reducing transaction costs in green finance.

5. **Global Availability:** Blockchain's borderless nature enables the seamless transfer of green finance across borders. This facilitates international cooperation and investment in global sustainability initiatives, serving both developed and developing economies.
6. **Increased Trust and Reduced Risk:** Blockchain's secure and tamper-resistant structure reduces counterparty risk and increases trust among stakeholders in green finance systems. This can attract more investors and financiers to participate in sustainable enterprises.
7. **Carbon Market Integration:** Blockchain can create decentralized carbon markets, allowing easy trading of carbon credits and promoting greater participation in carbon offset systems. This liquidity and availability can encourage more companies to offset their emissions and invest in sustainable practices.
8. **Enhanced Data Integrity:** Blockchain ensures the integrity of data related to environmental criteria and impact assessments. This reliability in data can strengthen the creditworthiness of green projects and attract more institutional investors.
9. **Green Supply Chain Finance:** Blockchain can facilitate green supply chain finance by providing a transparent and auditable record of sustainability efforts along the supply chain. This encourages companies to adopt environmentally friendly practices and opens up funding opportunities for sustainable suppliers.
10. **Carbon Accounting and Reporting:** Blockchain-based platforms can streamline carbon accounting and reporting processes for companies and governments, simplifying compliance with emission reduction commitments and easing emission trading.

By using blockchain technology, green finance can be made more inclusive, effective, and transparent, driving sustainable investments and promoting environmentally responsible practices across various sectors. The synergy between blockchain and green finance has the potential to create a significant positive impact on environmental preservation and sustainable development worldwide. Automation, quick execution, traceability, and transparency will all be made possible by the effective use of smart contracts in the processes which include:

1. **Trustworthy Transaction Recording:** One of the primary justifications for using blockchain technology in the financial industry is that it provides a productive and dependable platform for carrying out and accurately documenting transactions. It makes monitoring asset ownership possible. Energy trade is made more exact and very reputable by blockchain.
2. **The Removal of Middlemen:** Transactions in renewable energy finance are automatically validated by the smart contract on a blockchain. This gets rid of the necessity for intermediaries. The common trading platform for physical and financial trading with a variety of renewable energy commodities is made more efficient.
3. **Regulatory Reporting and Compliance:** At the moment, regulators in the industry expect that trading companies for renewable energy provide thorough data in order to spot noncompliance and other regulatory flaws. The data collection and cleansing process is laborious with the current tools and techniques, and there is a significant chance that the data may end up in the wrong hands. Blockchain is used to solve all of the aforementioned issues. It offers perfect ownership rights, unchangeable data, and maximum transparency.
4. **Global Supply Network:** With the existing setup, it is hard to have a comprehensive understanding of the sector's end-to-end procedures. DLTs give organizations the ability to exchange information quickly, productively, securely, and transparently on an open platform without worrying that their private and vitally important business information would be compromised.

5.7 FOSTERING GREEN INVESTMENT

Fostering green investment is indeed a great global ambition with far-reaching consequences for combating climate change, promoting sustainable development, and securing a greener and more flexible future for our Earth. This ambition is pivotal for addressing pressing environmental challenges, such as greenhouse gas emissions, biodiversity loss, and resource depletion, while ensuring economic viability and social well-being. Here are some crucial reasons why fostering green investment is a critical global ambition.

1. **Addressing Climate Change:** Green investments in renewable energy, energy efficiency, and low-carbon technologies are essential to reduce greenhouse gas emissions and limit global warming. By transitioning to cleaner and more sustainable energy sources, we can alleviate the impacts of climate change and work towards a carbon-neutral future.
2. **Promoting Sustainable Development:** Green investments support initiatives that contribute to sustainable development goals, such as poverty alleviation, clean water access, and biodiversity conservation. These investments prioritize social and environmental benefits, fostering equitable and inclusive development.
3. **Creating Green Jobs and Economic Growth:** Green investments can drive economic growth by creating new opportunities and employment in the renewable energy sector, sustainable infrastructure, and green technologies. This, in turn, boosts economic resilience and diversification.
4. **Enhancing Energy Security:** Investing in renewable energy and energy-efficient technologies reduces dependence on fossil fuels and enhances energy security. Diversifying energy sources helps countries become more resilient to energy price fluctuations and supply disruptions.
5. **Improving Public Health:** Transitioning to clean and sustainable energy sources improves air and water quality, leading to better public health outcomes and reducing healthcare costs associated with pollution-related illnesses.
6. **Encouraging Innovation and Technological Advancements:** Green investments drive innovation in renewable energy, green transportation, and sustainable technologies. This fosters technological advancements and competitiveness in global markets.
7. **Attracting Private Sector Engagement:** By providing conducive policy frameworks and financial incentives, green investments attract private sector engagement. This collaboration between governments, businesses, and investors accelerates the scale and impact of sustainable initiatives.

8. **Strengthening International Cooperation:** The shared ambition of fostering green investment promotes international cooperation in addressing climate change and other environmental challenges. Global collaboration is essential for achieving meaningful and lasting impact.

9. **Conserving Biodiversity and Ecosystems:** Green investments support conservation efforts, protecting critical ecosystems and conserving biodiversity. This is vital for maintaining ecosystem services that benefit human well-being.

10. **Empowering Original Communities:** Green investments often involve community engagement and empowerment. Local communities are encouraged to participate in sustainable initiatives, ensuring their voices are heard and their needs are met.

5.8 CHALLENGES OF NEW APPROACHES TO GREEN FINANCE

While new approaches to green finance hold great promise for driving sustainable development and combating climate change, they also face several challenges that need attention to achieve their full potential. Some of the key challenges are:

1. **The Lack of Standardized Definitions and Criteria:** The lack of standardized definitions for green finance can lead to confusion and greenwashing, where investments are falsely presented as environmentally friendly. The absence of clear and consistent criteria hinders the credibility and acceptance of green finance initiatives.

2. **Limited Track Record:** Some new green finance approaches, such as green bonds and sustainable investment funds, are relatively new and may lack a long-term track record of performance. This can discourage risk-averse investors from participating in these opportunities.

3. **High Initial Costs:** Investing in green technologies and initiatives may require higher upfront costs compared to traditional alternatives. While green investments often yield long-term benefits, the initial expenses can be a barrier for investors, especially in emerging markets.

4. **Risk Perception:** Investors may perceive green finance initiatives as risky due to uncertainties related to policy changes, technological advancements, and climate-related impacts. These perceived risks can affect the willingness to invest in green projects.
5. **Liquidity Issues:** Some green investments, such as renewable energy projects and carbon credits, may have limited liquidity, making it challenging for investors to exit their investments when needed.
6. **Regulatory and Policy Barriers:** Inconsistent or weak regulatory frameworks can hamper the growth of green finance initiatives. Clear and supportive policies are crucial for incentivizing investment and ensuring a level playing field for green projects.
7. **Data Limitations:** Reliable data on environmental impact and sustainability performance are essential for making informed green investment decisions. However, data availability and quality can be limited, particularly in developing regions.
8. **Capacity and Expertise Gaps:** Green finance requires technical knowledge and expertise in ESG (Environmental, Social, and Governance) considerations. A lack of skilled professionals and capacity can hinder the effective implementation of green finance strategies.
9. **Perception of Lower Returns:** Some investors may perceive green investments as offering lower returns compared to traditional investments. However, studies have shown that well-managed green investments can be financially competitive while providing positive environmental and social impacts.
10. **Transition Risks:** As economies move towards sustainability, some industries and assets may become stranded or obsolete, leading to transition risks for investors in traditional sectors.

Addressing these challenges requires collaborative efforts from governments, financial institutions, and other stakeholders. Policy and regulatory support, increased transparency and reporting standards, capacity building, and awareness campaigns are essential to overcome these hurdles and accelerate the adoption of green finance on a global scale. By inclusively addressing these challenges, we can unleash the full potential of green finance and drive meaningful positive change for the environment and society.

5.9 CONCLUSION

The integration of blockchain technology into green finance has the potential to revolutionize sustainable development efforts and pave the way for a greener, more resilient future. By leveraging the unique attributes of transparency, decentralization, and immutability, blockchain can address critical challenges and unlock new opportunities in the realm of environmental finance.

Applications of blockchain in sustainable finance enable increased trust, accountability, and efficiency. Transparent and traceable transactions allow investors and stakeholders to verify the impact of initiatives, promoting confidence and credibility in sustainable investments. Smart contracts automate compliance and disbursement of funds, ensuring that green objectives are met and resources are used responsibly. Additionally, blockchain's tokenization of green assets democratizes access to green investments, making it easier for investors of all sizes to participate in sustainability initiatives. This inclusivity fosters greater engagement from both public and private sectors, which drives the mobilization of green finance on a global scale.

The borderless nature of blockchain technology facilitates international collaboration in addressing climate change and environmental challenges. Through seamless cross-border transactions, countries can share knowledge, resources, and support to collectively address global sustainability goals.

However, challenges such as standardization, data limitations, and regulatory frameworks must be addressed to fully realize the potential of blockchain in green finance. Cooperative efforts from governments, financial institutions, technology providers, and other stakeholders are essential to create supportive policy environments, establish common standards, and bridge capacity gaps.

Embracing blockchain technology in green finance is a transformative step towards achieving sustainable development goals and combating climate change. By fostering innovation, empowering financial inclusion, and strengthening environmental stewardship, the integration of blockchain and green finance opens a world of possibilities for a more sustainable and equitable future.

As we continue on this journey, the integration of blockchain technology and green finance must be guided by a commitment to transparency, integrity, and inclusivity. Together, we can harness the full potential of blockchain to drive sustainable development, protect the environment, and secure a prosperous future for generations to come.

KEYWORDS

- **blockchain**
- **coin offerings**
- **data limitations**
- **digital payments**
- **green bonds**
- **green finance**
- **policy barriers**
- **smart contracts**
- **transition risks**

REFERENCES

Brown, L., & Alayza, N. (2021). Why the Green Climate Fund should give developing countries greater direct access to finance. *World Resources Institute*. Retrieved from: https://www.wri.org/insights/why-green-climate-fund-should-give-developing-countries-greater-direct-access-finance (accessed on 14 December 2024).

Centobelli, P., Cerchione, R., Vecchio, P. D., Oropallo, E., & Secundo, G. (2021). Blockchain technology for bridging trust, traceability, and transparency in circular supply chains. *Information & Management, 58*(8), 103508. https://doi.org/10.1016/j.im.2021.103508.

Chauhan, R., & Kaul, V. (2021). Blockchain technology: A panacea for medical distribution ailments. In *Blockchain for Healthcare Systems* (pp. 43–60). CRC Press.

Chauhan, R., & Kaul, V. (2022). Blockchain technology: Unlocking the business model maze for evolving businesses and start-ups. In *Transformations Through Blockchain Technology: The New Digital Revolution* (pp. 291–309). Cham: Springer International Publishing.

Climate Bonds Initiative. (2019). *Blockchain Gateway for Sustainability-Linked Bonds.* Retrieved from: https://www.climatebonds.net/ (accessed on 14 December 2024).

Crosby, M., Nachiappan, S., & Pattanayak, V. (2015). *Blockchain Technology*. Sutardja Center for Entrepreneurship & Technology. https://scet.berkeley.edu/wp-content/uploads/BlockchainPaper.pdf (accessed on 14 December 2024).

Cryptopedia Staff. (2021). What is tokenization? Blockchain token types. *Gemini*. Retrieved from: https://www.gemini.com/cryptopedia/what-is-tokenization-definition-crypto-token (accessed on 14 December 2024).

Dorfleitner, G., Muck, F., & Scheckenbach, I. (2021). Blockchain applications for climate protection: A global empirical investigation. *Renewable and Sustainable Energy Reviews, 149*, 111378. https://doi.org/10.1016/j.rser.2021.111378.

European Investment Bank. (n.d.). *Climate Awareness Bonds. EIB.org.* Retrieved from: https://www.eib.org/en/investor-relations/cab/index.htm (accessed on 14 December 2024).

Fleming, S. (2020). What is green finance and why is it important? *World Economic Forum*. Retrieved from: https://www.weforum.org/agenda/2020/11/what-is-green-finance (accessed on 14 December 2024).

HDFC Bank. (2022). What is DIY investing? *HDFC Bank Learning Center*. Retrieved from: https://www.hdfcbank.com/personal/resources/learning-centre/invest/what-is-diy-investing (accessed on 14 December 2024).

HM Government. (2021). *Greening Finance: A Roadmap to Sustainable Investing.*

HSBC & Green Digital Finance Alliance. (2019). Blockchain gateway for sustainability-linked bonds. Retrieved from: https://greendigitalfinancealliance.org/wp-content/uploads/2019/12/blockchain-gateway-for-sustainability.pdf (accessed on 14 December 2024).

Kenton, W. (2019). Disintermediation. *Investopedia*. Retrieved from: https://www.investopedia.com/terms/d/disintermediation.asp (accessed on 14 December 2024).

Kowalski, M., Lee, Z. W. Y., & Chan, T. K. H. (2021). Blockchain technology and trust relationships in trade finance. *Technological Forecasting and Social Change, 166*, 120641. https://doi.org/10.1016/j.techfore.2021.120641.

Lim, M. (2022). How blockchain can power the growth of green finance. *Forkast News*. Retrieved from: http://forkastnews/blockchain-green-finance-bonds-asia (accessed on 14 December 2024).

Morris, D. Z. (2016). Leaderless, blockchain-based venture capital fund raises $100 million, and counting. *Fortune.*

Naderi, N. (2021). Utilizing blockchain technology in international remittances for poverty reduction and inclusive growth. In F. Taghizadeh-Hesary, N. Panthamit, & N. Yoshino (Eds.), *Poverty Reduction for Inclusive Sustainable Growth in Developing Asia* (pp. 149–163). Springer. https://doi.org/10.1007/978-981-16-1107-0_7.

OECD. (2020). *Global Outlook on Financing for Sustainable Development.*

Pinna, A., & Ruttenberg, W. (2016). *Distributed Ledger Technologies in Securities Post-Trading: Revolution or Evolution? (Occasional Paper Series No. 172).* European Central Bank. https://www.ecb.europa.eu/pub/pdf/scpops/ecbop172.en.pdf (accessed on 14 December 2024).

Rozemberg, R., Barafani, M., & Garcia, P. (2021). Blockchain technology: A new opportunity for international trade beyond borders. *IDB Blogs*. Retrieved from: https://

blogs.iadb.org/integration-trade/en/blockchain-technology-a-new-opportunity-for-international-trade (accessed on 14 December 2024).

Schletz, M., Nassiry, D., & Lee, M. K. (2020). Blockchain and tokenized securities: The potential for green finance. *Asian Development Bank Institute Working Paper Series, 1079*. Retrieved from: https://www.adb.org/sites/default/files/publication/566271/adbi-wp1079.pdf (accessed on 14 December 2024).

Schulz, K., & Feist, M. (2021). Leveraging blockchain technology for innovative climate finance under the Green Climate Fund. *Earth System Governance, 7*, 100084. https://doi.org/10.1016/j.esg.2020.100084.

Taghizadeh-Hesary, F., & Yoshino, N. (2019). The way to induce private participation in green finance and investment. *Finance Research Letters, 31*, 98–103. https://doi.org/10.1016/j.frl.2019.04.016.

UN. (2019). Take urgent action to combat climate change and its impacts. *United Nations*. Retrieved from: https://www.un.org/sustainabledevelopment/climatechange (accessed on 14 December 2024).

United Nations Environment Programme. (2018). Green financing. *UNEP–UN Environment Programme.*

UNOPS. (2019). *Building for a Sustainable Future. The Economist.* https://content.unops.org/publications/UNOPS-Building-for-a-sustainable-future.pdf (accessed on 14 December 2024).

Wang, L., Luo, X., Lee, F., & Benitez, J. (2021). Value creation in blockchain-driven supply chain finance. *Information & Management, 58*(7), 103510. https://doi.org/10.1016/j.im.2021.103510.

World Bank. (2018). *World Bank marks 10th Green Bond Anniversary with Landmark Issuance: US$1.3 Billion Issuances Bring World Bank Green Bond Program to US$12.6 Billion.* Retrieved from: https://www.worldbank.org/en/news/press-release/2018/11/13/world-bank-marks-10-year-green-bond-anniversary-with-landmark-issuance-us-1-2-billion-issuances-bring-world-bank-green-bond-program-to-us-12-6-billion (accessed on 14 December 2024).

Yoshino, N., Schloesser, T., & Taghizadeh-Hesary, F. (2021). Social funding of green financing: An application of distributed ledger technologies. *International Journal of Finance & Economics, 26*(4), 6060–6073. https://doi.org/10.1002/ijfe.2108101121.

Zhou, Z., Liu, X., Zhong, F., & Shi, J. (2022). Improving the reliability of information disclosure in the supply chain based on blockchain technology. *Electronic Commerce Research and Applications, 52*. https://doi.org/10.1016/j.elerap.2022.101121.

CHAPTER 6

Enhancing Resilience and Visibility in Green Supply Chains with Blockchain Technology

TANUSHREE SANWAL,[1] SAPNA YADAV,[2] SANDHYA AVASTHI,[3] and MEENAKSHI TYAGI[1]

[1]*KIET School of Management, KIET Group of Institutions Delhi-NCR, Ghaziabad, Uttar Pradesh, India*

[2]*Jaypee Business School, Jaypee Institute of Information Technology, Noida, Uttar Pradesh, India*

[3]*ABES Engineering College, Ghaziabad, Uttar Pradesh, India*

*Corresponding author

ABSTRACT

Supply chain management (SCM) focuses on coordinating and regulating various processes to save expenses, improve operational effectiveness, and enhance product excellence. However, the effectiveness of SCM is hindered by a range of issues, such as disruptions, uncertainties in international trade, sustainability concerns, the digitalization process, shortages of a qualified workforce, and the necessity to adhere to regulatory obligations. The concepts of blockchain technology and green supply chain management (GSCM) are inherently interconnected, as they both contribute to the improvement of sustainability and transparency. The amalgamation of SCM, GSCM, and blockchain technology is crucial in India due to environmental concerns, regulatory mandates, resource availability, and the quest for sustainable economic advancement. Effectively managing data requires addressing and resolving a range of challenges,

Blockchain Technology: Unlocking the Value of 3Ts in the Business World.
Richa Chauhan & Vidhi Kaul (Eds.)

DOI: 10.1201/9781998511914-6

such as data quality, accuracy, energy usage, legal compliance, financial implications, industry norms, smart contracts, and environmental impacts. Implementing green supply chain management practices in India has the capacity to improve supply chain operations, foster cooperation, and support the development of sustainability and resilience in the economy.

6.1 INTRODUCTION

The growing phenomenon of globalization, the widespread use of e-commerce, and the heightened social consciousness among consumers have resulted in an upsurge in the need for a diverse range of products, competitive pricing, convenient purchasing options, prompt delivery, assurance of authenticity and origin, responsible sourcing and manufacturing practices, adherence to regulatory standards, and reliable post-sales services (Sanwal, Avasthi, and Saxena, 2016). Meeting the growing intricacy of consumer demands has necessitated the transformation of supply chains into multi-enterprise networks, encompassing multiple pathways for the flow of production, merchandising, and fulfillment. These networks involve numerous organizational and institutional transitions, enabling the efficient management of a vast array of complex products characterized by shorter life cycles and high transaction volumes.

The planning, coordination, and control of the transportation and storage of raw materials, components, completed items, and evidence from the point of origin to the site of consumption is known as supply chain management (SCM). To ensure the effective flow of products and services, it entails the integration and optimization of numerous operations and parties, including suppliers, manufacturers, distributors, retailers, and customers.

The key objectives of supply chain management are to curtail costs, improve efficiency, enhance product quality, reduce lead times, and enhance overall customer satisfaction. It necessitates effective collaboration and communication among all parties involved in the supply chain to streamline processes and eliminate inefficiencies (Yusuf and Soediantono, 2022).

6.1.1 IMPORTANCE OF SUPPLY CHAIN MANAGEMENT

Supply chain management is essential to a company's ability to succeed and remain competitive across a variety of industries. Its significance stems from the significant impact it has on various aspects of a company's operations, performance, and customer satisfaction. Here are some key reasons why supply chain management is essential (Meredith and Shafer, 2023):

1. **Cost Efficiency:** Effective supply chain management helps optimize costs at every stage, from procurement to supply. By minimizing waste, streamlining processes, and reducing inventory levels, companies can achieve cost savings and maintain a competitive edge.
2. **Customer Satisfaction:** Effective management of the supply chain ensures the consistent availability of products at the desired time and location, meeting the demands of customers. It improves order fulfillment rates, reduces lead times, and augments overall customer service, leading to higher buyer satisfaction and loyalty.
3. **Business Performance:** An efficient supply chain can improve overall business performance by ensuring that resources are utilized effectively, production is aligned with demand, and operational bottlenecks are addressed (Sanwal and Sareen, 2023).
4. **Flexibility and Responsiveness:** Companies may respond swiftly to shifting market conditions, customer requests, and unanticipated occurrences, such as supply outages or adjustments in consumer tastes, with the help of an agile supply chain.
5. **Inventory Management:** Supply chain management facilitates the attainment of an optimal equilibrium between maintaining an adequate inventory level to satisfy demand and preventing excessive inventory that would lead to capital immobilization and incur carrying costs.
6. **Risk Mitigation:** Effective supply chain management includes identifying and managing risks related to supply disruptions and other uncertainties that could affect the movement of goods and services (Avasthi et al., 2022).
7. **Innovation and Collaboration:** Collaboration among supply chain partners fosters innovation, sharing of best practices, and mutual problem-solving, leading to continuous improvement and better outcomes for all stakeholders.

8. **Global Reach:** In today's globalized economy, supply chain management enables companies to source materials and products from around the world, tapping into diverse markets and benefiting from competitive advantages.
9. **Sustainability and Corporate Social Responsibility:** Supply chain management can be instrumental in implementing sustainable practices, reducing environmental impact, and ensuring ethical sourcing and manufacturing (Avasthi et al., 2022).
10. **Data-Driven Decision-Making:** With the increasing use of technology and data analytics, supply chain managers can make decisions grounded on real-time insights, leading to more efficient and informed business strategies (Avasthi et al., 2023).
11. **Compliance and Regulation:** An efficient supply chain helps ensure compliance with various regulations and standards, preventing potential legal and reputational risks.

In summary, supply chain management is crucial for improving operations, reducing costs, meeting customer demands, and gaining a competitive edge in today's dynamic business environment. Companies that invest in supply chain management strategies and technologies are better equipped to adapt to market changes, deliver superior customer experiences, and drive sustainable growth.

6.1.2 COMPONENTS OF SUPPLY CHAIN MANAGEMENT

The components of SCM include:

1. **Planning and Forecasting:** Anticipating demand, devising production plans, and determining inventory levels to meet customer needs while minimizing excess inventory.
2. **Sourcing:** The procurement process encompasses the identification and selection of suppliers, the negotiation of contracts, and the management of relationships to facilitate the timely and cost-effective acquisition of products and services.
3. **Manufacturing:** Coordinating production processes to meet demand while maintaining quality standards and efficient manufacturing operations.

4. **Logistics and Transportation:** It entails the management of the flow of products from suppliers to manufacturers, distribution centers, retailers, and end customers.
5. **Inventory Management:** It entails monitoring and controlling inventory levels to prevent stock-outs while minimizing transportation costs and excess inventory.
6. **Information Technology:** Using sophisticated software and systems to acquire, analyze, and share data across the supply chain, thereby facilitating better decision-making and performance.
7. **Collaboration and Communication:** Promoting effective collaboration between supply chain stakeholders in order to share information, manage risks, and coordinate activities.

With effective supply chain management, businesses can achieve competitive advantages by optimizing their operations, reducing costs, improving customer service, and adapting quickly to changes in market conditions. It has become increasingly important in a globalized and interconnected world.

6.1.3 CHALLENGES IN SUPPLY CHAIN MANAGEMENT

The field of supply chain management faces several ongoing and emerging challenges. Some of these challenges are (Kumar, 2020; Mukwarami, Nkwaira, and van der Poll, 2023):

1. **Supply Chain Disruptions and Resilience:** The COVID-19 pandemic has brought attention to the inherent weaknesses present within global supply systems. The disruptions resulting from the implementation of lockdown measures, travel restrictions, and staff shortages have highlighted the imperative for the development of supply chains that possess enhanced resilience, enabling them to promptly and effectively respond to unforeseen circumstances.
2. **Global Trade Uncertainties:** Geopolitical tensions, trade wars, and shifts in trade policies between countries have created uncertainties in supply chain operations. Companies have to navigate

changing tariffs, export/import regulations, and potential disruptions to supply chain flows.

3. **Sustainability and Environmental Concerns:** Consumers and stakeholders are increasingly demanding sustainable practices from companies. Supply chain managers face pressure to reduce carbon footprints, minimize waste, and ensure ethical sourcing of materials.

4. **Supply Chain Digitization and Technology Integration:** The implementation of digital technologies, such as AI, blockchain, IoT, and data analytics, brings opportunities for better visibility and efficiency in supply chains. However, integrating these technologies and managing the vast amount of data can be challenging.

5. **E-Commerce and Omni-Channel Distribution:** The rise of e-commerce and the expectation for seamless omni-channel experiences require supply chains to be agile and capable of fulfilling orders from multiple channels effectively.

6. **Shortage of Skilled Talent:** Supply chain management is becoming increasingly complex, requiring professionals with diverse skill sets. The demand for supply chain talent often outpaces the availability of skilled individuals.

7. **Supplier Risk Management:** Relying on a complex network of suppliers exposes businesses to risks related to financial stability, quality issues, natural disasters, or geopolitical events in various parts of the world.

8. **Demand Volatility and Forecasting Accuracy:** The rapidly changing market conditions and consumer behavior make demand forecasting more challenging. Accurate forecasting is crucial to optimize inventory levels and prevent stockouts or excess inventory.

9. **Regulatory Compliance:** Meeting various regulatory requirements related to product safety, transportation, data privacy, and labor practices can pose significant challenges, especially for global supply chains.

10. **Reverse Logistics and Waste Management:** Efficiently managing product returns, recycling, and disposal is critical

for sustainability and cost reduction, but it can be complex and resource-intensive.

To address these challenges, supply chain professionals are increasingly focusing on building resilience, leveraging data and technology, fostering transparency and collaboration in the supply chain, and embracing sustainable and responsible practices. As the business landscape continues to evolve, new challenges and opportunities will emerge, shaping the future of supply chain management.

6.2 GREEN SUPPLY CHAIN MANAGEMENT (GSCM)

Green supply chain management (GSCM), also known as sustainable supply chain management, focuses on integrating environmentally friendly practices throughout the entire supply chain. It involves minimizing the environmental effects of the supply chain by reducing waste and other negative ecological impacts. The key principles of GSCM include choosing suppliers that adhere to environmentally responsible practices, such as using renewable materials and minimizing their carbon footprint. This is known as sustainable sourcing. Green logistics and transportation is also one of the principles of GSCM, which includes optimizing transportation routes, adopting fuel-efficient vehicles, and reducing emissions in distribution processes. Following energy-efficient operations, such as implementing energy-saving measures in manufacturing and warehouse facilities to reduce overall energy consumption, is also a primary concern of GSCM.

Furthermore, modern models of green supply chains work on waste reduction and recycling, promoting waste reduction strategies and recycling initiatives to minimize the environmental impact of supply chain activities. The models also focus on lifecycle assessment, evaluating the ecological influence of products during their entire lifecycle, from raw material extraction to end-of-life disposal or recycling.

6.2.1 IMPORTANCE OF GREEN SUPPLY CHAIN MANAGEMENT

The significance of GSCM has experienced notable growth in recent years as a result of heightened recognition of environmental concerns and the imperative for sustainable business strategies (Gu et al., 2022). The field of GSCM extends beyond conventional supply chain management by integrating environmentally sustainable practices throughout all stages of the supply chain. Several significant factors highlight the importance of GSCM.

Environmental conservation is a critical aspect in which GSCM plays a noteworthy role by effectively mitigating the environmental consequences associated with supply chain operations. Businesses have the potential to contribute to the preservation of natural resources and the reduction of pollution by implementing sustainable practices, which include energy efficiency, waste reduction, and responsible sourcing. Numerous countries and regions have established environmental policies and standards that necessitate companies' adherence. The implementation of GSCM practices guarantees that firms comply with these specified requirements, thereby mitigating the potential for legal sanctions and safeguarding their reputation. One of the benefits of adopting environmentally friendly practices is the potential for long-term cost reductions. One illustration of this notion is that the application of energy-efficient practices can lead to a decrease in utility expenses, whereas the adoption of waste reduction measures can result in the mitigation of disposal costs. Additionally, in light of increasing consumer environmental awareness, enterprises that exhibit a dedication to sustainability through GSCM practices can gain a competitive edge. The implementation of green initiatives has the potential to appeal to customers who prioritize environmental concerns and foster a sense of loyalty toward the company (Al-Khawaldah et al., 2022; Gawusu et al., 2022; Sheng et al., 2023).

Risk management is a vital aspect of GSCM since it enables organizations to effectively detect and mitigate supply chain risks that are linked to environmental concerns. This encompasses the potential hazards associated with climate change, occurrences of natural disasters, and interruptions in supply chains caused by the limited availability of resources. Companies that place a high emphasis on sustainability and green practices through their GSCM strategies frequently experience a favorable brand image and reputation. The establishment of a robust reputation in terms of environmental stewardship has the potential to attract customers, investors, and partners who share similar values.

The adoption of GSCM practices can foster enhanced collaboration and trust among firms and their suppliers. Businesses that prioritize sustainable practices are more likely to be preferred by environmentally conscientious companies. The pursuit of green supply chain strategies frequently fosters innovation and enhances operational efficiency. Organizations are incentivized to actively pursue novel technologies and methodologies that demonstrate dual attributes of environmental sustainability and operational efficiency. The rising environmental consciousness among customers has led to a growing demand for sustainable products and services, hence shaping customer expectations. The application of GSCM enables firms to effectively fulfill customer expectations and effectively target an expanding market segment. The long-term sustainability of enterprises and industries relies on the implementation of sustainable practices. The field of GSCM plays a crucial role in ensuring the resilience and adaptability of supply chains in response to environmental problems.

In general, the application of green supply chain management is vital in cultivating a sustainable business milieu and effectively tackling the worldwide predicaments presented by climate change and the reduction of resources. By including environmental factors into supply chain operations, companies have the potential to not only contribute to a more sustainable future but also enhance their competitive advantage and financial performance in a dynamic market environment.

6.3 BLOCKCHAIN TECHNOLOGY IN SUPPLY CHAIN

Blockchain technology is a system that is decentralized and provides a secure and transparent way to record and validate transactions across a system of computers. Each transaction (block) is linked to the previous one, creating an immutable chain of data blocks (Guo and Yu, 2022). Concerning supply chain management, blockchain offers several benefits as depicted in Figure 6.1.

The utilization of blockchain technology ensures a high level of transparency and immutability in documenting all transactions and the distribution of items across the entirety of the supply chain. This transparency can help identify inefficiencies, improve traceability, and ensure compliance with sustainability standards. Blockchain also enables end-to-end traceability of products, allowing customers and businesses to track the origin, manufacturing process, and journey of products, ensuring authenticity and ethical sourcing (Risso et al., 2023).

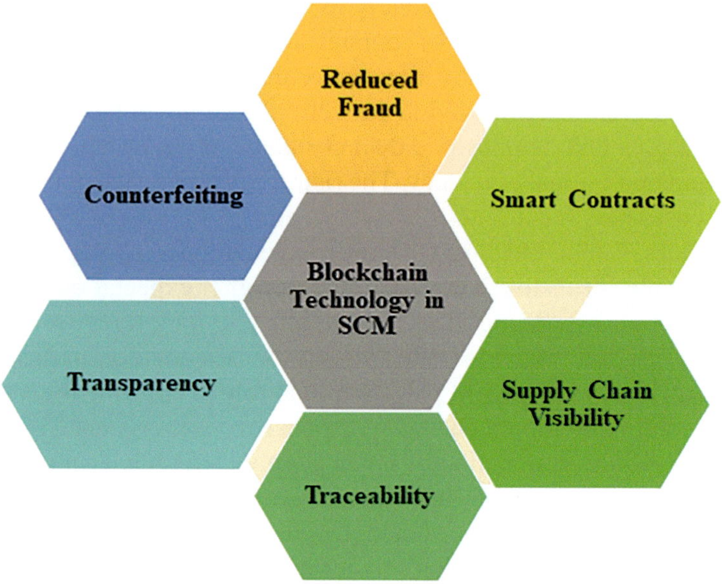

FIGURE 6.1 Blockchain technology in supply chain.

Blockchain enables complete traceability of products and commodities throughout the supply chain. Each transaction or transfer is recorded as a block, establishing an immutable and transparent chain of custody. This allows stakeholders to track the origin, production, and distribution of products, ensuring authenticity and verifying claims of ethical and sustainable practices. Each transaction or movement is recorded as a block, creating an immutable and transparent chain of custody. This allows stakeholders to track the origin, production, and distribution of products, ensuring authenticity and verifying claims of ethical and sustainable practices.

Blockchain further enhances visibility by providing real-time updates on the status and location of goods at every stage of the supply chain, improving coordination and reducing delays.

Smart contracts are contracts that are capable of executing themselves based on predetermined rules. They can automate various processes in the supply chain, such as payments, quality control, and compliance verification. This automation reduces administrative overhead and ensures that all parties involved adhere to agreed-upon terms (Kumar et al., 2023).

Blockchain's tamper-resistant nature reduces the risk of fraud and counterfeiting, ensuring the authenticity and integrity of products. With blockchain, companies can verify the authenticity and origin of raw materials and components. This is especially crucial for industries like food and luxury goods, where ensuring the integrity of the supply chain is essential to prevent counterfeiting and fraud.

6.3.1 BLOCKCHAIN APPLICATION IN SUPPLY CHAIN MANAGEMENT

Blockchain technology has several valuable applications in supply chain management, offering benefits such as enhanced transparency, traceability, security, and efficiency (Gadekallu et al., 2022; Sangari and Mashatan, 2022). Here are some key applications of blockchain in supply chain management as stated in Figure 6.2.

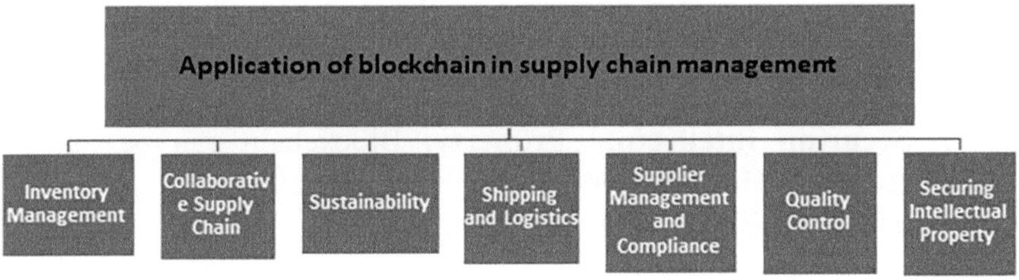

FIGURE 6.2 Application of blockchain in supply chain management.

1. **Inventory Management:** Blockchain can provide real-time visibility of inventory levels at various stages of the supply chain. This visibility helps optimize inventory management, decreasing the risk of stockouts and excess inventory, leading to cost savings and better customer service.
2. **Supplier Management and Compliance:** Blockchain facilitates secure sharing of information and documents among suppliers, manufacturers, and other stakeholders. It enables verification of supplier qualifications, certifications, and compliance with regulatory standards, fostering trust and accountability.
3. **Shipping and Logistics:** Blockchain can streamline shipping and logistics operations by providing a tamper-proof record of the movement of goods and optimizing routes based on real-time data. This enhances efficiency and reduces delays in the supply chain.

4. **Quality Control and Recall Management:** Blockchain's transparency allows for more effective quality control measures. In the event of a product recall, blockchain can help quickly identify affected batches and trace them back to the source, preventing further distribution of defective products.
5. **Sustainability and Green Supply Chain:** Blockchain can track environmental metrics, such as carbon emissions and waste generation, throughout the supply chain (Vu, Ghadge, and Bourlakis, 2023). This data helps businesses assess their environmental impact and implement more sustainable practices.
6. **Collaborative Supply Chain Networks:** Blockchain allows multiple stakeholders to collaborate on a single, shared platform with secure access to data. This fosters trust, improves communication, and facilitates collaboration among suppliers, manufacturers, distributors, and retailers.
7. **Securing Intellectual Property:** Blockchain can be used to record intellectual property rights, protecting companies from unauthorized use or infringement.

By integrating blockchain technology into supply chain management, businesses can streamline processes, improve trust and transparency, reduce costs, and enhance overall efficiency, contributing to a more resilient and sustainable supply chain ecosystem.

6.4 THE INTEGRATION OF GSCM AND BLOCKCHAIN

GSCM and blockchain technology are two concepts that are increasingly becoming intertwined as businesses seek to enhance their sustainability efforts and improve transparency in their supply chain operations. By using blockchain to record and track sustainable practices throughout the supply chain, companies can enhance transparency and accountability. Consumers and stakeholders can be assured that environmentally friendly practices are being followed, and sustainable claims can be verified.

Blockchain, for example, can be used to monitor the origin of raw commodities, validate certifications, monitor carbon emissions, and ensure responsible sourcing practices are implemented. This transparency fosters consumer trust and encourages businesses to adopt more sustainable practices throughout their supply chains.

Overall, the integration of green supply chain management and blockchain can pave the way for more sustainable, efficient, and ethical supply chain operations, contributing to a greener and more responsible future for businesses and the planet.

6.4.1 IMPORTANCE OF BLOCKCHAIN IN GREEN SUPPLY CHAIN MANAGEMENT

The utilization of blockchain technology within the context of green supply chain management presents a multitude of advantageous outcomes (Elhidaoui et al., 2022; Han and Rani, 2022; Munir et al., 2022; Sahoo et al., 2022). The provision of transparency and traceability enables stakeholders and consumers to effectively monitor and assess the environmental impact associated with various products. Additionally, it facilitates the process of validating sustainable assertions, enabling businesses to furnish substantiation of their environmentally conscious initiatives. The utilization of blockchain technology enhances the level of transparency within supply chains, facilitates effective collaboration, and streamlines the process of verifying compliance and meeting sustainability standards by means of smart contracts. The implementation of this measure decreases the likelihood of environmental fraud occurrences and streamlines the auditing procedure. Furthermore, the implementation of blockchain technology not only strengthens consumer confidence but also facilitates the efficient monitoring and implementation of carbon emission tracking and offset initiatives. In general, the utilization of blockchain technology serves as a motivating factor for enterprises to embrace and integrate more environmentally sustainable practices, hence engendering a beneficial cycle of competition.

Blockchain technology brings several crucial advantages to green supply chain management, enhancing its effectiveness and impact. Here are the key reasons why blockchain is important in green supply chain management (Esmaeilian et al., 2020; Khan et al., 2022; Kshetri, 2021; Li et al., 2022; Mangla et al., 2022):

1. **Transparency and Traceability:** The decentralized and unchangeable nature of blockchain enables transparent and tamper-proof tracking of all transactions and movements of items across the supply chain. This transparency allows stakeholders

and consumers to follow a product's environmental impact from raw material procurement to final disposal, assuring compliance with sustainable practices and certifications.

2. **Verification of Sustainable Claims:** Blockchain can be used to verify and authenticate sustainability claims made by suppliers or manufacturers. By recording certifications, environmental data, and compliance information on the blockchain, businesses can provide evidence of their eco-friendly practices and responsible sourcing.
3. **Supply Chain Visibility:** Blockchain's real-time tracking and data-sharing capabilities improve supply chain visibility. This visibility is essential for identifying inefficiencies, reducing waste, and optimizing resource utilization, leading to a more eco-efficient supply chain.
4. **Efficient Collaboration:** Blockchain allows multiple stakeholders to share data securely on a distributed platform. This fosters collaborative efforts to implement green initiatives, share best practices, and collectively work towards sustainability goals.
5. **Smart Contracts for Sustainable Practices:** Smart contracts on the blockchain can automate compliance checks and sustainability requirements. For example, these contracts can automatically verify that suppliers adhere to environmental standards before conducting transactions, promoting responsible sourcing.
6. **Reducing Environmental Fraud:** Blockchain's tamper-resistant nature reduces the risk of fraudulent or false claims regarding sustainability efforts. Companies can rely on verified data on the blockchain to ensure the authenticity of eco-friendly practices.
7. **Auditing and Reporting:** Blockchain simplifies the auditing process for green supply chains. Instead of relying on complex and time-consuming audits, auditors can access the blockchain to verify environmental data and compliance records efficiently.
8. **Enhancing Consumer Trust:** By leveraging blockchain technology, businesses can build trust and credibility with environmentally conscious consumers. Transparently sharing sustainability data fosters customer confidence in the genuineness of green products and ethical business practices.

9. **Carbon Emission Tracking and Offset Programs:** Blockchain can record and track carbon emissions along the supply chain. This data enables businesses to implement carbon offset programs effectively, taking necessary actions to reduce their carbon footprint.
10. **Encouraging Sustainability Adoption:** The implementation of blockchain in GSCM incentivizes companies to adopt more sustainable practices. This can trigger a positive competitive cycle where businesses strive to enhance their environmental performance to maintain a competitive edge.

In summary, blockchain technology empowers green supply chain management with increased transparency, traceability, collaboration, and credibility (Paliwal, Chandra, and Sharma, 2020). It enhances the ability of businesses to monitor, validate, and improve their sustainability efforts, contributing to a more environmentally responsible and strong supply chain ecosystem.

6.4.2 CHALLENGES OF BLOCKCHAIN IN GREEN SUPPLY CHAIN MANAGEMENT (GSCM)

The potential application of blockchain technology in the realm of GSCM is considerable; yet, several issues necessitate resolution before its widespread adoption. This list includes data quality, accuracy, interoperability, scalability, energy consumption, compliance with regulations, adoption and education, costs and investments, the lack of industry standards, the difficulty of smart contracts, and the effects of mining on the environment (Bag et al., 2020). For projects to go well, it's important to ensure that environmental information from different sources is accurate and reliable. Blockchain networks, especially those that use proof of work (PoW) consensus methods, have been seen to use a lot of energy. This means that we need to find new consensus methods that use less energy. To protect data privacy and keep private information safe, you have to follow the rules and laws. For blockchain adoption to work, it's important to think carefully about methods for adoption and educational programs. Companies should think carefully about the return on their investment and the long-term benefits of using blockchain technology. Even with these problems, blockchain can be used to its fullest potential in the field of environmentally sustainable supply chain management if there is continued academic research, technological progress, and cooperation between the right people.

Even though blockchain offers significant potential in GSCM, there exist a multitude of problems that necessitate attention to achieve successful implementation. Some of the key challenges include (Caldarelli, Zardini, and Rossignoli, 2021; Kouhizadeh, Saberi, and Sarkis, 2021):

1. **Data Quality and Accuracy:** The effectiveness of blockchain in GSCM relies on accurate and reliable data. Ensuring the accuracy of environmental data from various sources, including suppliers, requires robust data verification processes and standardized metrics.
2. **Interoperability:** Achieving seamless data sharing and collaboration across multiple supply chain stakeholders can be challenging, especially when different organizations use different blockchain platforms or protocols. Ensuring interoperability between various blockchain networks is essential for effective communication.
3. **Scalability:** As the number of transactions and participants in a green supply chain increases, scalability becomes a concern. Current blockchain networks may face performance issues when processing a large volume of data and transactions, leading to delays and increased costs.
4. **Energy Consumption:** Blockchain networks, particularly those employing PoW consensus algorithms, can consume a significant amount of energy. The energy consumption associated with blockchain operations might counteract the sustainability goals of green supply chains, necessitating the adoption of more energy-efficient consensus mechanisms.
5. **Regulatory and Legal Compliance:** Blockchain's transparency can create challenges in meeting data privacy regulations and protecting sensitive information. Companies must navigate legal frameworks and ensure that personal and confidential data remains secure and compliant.
6. **Adoption and Education:** Implementing blockchain technology requires a change in organizational processes and may be met

with resistance from stakeholders who are unfamiliar with the technology. Educating and gaining buy-in from all participants in the supply chain is crucial for successful adoption.

7. **Costs and Investment:** Integrating blockchain into existing supply chain systems can involve significant upfront costs, including technology infrastructure, training, and development. Companies must carefully assess the return on investment (ROI) and long-term benefits of blockchain adoption.
8. **Lack of Industry Standards:** The absence of standardized approaches and protocols for implementing blockchain in GSCM can lead to fragmentation and hinder broader adoption across industries.
9. **Smart Contract Complexity:** While smart contracts offer automation and efficiency benefits, their development and deployment can be complex and require thorough testing to avoid potential vulnerabilities and errors.
10. **Environmental Impact of Mining:** In some blockchain networks that use PoW, the mining process consumes a considerable amount of energy, raising concerns about its environmental impact. The sustainability of mining activities must be addressed in green supply chains.

Despite these challenges, ongoing research, technological advancements, and collaborative efforts among stakeholders can help address the barriers and maximize the potential of blockchain in GSCM. By tackling these challenges, businesses can unlock the benefits of blockchain technology to enhance transparency, traceability, and sustainability in their supply chain operations.

6.5 IMPORTANCE OF GREEN SUPPLY CHAIN MANAGEMENT (GSCM) IN INDIA

GSCM holds considerable significance in the context of India for various reasons. India is currently confronted with a range of environmental issues, encompassing air and water pollution, deforestation, and garbage generation. GSCM possesses the potential to assume an essential role in the reduction of environmental consequences associated with supply chain operations and the effective management of these difficulties.

Regulatory compliance in India has witnessed a notable surge as the government has taken proactive measures to enforce stringent environmental regulations and standards, intending to effectively tackle sustainability concerns. The application of GSCM techniques enables organizations to adhere to relevant legislation and mitigate the risk of incurring legal and financial consequences.

India is currently experiencing mounting constraints on its natural resources, encompassing water, electricity, and raw minerals, due to resource scarcity. The field of Global Supply Chain Management is dedicated to the promotion of resource efficiency and waste reduction, with the ultimate goal of assuring the responsible utilization of limited resources.

6.5.1 THE IMPACT OF PUBLIC PERCEPTION AND REPUTATION

With the increasing knowledge of environmental concerns, there is a noticeable shift in the mindset of Indian customers towards more environmental consciousness. The usage of GSCM methods has the potential to augment a company's standing, entice customers with a pro-environmental mindset, and fortify brand allegiance.

India exhibits vulnerability to the adverse effects of climate change, encompassing phenomena such as extreme weather occurrences and interruptions in agricultural activities. The field of GSCM offers organizations the opportunity to enhance their resilience by effectively detecting climate-related risks and using adaptive methods to navigate evolving environmental conditions.

The addition of environmental sustainability into supply chain management contributes to sustainable economic growth through cost reduction, improved resource efficiency, and the stimulation of innovation in eco-friendly technology and practices.

India's agricultural sector encounters various issues pertaining to water constraints, soil degradation, and excessive utilization of chemicals, hence necessitating the adoption of sustainable farming practices. The use of GSCM strategies can contribute to the advancement of sustainable agriculture practices through the facilitation of responsible sourcing of inputs and the mitigation of the environmental impact associated with agricultural supply chains.

India is currently undergoing a significant shift towards the adoption of cleaner energy sources. The use of GSCM practices has the potential to support the integration of renewable energy solutions within supply chain operations, thereby making a significant contribution towards India's objectives of achieving a cleaner and more sustainable energy future.

The promotion of green technology and developments in many industries is facilitated by the integration of GSCM practices. This cultivates an environment that prioritizes sustainability and encourages the advancement of environmentally conscious goods and services.

India's involvement in international trade necessitates adherence to worldwide sustainability norms. The application of GSCM techniques in India contributes to the enhancement of the country's reputation as a responsible global partner and facilitates access to sustainable markets.

GSCM facilitates the shift towards a circular economy paradigm in India, wherein the objective is to limit waste generation and optimize resource utilization through the implementation of recycling and reuse practices.

The implementation of GSCM programs frequently results in favorable social outcomes, including the advancement of impartial labor practices, the facilitation of assistance to local communities, and the improvement of the economic conditions for small-scale producers.

The adoption of GSCM strategies by Indian enterprises has the potential to make a noteworthy contribution towards developing a more sustainable and resilient economy, while simultaneously addressing urgent environmental concerns. The aforementioned initiative is in accordance with India's dedication to attaining the United Nations Sustainable Development Goals, thereby bolstering the nation's competitiveness in the international market, given the growing significance of sustainability as a determinant of corporate prosperity.

6.5.2 BLOCKCHAIN TECHNOLOGY AND SUPPLY CHAIN MANAGEMENT (SCM) IN INDIA

The implementation of blockchain technology in SCM in India has the potential to yield numerous benefits, including enhanced transparency, efficiency, and confidence among various stakeholders involved in the process. There exist several mechanisms through which blockchain technology can augment the efficiency of SCM within the context of India (Saberi et al., 2019).

Blockchain technology offers a durable and transparent ledger that maintains an unalterable record of transactions and the flow of commodities throughout the entirety of the supply chain. The level of visibility provided enables stakeholders to effectively monitor the source, manufacturing process, and supply chain of goods, thereby guaranteeing their genuineness and validating adherence to ethical and sustainable standards.

Counterfeit prevention is a matter of great concern in India, particularly within industries such as pharmaceuticals and luxury goods. The use of blockchain technology enables the verification of product authenticity and serves as a deterrent against the proliferation of counterfeit merchandise.

Supplier management is facilitated by blockchain technology, which allows for the secure exchange of information and documentation among suppliers. This results in the optimization of supplier verification procedures and the guarantee of adherence to environmental and social regulations.

Smart contracts can automate several aspects of supply chain operations, including payment dispensation, value control, and compliance verification. This automation minimizes administrative burdens and guarantees that all involved parties comply with predetermined norms and agreements.

The digitized nature of blockchain technology leads to a reduction in chapter work and manual record-keeping, resulting in streamlined administrative operations and a decreased probability of errors.

Efficient logistics operations can be improved through the utilization of blockchain technology, which offers real-time tracking capabilities. This enables improved visibility within the supply chain, facilitating route optimization and ensuring the timely delivery of goods.

Blockchain technology has the potential to improve the efficiency of payment systems by enabling expedited and highly secure cross-border transactions. This has the advantage of minimizing the time and expenses typically incurred in international trade.

The use of blockchain technology enables the recording and monitoring of environmental indicators, such as carbon emissions and waste production, across the entirety of the supply chain. The aforementioned data facilitates the evaluation of organizations' environmental footprint and the adoption of more ecologically sound strategies. Trust and collaboration are facilitated by the decentralized structure of blockchain technology, which promotes a sense of trust among participants in the supply chain.

This, in turn, encourages collaboration and the exchange of information, all while ensuring the protection of data.

The use of blockchain technology has the potential to facilitate adherence to regulatory requirements and standards of environmental, health, safety, and labor policies. This enables enterprises to ensure compliance with both local and international legislation. Furthermore, the implementation of blockchain technology ensures the safeguarding of sensitive supply chain data through the utilization of cryptographic security features, hence preventing unauthorized access and alteration. Blockchain technology also facilitates the auditing process by offering a visible and traceable record of transactions and occurrences across the whole supply chain, hence streamlining the process.

Blockchain technology augments the traceability of agricultural produce in India, hence facilitating equitable pricing for farmers and enhancing the safety of food for consumers. This technology can bring about notable improvements in transparency, efficiency, and sustainability, while fostering collaboration and trust among the various stakeholders involved in the supply chain. The continuous evolution and increasing usage of technology are anticipated to have a substantial positive influence on India's supply chain ecology.

6.6 DISCUSSION

Supply chain management (SCM) refers to the organized coordination and oversight of many operations encompassing the planning, organization, and administration of the movement and storage of goods and information over the whole supply chain, spanning from the initial source to the final destination of consumption. The objective is to decrease costs, enhance efficiency, elevate product quality, reduce production time, and enhance customer satisfaction. SCM encounters various obstacles, including interruptions, uncertainties in global trade, concerns regarding sustainability, the process of digitization, lack of skilled labor, fluctuating demand, adherence to regulatory requirements, and the management of reverse logistics (Chauhan and Kaul, 2021, 2022). Supply chain professionals prioritize the growth of resilience, leveraging data and technology, fostering transparency and collaboration, and embracing sustainable practices to address these challenges.

The concepts of blockchain technology and GSCM are interconnected and contribute to enhancing sustainability and transparency. The utilization of blockchain technology has numerous benefits, including increased transparency, traceability, the implementation of smart contracts, reduced instances of fraudulent activities, and improved inventory management. On the other hand, GSCM primarily emphasizes the adoption of ecologically sustainable practices and procedures. The integration of blockchain technology into the domain of GSCM offers businesses the potential to enhance transparency, accountability, and trustworthiness, while also assuring compliance with sustainable practices. Nevertheless, it is imperative for blockchain technology to effectively tackle concerns pertaining to data accuracy, scalability, energy consumption, and adherence to regulatory standards.

In the milieu of India, the acceptance of GSCM is imperative owing to a range of factors including environmental concerns, adherence to regulatory requirements, limitations in resource availability, and the pursuit of sustainable economic development. The integration of GSCM with blockchain technology can increase several aspects of supply chain operations in Indian enterprises. This integration can lead to improvements in transparency, traceability, and sustainability within the supply chains. Additionally, it can contribute to the enhancement of transparency, efficiency, and sustainability specifically within the agricultural supply chain. In general, the integration of supply chain management, GSCM, and blockchain technology synergistically contributes to the enhancement of sustainability, transparency, and efficiency in supply chain management practices. The utilization of blockchain technology in supply chain management has a multitude of benefits, encompassing sharp levels of efficiency, traceability, security, and transparency. The primary applications encompass several aspects of supply chain management, such as inventory management, supplier management, shipping and logistics, quality control and recall management, sustainability and green supply chain practices, collaborative supply chain networks, and intellectual property security. The integration of blockchain technology in enterprises offers the potential to optimize operational procedures, promote trust and transparency, mitigate expenses, and bolster overall efficacy, hence fostering a more robust and sustainable supply chain ecosystem.

The integration of GSCM and blockchain technology has become more prevalent as organizations strive to better their sustainability initiatives and enhance the visibility of their supply chain activities. The application of blockchain technology for documenting and monitoring sustainable activities across the supply chain can bolster transparency and accountability, hence fostering consumer confidence and incentivizing enterprises to embrace further sustainable practices. The utilization of blockchain technology enables the tracing of the provenance of raw materials, verification of certifications, surveillance of carbon emissions, and enforcement of responsible sourcing protocols.

The amalgamation of GSCM and blockchain technology has the potential to facilitate the development of supply chain operations that are more sustainable, efficient, and ethical. This integration holds promise for fostering a future in which businesses and the environment may coexist in a greener and more responsible manner. Nevertheless, it is imperative to address a multitude of obstacles in order to effectively navigate the landscape of data management. These challenges encompass aspects such as the quality and accuracy of data, the ability to seamlessly integrate and exchange information across various systems, the capacity to handle increasing volumes of data, the energy consumption associated with data processing, adherence to legal requirements, the need for widespread acceptance and understanding of data management practices, the financial implications and investments required, the absence of universally recognized industry standards, the intricacies involved in implementing smart contracts, and the environmental consequences associated with data mining activities. The execution of GSCM techniques in India has the promise of enhancing supply chain management practices, facilitating cooperation and confidence among various actors, and advancing the cause of sustainability and resilience within the economy.

6.7 CONCLUSION

Supply chain management encompasses the methodical coordination and oversight of several operations pertaining to the strategic planning, organizational structuring, and operational administration of the movement and warehousing of goods and information, commencing from their initial source and culminating at their final destination for consumption. The objective is to decrease costs, enhance efficiency, elevate product quality, cut production time, and augment customer satisfaction. Supply chain management encounters various obstacles, including interruptions, uncertainties in global trade, concerns regarding sustainability, the process of digitization, lack of skilled labor, unpredictable fluctuations in demand, adherence to compliance regulations, and the management of reverse logistics. Supply chain professionals prioritize the development of resilience, the use of data and technology, the promotion of transparency and collaboration, and the adoption of sustainable practices. The concepts of blockchain technology and GSCM are interconnected and contribute to the enhancement of sustainability and transparency. The integration of blockchain technology into the field of GSCM has the potential to enhance transparency, accountability, and trust within enterprises, while also promoting the adoption of sustainable practices. Nevertheless, it is imperative for blockchain technology to effectively tackle challenges related to data accuracy, scalability, energy consumption, and adherence to regulatory standards. In the context of India, the adoption of GSCM is imperative due to many factors such as environmental concerns, adherence to legislative requirements, limitations in resources, and the pursuit of sustainable economic development. The integration of GSCM and blockchain technology has the potential to augment transparency, traceability, and sustainability within the supply chains of Indian enterprises, including those operating in the agriculture sector. In general, the integration of supply chain management, GSCM, and blockchain technology synergistically contributes to the enhancement of sustainability, transparency, and efficiency within the domain of supply chain management.

KEYWORDS

- **blockchain**
- **cost reduction**
- **digitization**
- **green supply chain management**
- **information flow**
- **operational administration**
- **supply chain management**
- **sustainable development**
- **warehousing**

REFERENCES

Al-Khawaldah, R., Al-Zoubi, W., Alshaer, S., Almarshad, M., AlShalabi, F., Altahrawi, M., & Al-Hawary, S. (2022). GSCM and competitive advantage: The mediating role of organizational ambidexterity. *Uncertain Supply Chain Management, 10*(3), 961–972.

Avasthi, S., Chauhan, R., Tripathi, S. L., & Sanwal, T. (2022). COVID-19 research: Open data resources and challenges. In *Biomedical Engineering Applications for People with Disabilities and the Elderly in the COVID-19 Pandemic and Beyond* (pp. 93–104). Academic Press.

Avasthi, S., Sanwal, T., Sareen, P., & Tripathi, S. L. (2022). Augmenting mental healthcare with artificial intelligence, machine learning, and challenges in telemedicine. In *Handbook of Research on Lifestyle Sustainability and Management Solutions Using AI, Big Data Analytics, and Visualization* (pp. 75–90). IGI Global.

Avasthi, S., Sanwal, T., Sharma, S., & Roy, S. (2023). VANETs and the use of IoT: Approaches, applications, and challenges. In *Revolutionizing Industrial Automation Through the Convergence of Artificial Intelligence and the Internet of Things* (pp. 1–23).

Bag, S., Viktorovich, D. A., Sahu, A. K., & Sahu, A. K. (2020). Barriers to adoption of blockchain technology in GSCM. *Journal of Global Operations and Strategic Sourcing, 14*(1), 104–133.

Caldarelli, G., Zardini, A., & Rossignoli, C. (2021). Blockchain adoption in the fashion sustainable supply chain: Pragmatically addressing barriers. *Journal of Organizational Change Management, 34*(2), 507–524.

Chauhan, R., & Kaul, V. (2021). Blockchain technology: A panacea for medical distribution ailments. In *Blockchain for Healthcare Systems* (pp. 43–60). CRC Press.

Chauhan, R., & Kaul, V. (2022). Blockchain technology: Unlocking the business model maze for evolving businesses and start-ups. In *Transformations Through Blockchain Technology: The New Digital Revolution* (pp. 291–309). Cham: Springer International Publishing.

Elhidaoui, S., Benhida, K., El Fezazi, S., Kota, S., & Lamalem, A. (2022). Critical success factors of blockchain adoption in GSCM: Contribution through an interpretive structural model. *Production & Manufacturing Research, 10*(1), 1–23.

Esmaeilian, B., Sarkis, J., Lewis, K., & Behdad, S. (2020). Blockchain for the future of sustainable supply chain management in Industry 4.0. *Resources, Conservation and Recycling, 163,* 105064.

Gadekallu, T. R., Huynh-The, T., Wang, W., Yenduri, G., Ranaweera, P., Pham, Q. V., & Liyanage, M. (2022). Blockchain for the metaverse: A review. *Future Generation Computer System, 143, 401–419.*

Gawusu, S., Zhang, X., Jamatutu, S. A., Ahmed, A., Amadu, A. A., & Djam Miensah, E. (2022). The dynamics of GSCM within the framework of renewable energy. *International Journal of Energy Research, 46*(2), 684–711.

Gu, Z., Malik, H. A., Chupradit, S., Albasher, G., Borisov, V., & Murtaza, N. (2022). GSCM with sustainable economic growth by CS-ARDL technique: Perspective to blockchain technology. *Frontiers in Public Health, 9,* 818614.

Guo, H., & Yu, X. (2022). A survey on blockchain technology and its security. *Blockchain: Research and Applications, 3*(2), 100067.

Han, X., & Rani, P. (2022). Evaluate the barriers to blockchain technology adoption in sustainable supply chain management in the manufacturing sector using a novel Pythagorean fuzzy-CRITIC-CoCoSo approach. *Operations Management Research, 15*(3, 4), 725–742.

Khan, S. A., Mubarik, M. S., Kusi-Sarpong, S., Gupta, H., Zaman, S. I., & Mubarik, M. (2022). Blockchain technologies as enablers of supply chain mapping for sustainable supply chains. *Business Strategy and the Environment, 31*(8), 3742–3756.

Kouhizadeh, M., Saberi, S., & Sarkis, J. (2021). Blockchain technology and the sustainable supply chain: Theoretically exploring adoption barriers. *International Journal of Production Economics, 231,* 107831.

Kshetri, N. (2021). Blockchain and sustainable supply chain management in developing countries. *International Journal of Information Management, 60,* 102376.

Kumar, R. (2020). Sustainable supply chain management in the era of digitalization: Issues and challenges. In *Handbook of Research on Social and Organizational Dynamics in the Digital Era* (pp. 446–460).

Kumar, R., Badwal, L., Avasthi, S., & Prakash, A. (2023). A secure decentralized e-voting with blockchain & smart contracts. In *2023 13th International Conference on Cloud Computing, Data Science & Engineering (Confluence)* (pp. 419–424). IEEE.

Li, Q., Ma, M., Shi, T., & Zhu, C. (2022). Green investment in a sustainable supply chain: The role of blockchain and fairness. *Transportation Research Part E: Logistics and Transportation Review, 167,* 102908.

Mangla, S. K., Kazançoğlu, Y., Yıldızbaşı, A., Öztürk, C., & Çalık, A. (2022). A conceptual framework for blockchain-based sustainable supply chain and evaluating implementation barriers: A case of the tea supply chain. *Business Strategy and the Environment, 31*(8), 3693–3716.

Meredith, J. R., & Shafer, S. M. (2023). *Operations and Supply Chain Management for MBAs.* John Wiley & Sons.

Mukwarami, S., Nkwaira, C., & van der Poll, H. M. (2023). Environmental management accounting implementation challenges and supply chain management in emerging economies' manufacturing sector. *Sustainability, 15*(2), 1061.

Munir, M. A., Habib, M. S., Hussain, A., Shahbaz, M. A., Qamar, A., Masood, T., & Salman, C. A. (2022). Blockchain adoption for sustainable supply chain management: Economic, environmental, and social perspectives. *Frontiers in Energy Research, 10,* 899632.

Paliwal, V., Chandra, S., & Sharma, S. (2020). Blockchain technology for sustainable supply chain management: A systematic literature review and a classification framework. *Sustainability, 12*(18), 7638.

Risso, L. A., Ganga, G. M. D., Godinho Filho, M., de Santa-Eulalia, L. A., Chikhi, T., & Mosconi, E. (2023). Present and future perspectives of blockchain in supply chain management: A review of reviews and research agenda. *Computers & Industrial Engineering,* 109195.

Saberi, S., Kouhizadeh, M., Sarkis, J., & Shen, L. (2019). Blockchain technology and its relationships to sustainable supply chain management. *International Journal of Production Research, 57*(7), 2117–2135.

Sahoo, S., Kumar, S., Sivarajah, U., Lim, W. M., Westland, J. C., & Kumar, A. (2022). Blockchain for sustainable supply chain management: Trends and ways forward. *Electronic Commerce Research,* 1–56.

Sangari, M. S., & Mashatan, A. (2022). A data-driven, comparative review of the academic literature and news media on blockchain-enabled supply chain management: Trends, gaps, and research needs. *Computers in Industry, 143,* 103769.

Sanwal, T., & Sareen, P. (2023). Higher employee engagement through social intelligence: A perspective of the Indian scenario. *Employee Responsibilities and Rights Journal, 35*(1), 111–126.

Sanwal, T., Avasthi, S., & Saxena, S. (2016). E-commerce and its sway on the minds of the young generation. *International Journal of Scientific and Research Publications, 6*(3), 112–117.

Sheng, X., Chen, L., Yuan, X., Tang, Y., Yuan, Q., Chen, R., & Liu, H. (2023). GSCM for a more sustainable manufacturing industry in China: A critical review. *Environment, Development and Sustainability, 25*(2), 1151–1183.

Vu, N., Ghadge, A., & Bourlakis, M. (2023). Blockchain adoption in food supply chains: A review and implementation framework. *Production Planning & Control, 34*(6), 506–523.

Yusuf, A., & Soediantono, D. (2022). Supply chain management and recommendations for implementation in the defense industry: A literature review. *International Journal of Social and Management Studies, 3*(3), 63–77.

CHAPTER 7

Blockchain in Tourism: Revolutionizing Travel and Hospitality

RAJEEV SEMWAL,[1,*] NANDITA TRIPATHI,[1] ADITI DHAMA,[2] and VIPIN NADDA[3]

[1]*Amity Institute of Psychology and Allied Sciences, Amity University, Greater Noida, Uttar Pradesh, India*

[2]*Amity University Uttar Pradesh, Greater Noida Campus, Greater Noida, Uttar Pradesh, India*

[3]*University of Sunderland, London, United Kingdom*

*Corresponding author

ABSTRACT

The worldwide tour and hospitality industry is getting ready for an innovative transformation, and at the leading edge of this paradigm shift stands blockchain generation. This summary offers an outline of an in-depth exploration of the various packages of blockchain inside the tourism sector. From redefining supply chain dynamics to reshaping tour guide systems, promoting sustainable tourism practices, safeguarding record privacy, and mitigating fraud in journey insurance, blockchain emerges as a disruptive pressure with the potential to redefine the very fabric of the industry. The introductory phase sets the tone by elucidating the essential ideas of blockchain and its inherent ability to introduce transparency, safety, and performance into conventional tourism systems. As a decentralized and tamper-resistant ledger, blockchain lays the groundwork for a transformative adventure, challenging traditional norms and offering modern solutions. The subsequent chapters of the exploration resolve the effect of

Blockchain Technology: Unlocking the Value of 3Ts in the Business World.
Richa Chauhan & Vidhi Kaul (Eds.)

DOI: 10.1201/9781998511914-7

blockchain across various facets of tourism. Supply chain management takes center stage, showcasing how blockchain tracks and verifies the origin, authenticity, and excellence of products and services in the journey industry. From ensuring the integrity of hospitality offerings to selling sustainable practices, blockchain acts as a catalyst for superb alternate, fostering trust among stakeholders. Decentralized tour manual platforms emerge as a disruptive pressure in the tourism industry, empowering local communities and redefining the dynamics of vacationer-manual interactions. Smart contracts play a pivotal role in automating strategies and improving the efficiency and authenticity of excursion-associated transactions. The exploration then ventures into the world of sustainable tourism, elucidating how blockchain allows transparent projects, monitors and verifies eco-friendly practices, and encourages vacationers to make informed selections aligned with environmental dreams. Through the tokenization of carbon offset applications, blockchain affords a secure and traceable platform, ensuring the credibility and effectiveness of sustainability efforts. The recognition then shifts to facts and privateness, a critical concern in the virtual age. Blockchain's encryption mechanisms and permissioned right of entry redefine how traveler records are handled, ensuring compliance with information safety guidelines and fostering a heightened sense of trust between travelers and provider carriers. Fraud mitigation in journey insurance emerges as a vital utility of blockchain in the very last chapters. By developing transparent and automatic claims approaches, blockchain minimizes the dangers associated with fraudulent sports, constructing trust among insurers and policyholders. In the end, this exploration underscores the transformative capability of blockchain in revolutionizing journeys and hospitality. As stakeholders across the enterprise embrace blockchain's abilities, the panorama of tourism is poised for unheard-of efficiency, transparency, and consideration. The summary encapsulates the essence of this transformative adventure, inviting in addition studies, implementation, and collaboration to release the total spectrum of opportunities that blockchain gives to the dynamic global of tourism and hospitality.

7.1 INTRODUCTION

The international tourism industry stands at the crossroads of evolution, encountering unprecedented growth and complicated challenges. In recent years, the surge in international travel and the digitization of offerings have propelled the industry into new dimensions. However, the traditional processes governing the tourism sector grapple with inherent inefficiencies and vulnerabilities. It is within this panorama of opportunities and demanding situations that the modern capacity of blockchain technology emerges.

7.1.1 BACKGROUND OF THE TOURISM INDUSTRY

The tourism industry, a dynamic and multifaceted zone, performs a pivotal role in the international economy. According to the World Travel and Tourism Council (WTTC), the industry contributed over 10% to the global GDP during the pre-pandemic era, employing tens of millions of people worldwide. From leisure trips and hospitality services to business meetings and cultural exchanges, the scope of the tourism industry is vast and varied.

As international tourism continues to rise, traditional tourism approaches find themselves strained under the weight of increasing demands. Booking systems, identity verification, and supply chain management within the hospitality sector face challenges that impact efficiency, transparency, and, most crucially, trust. These challenges have spurred the exploration of innovative solutions, and one technology, in particular, has emerged as a beacon of transformation: blockchain.

7.1.2 CHALLENGES FACED BY TRADITIONAL TOURISM PROCESSES

Despite its importance, the tourism industry grapples with several challenges that obstruct its seamless functioning. One of the primary challenges is the lack of transparency in reservation systems. Centralized systems often result in discrepancies, incorrect information, or even fraudulent activities, eroding trust among providers and travelers.

Identity management and security concerns represent another essential hurdle. Traditional methods of verifying identities and handling sensitive information pose risks of data breaches and identity theft. Moreover, the reliance on intermediaries in the booking process introduces inefficiencies, delays, and additional costs.

These challenges are compounded within the supply chain of the hospitality sector. From ensuring the authenticity of hotels to monitoring the origin of services, the existing systems lack the transparency and traceability required for the modern traveler's expectations.

7.1.3 THE ROLE AND SIGNIFICANCE OF BLOCKCHAIN IN ADDRESSING CHALLENGES

Blockchain technology, often associated with cryptocurrencies, provides a paradigm shift in addressing the challenges typical of the tourism industry (Dadkhah et al., 2022). At its core, blockchain is a decentralized and distributed ledger technology that ensures transparency, security, and efficiency in data management and transactions.

In the context of tourism, blockchain can revolutionize booking systems by providing a decentralized platform for transparent and automated transactions. Smart contracts, self-executing contracts with the terms of the agreement directly written into code, eliminate the need for intermediaries, reducing costs and minimizing the risk of fraud (Dutta, 2020). Identity verification and security receive a significant boost through blockchain's cryptographic features (Velmurugadass et al., 2021). Immutable data on the blockchain ensures that personal information remains secure and can be selectively shared, enhancing privacy and reducing the risk of identity-related crimes.

The supply chain within the hospitality industry can greatly benefit from blockchain's transparency. Tracking the origin and authenticity of services becomes streamlined, fostering trust among travelers and ensuring the quality and safety of hotels. As we delve deeper into the transformative potential of blockchain in different aspects of the tourism industry, the subsequent sections will explore how this technology reshapes travel security, booking processes, loyalty programs, and the overall travel experience. Through real-world examples and in-depth analyses, we will explore the complex ways in which blockchain is poised to become the cornerstone of a new era in global tourism.

7.2 ENHANCING TRAVEL SECURITY AND IDENTITY MANAGEMENT

In a era defined by rapid technological advancements, the travel industry finds itself at the forefront of innovation, seeking solutions to age-old challenges. As the digitalization of services accelerates, so do the vulnerabilities associated with traditional security and identity verification methods. It is within this landscape of opportunities and risks that blockchain technology emerges as a powerful force, poised to redefine how we secure and verify identities within the dynamic world of travel.

7.2.1 BLOCKCHAIN SOLUTIONS FOR SECURE AND VERIFIABLE IDENTITY

Identity theft and fraudulent activities have long plagued the travel industry, prompting the exploration of more robust and secure verification methods. Blockchain, characterized by its decentralized and cryptographic structure, represents a paradigm shift in how we approach identity management in the digital age (Zhu and Badr, 2018).

On the blockchain, personal identity information is not stored in a centralized repository vulnerable to breaches. Instead, each traveler is assigned a unique cryptographic key, ensuring the security and privacy of their data. This decentralized approach minimizes the risk associated with a single point of failure, presenting a resilient system against large-scale data breaches.

In addition, the integration of smart contracts enhances the security of identity verification. These self-executing contracts facilitate automated and secure identity checks, reducing reliance on traditional, often manual verification methods. Travelers benefit from a more efficient onboarding experience while maintaining control over the sharing of their personal information.

7.2.2 STREAMLINING AIRPORT SECURITY AND IMMIGRATION

Airports, often characterized by long queues and stringent security measures, stand as gatekeepers to the travel experience. Blockchain's capacity to streamline airport security processes offers a glimpse into a future where efficiency and security coexist seamlessly (Zorlu et al., 2023).

In a blockchain-based airport security model, travelers' identity information is securely stored on a decentralized ledger. Authorized personnel at various checkpoints can access this information using cryptographic keys, allowing for a smoother and more secure verification process. This decentralized approach reduces redundancy in identity checks, ultimately expediting overall security procedures within the airport.

Immigration processes, historically marked by paperwork and manual verification, can benefit significantly from blockchain's efficiency (Shinde and Gurrala, 2023). The decentralized and secure nature of blockchain ensures that immigration officials have access to accurate and up-to-date information, reducing the time and resources required for manual checks. This not only enhances operational efficiency but also contributes to a positive and seamless travel experience for passengers.

7.2.3 IMMUTABLE TRAVEL RECORDS AND DIGITAL PASSPORTS

The immutability of blockchain, a fundamental feature, holds transformative implications for the integrity of travel information and digital passports. Traditional paper passports, prone to loss or damage, are steadily giving way to digital counterparts securely stored on the blockchain.

Digital passports, housed within a traveler's digital wallet, contain verified information about their identity, travel history, and vaccination records (Cook, 2020). The immutability of the blockchain ensures that this information remains accurate and tamper-proof, providing an undeniable record of the traveler's journey.

In the context of global health considerations, blockchain-based digital passports play a pivotal role in verifying vaccination status and ensuring compliance with health regulations. This not only contributes to public health and safety but also instills confidence in global travel, facilitating the industry's recovery from the effects of global health crises.

In conclusion, the application of blockchain technology in travel security and identity management is a journey toward a more secure, efficient, and consumer-centric travel experience. From secure and verifiable identity solutions to streamlined airport processes and immutable travel records, blockchain's impact on industries is multifaceted. As we navigate this landscape of innovation, it becomes increasingly clear that blockchain is not just a technological advancement but a foundational shift in how we approach security and identity within the dynamic global context of world tourism.

7.3 SMART CONTRACTS FOR TRANSPARENT AND EFFICIENT BOOKING SYSTEMS

In the ever-evolving landscape of the travel enterprise, conventional reserving structures have faced challenges, ranging from a lack of transparency to susceptibility to fraudulent activities. The integration of blockchain technology, particularly through the implementation of smart contracts, promises to revolutionize booking systems, making them more transparent, efficient, and resistant to fraudulent practices (Figure 7.1).

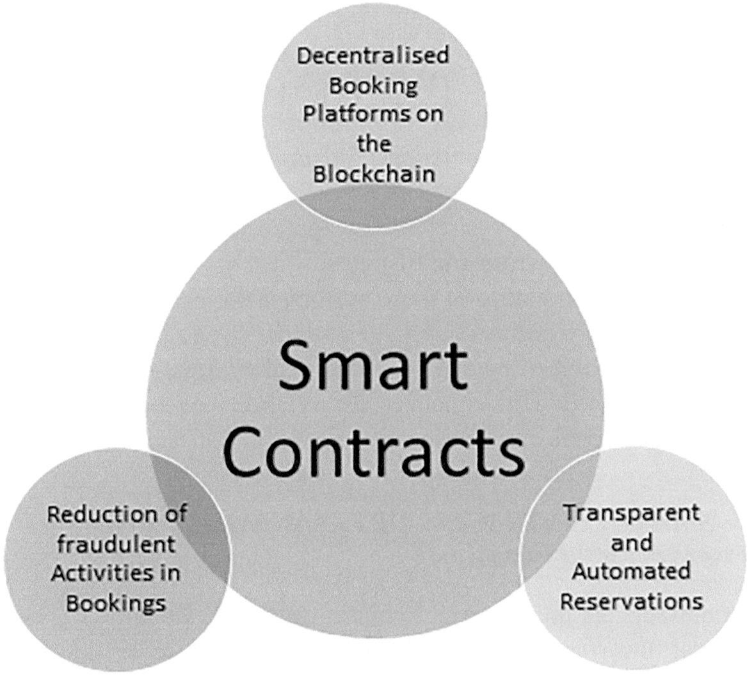

FIGURE 7.1 .Smart contracts for transparent and efficient booking systems.

Source: Authors own compilation.

7.3.1 DECENTRALIZED BOOKING PLATFORMS ON THE BLOCKCHAIN

Traditional booking systems often operate through centralized structures, where intermediaries play a significant role in facilitating transactions between travelers and service providers (Rashideh, 2020). This centralized model, while common, introduces inefficiencies, increased costs, and a lack of transparency. Blockchain technology offers a decentralized alternative, providing a transformative shift in how bookings are made in the travel industry.

On a decentralized booking platform powered by blockchain, the middleman function diminishes, and transactions occur directly between the traveler and the service provider. The decentralized ledger ensures that the booking history and terms of agreements are securely recorded and transparently accessible to all relevant parties.

Smart contracts, self-executing contracts with predefined rules and conditions, play an important role in decentralized booking platforms (John William et al., 2022). These contracts automate the execution of agreements, from confirming reservations to processing payments, without the need for intermediaries. The terms of the contract are written into code and executed automatically when predefined conditions are met. This automation not only streamlines the reservation process but also minimizes the risk of errors and disputes.

The decentralized nature of these systems also means that they are less susceptible to system failures or targeted attacks (Chauhan and Kaul, 2021). With no central point of control, the risk of downtime or manipulation is significantly reduced. This enhances the reliability of the booking system, ensuring that travelers and service providers can transact with confidence.

7.3.2 SMART CONTRACTS FOR TRANSPARENT AND AUTOMATED RESERVATIONS

Smart contracts, as they are the linchpin of the blockchain era, introduce exceptional transparency and efficiency into the reservation process (Lee, 2022). In the context of travel bookings, smart contracts automate numerous stages of the reservation lifecycle, providing benefits for both travelers and service providers.

When a traveler initiates a booking, a smart contract is created, containing all of the relevant information, including check-in and check-out dates, accommodation options, and payment details. This contract is then broadcasted to the decentralized network for validation. Once verified, the contract is added to the blockchain, creating an immutable record of the reservation.

The transparency of this process ensures that each party has visibility into the terms of the reservation. Travelers can verify the conditions of their booking, and service providers can access accurate and unalterable records. This transparency facilitates agreement between parties, mitigating disputes or misunderstandings.

Automation through smart contracts also extends to payment processing (Wang et al., 2019). As predefined conditions are met, such as the successful completion of a stay, the smart contract automatically executes the payment. This not only eliminates the need for intermediaries to facilitate transactions but also ensures that payments are processed directly and without the risk of delays.

7.3.3 REDUCTION OF FRAUDULENT ACTIVITIES IN BOOKINGS

Fraudulent activities, such as identity theft, fake bookings, and chargebacks, have been persistent challenges within the travel industry. Blockchain's inherent security features, combined with the automation of smart contracts, offer robust solutions to mitigate these risks and reduce fraudulent activity (Riesco et al., 2020). The decentralized and immutable nature of the blockchain ensures that once a booking record is added to the ledger, it cannot be altered or tampered with. This provides a transparent and tamper-proof audit trail of all transactions, lowering the risk of fake bookings or fraudulent alterations to reservation information.

Smart contracts contribute significantly to fraud prevention. Identity verification procedures can be embedded into smart contracts, ensuring that only legitimate customers can make bookings. This reduces the likelihood of identity theft and ensures that the reservation system operates with the utmost security. Chargeback fraud, a common problem for service providers, is also mitigated through smart contracts. Since the terms and conditions of the reservation are predefined in the contract, disputes arising from misunderstandings or false claims are minimized. The automated execution of payments upon successful completion of a stay further reduces the opportunity for chargeback fraud.

In the end, the combination of smart contracts with booking systems powered by blockchain represents a pivotal moment in the evolution of the travel industry (Chauhan and Kaul, 2022). By moving towards decentralized booking platforms and leveraging the transparency and automation provided by smart contracts, the industry stands to benefit not only in terms of efficiency but also in building trust and reducing fraudulent activities. As we explore the real-world applications and the potential impact of this technological advancement, it becomes clear that the future of travel bookings lies within the secure, transparent, and automated realm of blockchain-powered smart contracts.

7.4 IMPROVING LOYALTY PROGRAMS WITH BLOCKCHAIN

In the fiercely competitive landscape of the tourism industry, customer loyalty is a coveted asset. Traditional loyalty programs, while widely adopted, often face challenges related to transparency, security, and user engagement. The introduction of blockchain technology provides the opportunity to revolutionize loyalty programs, making them more efficient, transparent, and straightforward (Figure 7.2).

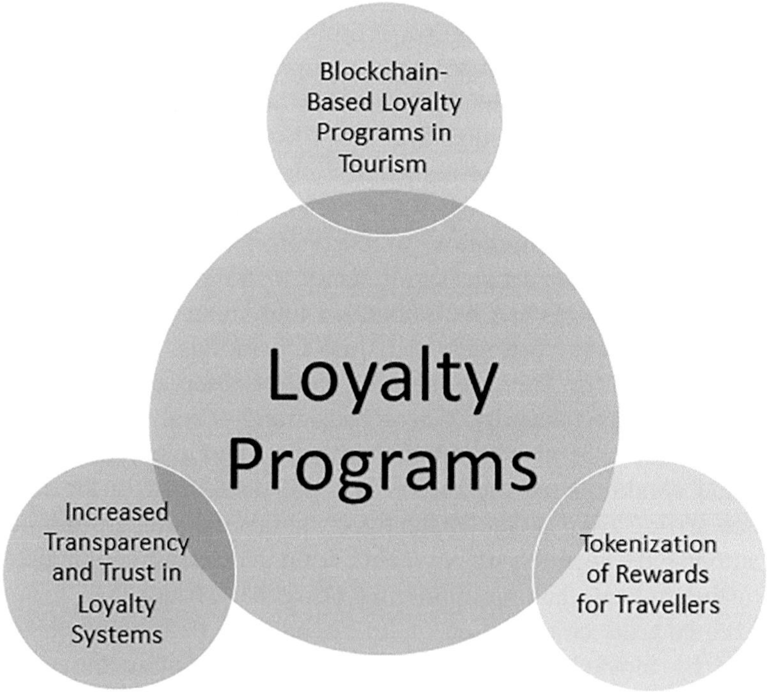

FIGURE 7.2 Improving loyalty programs with blockchain.

Source: Authors own compilation.

7.4.1 BLOCKCHAIN-BASED LOYALTY PROGRAMS IN TOURISM

Loyalty programs are imperative to retaining customers and fostering a sense of brand allegiance (Morais et al., 2005). However, the traditional models of loyalty programs can be marred by complexities, delays in reward distribution, and a lack of transparency. Blockchain, with its decentralized and transparent nature, offers a paradigm shift in loyalty program management. In a blockchain-based loyalty program, the entire rewards ecosystem is decentralized and operates on a transparent ledger. Each customer's loyalty points or rewards are securely recorded on the blockchain, creating an immutable record of their engagement with the brand. This transparency ensures that both clients and service providers have visibility into the accumulation, redemption, and expiration of loyalty points.

Smart contracts play a pivotal role in automating the operations of loyalty programs (Khan et al., 2021). These contracts are programmed to execute predefined rules and conditions automatically. For example, when a customer achieves a certain level of engagement or makes a qualifying purchase, the smart contract triggers the issuance of loyalty points. This automation streamlines the reward distribution process, eliminating delays and potential errors related to manual intervention.

7.4.2 TOKENIZATION OF REWARDS FOR TRAVELERS

Tokenization is a key innovation introduced by blockchain technology, and it has profound implications for loyalty programs in the tourism sector (Treiblmaier, 2022). Instead of traditional loyalty points, which are often specific to particular programs and lack interoperability, blockchain enables the tokenization of rewards.

Tokens are digital assets representing a certain value or right within specified environments (Westerkamp et al., 2020). In the context of loyalty programs, tokens can be seamlessly transferred and exchanged across different travel services and partners. This interoperability enhances the power and utility of loyalty rewards, providing travelers with a more versatile and valuable incentive.

Moreover, tokenization reduces the friction associated with redemption. Traditional loyalty programs may have restrictions on how and where rewards can be redeemed. With tokenized rewards on the blockchain, travelers can use their loyalty tokens for various offerings, from booking accommodations to accessing exclusive experiences, creating a more rewarding and personalized loyalty program.

7.4.3 INCREASED TRANSPARENCY AND TRUST IN LOYALTY SYSTEMS

Transparency is an important factor in fostering acceptance as true within loyalty packages (Lin et al., 2017). Customers want a warranty that their loyalty factors are recorded as they should be and that the terms and conditions of the program are upheld (Ozuem et al., 2016). Blockchain's obvious and decentralized ledger addresses those worries, offering an immutable report of all transactions related to the loyalty program.

Travelers can, without problems, verify their accrued loyalty points and the guidelines governing their redemption via the decentralized ledger. Any adjustments to the loyalty software, such as changes in redemption thresholds or expiration policies, can be transparently communicated through the blockchain. This expanded transparency instills self-assurance in customers, reinforcing the integrity of the loyalty program. Additionally, the decentralized nature of blockchain mitigates the threat of fraud within loyalty applications. Each transaction is securely recorded on the blockchain, reducing the probability of unauthorized alterations or manipulations. This complements the safety and reliability of loyalty programs, assuring vacationers that their engagement and loyalty are valued and guarded.

The blockchain era brings a fresh attitude to loyalty programs within the tourism enterprise. By decentralizing and tokenizing rewards and leveraging clever contracts for automation, the enterprise can create loyalty applications that are not only green but also extra obvious and sincere. As blockchain-based loyalty programs continue to gain momentum, tourists can expect a future wherein their loyalty is rewarded seamlessly and with extra versatility, fostering lasting connections between customers and the manufacturers they price.

7.5 SUPPLY CHAIN TRANSPARENCY IN THE HOSPITALITY SECTOR

The hospitality quarter is predicated on elaborate and multifaceted deliver chains to supply satisfactory services to vacationers. However, ensuring transparency, authenticity, and safety within these supply chains has been a longstanding challenge. The integration of blockchain technology emerges as a transformative answer, offering remarkable visibility and believe within the origin, first-class, and protection of hospitality services.

7.5.1 TRACKING ORIGIN AND AUTHENTICITY OF HOSPITALITY SERVICES

In an enterprise where the provenance of services is paramount, blockchain affords an immutable and transparent ledger for monitoring the origin and authenticity of hospitality services. From the sourcing of materials for room fixtures to the starting place of meals served in eating places, each stage of the supply chain may be recorded at the blockchain. For instance, bear in mind the journey of top rate linens used in inn rooms. Traditionally, tracing the foundation and authenticity of these linens will be a complex and regularly opaque procedure. With blockchain, each step within the deliver chain, from the producer to the distributor and subsequently to the lodge, is recorded on a decentralized ledger. This guarantees that the linens are ethically sourced, meet first-rate requirements, and are actual, offering transparency for both the inn and the environmentally aware traveler.

Moreover, Blockchain's decentralized nature gets rid of the threat of fraudulent claims regarding the beginning or authenticity of products and services. Travelers can get entry to this information, regularly presented in a user-friendly interface, to make knowledgeable choices that align with their values and possibilities.

7.5.2 ENSURING THE QUALITY AND SAFETY OF ACCOMMODATIONS

Quality and safety are non-negotiable factors in the hospitality sector (Vasilakakis and Sdrali, 2023). From the cleanliness of rooms to adherence to safety policies, retaining high standards is imperative. Blockchain generation introduces a unique approach to ensuring and verifying the quality and protection of accommodations. Blockchain allows the advent of a tamper-proof report detailing the renovation history and cleanliness standards of resort rooms (Jain et al., 2023; Nguyen et al., 2019). Smart contracts may be programmed to trigger alerts when it's time for recurring inspections or protection responsibilities. This no longer only guarantees that hotels are consistently up to standard, but additionally streamlines the auditing process for regulatory compliance.

Consider the state of affairs in which a hotel undergoes habitual hearth safety inspections. The consequences of these inspections, along with any vital corrective moves taken, are recorded on the blockchain. Travelers can get admission to these records to make informed decisions about their lives, and regulatory authorities can successfully verify compliance without relying on cumbersome chapter work. Furthermore, in the context of fitness and protection, especially within post-pandemic technology, blockchain can play a vital role. Records of sanitation protocols, air quality measures, and fitness and safety certifications may be securely saved on the blockchain, assuring tourists of a dedication to their well-being.

7.5.3 USE CASES OF BLOCKCHAIN IN SUPPLY CHAIN MANAGEMENT

Blockchain's ability to deliver chain management extends past tracking starting place and ensuring pleasant it additionally introduces efficiencies and fee financial savings. One compelling use case is within the procurement of goods and offerings for hospitality institutions.

Traditional procurement methods often contain several intermediaries, leading to extended charges, delays, and potential errors. Blockchain streamlines this system by presenting a decentralized and obvious platform for coping with procurement contracts. Smart contracts can automate various levels of the procurement lifecycle, from order placement to bill settlement, lowering the need for guidance intervention. For example, while a hotel places an order for sparkling produce from a supplier, a clever agreement can routinely execute the charge upon a successful transport, as verified by using predefined situations that include temperature management for the duration of transportation. This not only hurries up the procurement process but additionally minimizes the danger of mistakes and disputes.

Blockchain's decentralized ledger ensures that each applicable party, along with suppliers, distributors, and hospitality institutions, has real-time access to the reputation of orders, payments, and inventory stages.

This transparency complements collaboration, reduces inefficiencies, and fosters an extra resilient and agile delivery chain atmosphere. Blockchain technology is poised to revolutionize supply chain transparency within the hospitality sector. By monitoring the starting place and authenticity of offerings, ensuring the pleasantness and safety of resorts, and introducing efficiencies in delivery chain management, blockchain enhances the overall trust and performance of the enterprise. As the hospitality zone continues to include blockchain solutions, travelers can count on a future where their reviews aren't just high-priced but additionally underpinned by a dedication to transparency, authenticity, and safety.

7.6 STREAMLINING CROSS-BORDER PAYMENTS FOR INTERNATIONAL TRAVEL

International travel, marked by its international nature, is regularly hindered by the complexities and inefficiencies inherent in traditional cross-border payment systems. The elaborate web of intermediaries, high transaction prices, and delays pose demanding situations for both vacationers and service providers. The blockchain era emerges as a disruptive force, offering solutions to streamline cross-border bills, making them faster, more price-powerful, and handier.

7.6.1 CHALLENGES IN TRADITIONAL CROSS-BORDER PAYMENT SYSTEMS

Traditional pass-border charge structures have long been characterized by sluggish transaction processing instances, high fees, and a lack of transparency (Veljan, 2021). When travelers make global bills, their money usually goes via a convoluted chain of banks and monetary institutions, each taking a cut and introducing delays within the technique.

The forex market, in which currencies are offered and sold, similarly complicates cross-border transactions (Wojcik et al., 2017). Fluctuating trade quotes and additional expenses for foreign currency conversion contribute to the overall value of international payments. These challenges now not only impact the cost-effectiveness of international travel but additionally introduce uncertainty and frustration for travelers.

Moreover, the traditional banking infrastructure is often inaccessible to people in certain areas or those without a proper banking relationship, compromising the inclusivity of cross-border payments.

7.6.2 BLOCKCHAIN SOLUTIONS FOR FASTER AND COST-EFFECTIVE TRANSACTIONS

Blockchain technology, with its decentralized and distributed ledger, offers a transformative approach to the challenges posed by traditional cross-border payment systems. By cutting out intermediaries and providing a transparent and secure platform, blockchain accelerates transaction processing times and reduces costs. Smart contracts, a key feature of blockchain, can be leveraged to automate and streamline cross-border payments (Li and Kassem, 2021). These self-executing contracts include predefined rules and conditions, allowing for the automatic execution of transactions when certain criteria are met. For instance, a smart contract could facilitate the instant release of funds upon the successful verification of travel services, eliminating the need for time-consuming manual processes.

The decentralized nature of blockchain ensures that cross-border payments can occur directly between the payer and the recipient without the need for multiple intermediaries. This not only speeds up the transaction process but also minimizes the costs associated with each intermediary, resulting in a more cost-effective solution for both travelers and service providers. Blockchain also addresses the challenges associated with the foreign exchange market. Stablecoins, which are digital currencies pegged to the value of a fiat currency, can be used to reduce volatility and eliminate the need for traditional currency conversion. This approach ensures that the recipient receives the intended value without the uncertainty of fluctuating exchange rates.

7.6.3 CRYPTOCURRENCY ADOPTION IN THE TRAVEL INDUSTRY

Cryptocurrencies, including Bitcoin and Ethereum, are gaining traction within the travel industry as an alternative method of payment. By facilitating peer-to-peer transactions without the need for traditional banking channels, cryptocurrencies provide a decentralized and borderless solution for cross-border payments. Travel companies and service providers are increasingly accepting cryptocurrencies as a form of payment for bookings and reservations (Irannezhad and Mahadevan, 2021; Li and Kassem, 2021). This adoption not only caters to the growing number of crypto-savvy travelers but also aligns with the principles of decentralization and financial inclusivity.

Cryptocurrencies provide travelers with the ability to make payments directly from their digital wallets, bypassing the delays and costs associated with traditional payment methods. Additionally, the transparency and security features of blockchain ensure that transactions are recorded on an immutable ledger, reducing the risk of fraud and providing a traceable record of payments. Furthermore, as central bank digital currencies (CBDCs) gain prominence, they offer governments the opportunity to explore more efficient and transparent cross-border payment systems. CBDCs, built on blockchain technology, could potentially streamline global transactions while maintaining regulatory oversight.

Blockchain technology is redefining the landscape of cross-border payments within the realm of global travel. By addressing the inefficiencies of traditional systems, offering faster and more cost-effective solutions through smart contracts, and fostering cryptocurrency adoption, blockchain is paving the way for a more seamless and inclusive experience for travelers. As the travel industry continues to embrace blockchain innovations, the future holds the promise of a global payment environment that is faster, more accessible, and more aligned with the dynamic nature of international travel.

7.7 PRESERVING CULTURAL HERITAGE THROUGH BLOCKCHAIN

Cultural heritage, encompassing artifacts, monuments, and traditions, is a testament to the rich tapestry of human history. Unfortunately, these treasures often face threats from illegal trade, vandalism, and environmental factors. Blockchain technology emerges as an effective tool for the preservation of cultural assets and heritage sites, providing solutions in digital provenance, authentication, and community involvement.

7.7.1 BLOCKCHAIN FOR PROTECTION OF CULTURAL ASSETS AND HERITAGE SITES

Cultural assets, whether housed in museums or present as historical sites, are prone to various threats. The illegal trade of artifacts, fueled by the demand for rare and valuable objects, poses a significant challenge to cultural heritage preservation. Blockchain's decentralized and tamper-resistant ledger provides a revolutionary approach to combat these threats (Hari and Lakshman, 2016; Ichikawa et al., 2017). By recording the provenance of cultural property on the blockchain, a transparent and unalterable record of their ownership and origin is established. This digital provenance becomes an effective tool in verifying the authenticity of artifacts and preventing the trade of stolen cultural objects. When transactions concerning cultural assets are recorded on the blockchain, potential buyers can trace the history of the item, ensuring its legitimacy and ethical acquisition.

For historical sites, blockchain can play a crucial role in safeguarding against unauthorized alterations or developments. Land registries, architectural plans, and conservation efforts can be securely stored on the blockchain, providing an immutable record of the site's history and preservation initiatives. This transparency enhances accountability and reduces the risk of unauthorized activities that could compromise the integrity of cultural heritage sites.

7.7.2 DIGITAL PROVENANCE AND AUTHENTICATION OF ARTIFACTS

The authentication of artifacts is a cornerstone of cultural heritage preservation. Blockchain technology facilitates the creation of digital certificates of authenticity, securely stored on the blockchain (Hasan et al., 2020). These certificates offer a verifiable and permanent record of an artifact's origin, ownership history, and any relevant information about its cultural significance. When museums, collectors, or heritage organizations create a record of an artifact on the blockchain, it becomes a digital twin of the physical object. This digital representation includes details such as the artifact's age, materials, and historical context. As the artifact changes hands or undergoes conservation efforts, these transactions are recorded on the blockchain, ensuring a transparent and accessible history. This digital provenance not only helps combat the illicit trade of cultural artifacts but also serves as a valuable resource for researchers, historians, and the public. Accessible via blockchain-enabled systems, the digital provenance of artifacts becomes a tool for education and a means of fostering a deeper appreciation for cultural heritage.

7.7.3 CROWDSOURCING AND COMMUNITY INVOLVEMENT IN HERITAGE PRESERVATION

Preserving cultural heritage is not solely the responsibility of institutions or governments; it is a collective endeavor that involves communities and individuals. Blockchain technology enables crowdsourcing initiatives for heritage preservation, encouraging community involvement in the documentation and protection of cultural assets.

Through blockchain-enabled platforms, communities can contribute information about local heritage sites, share stories about cultural traditions, or even report instances of vandalism or illegal activities. These contributions, once verified and recorded on the blockchain, create a decentralized repository of community knowledge and efforts in heritage preservation.

Smart contracts can be employed to incentivize community participation. For instance, community members who make valuable contributions or actively participate in preservation activities can be rewarded with tokens or other forms of recognition. This gamification of heritage preservation encourages a sense of ownership and pride among community members, fostering a sustainable and collaborative approach to cultural heritage protection.

In conclusion, blockchain technology offers a multifaceted approach to preserving cultural heritage. From protecting artifacts against illegal trade through digital provenance to involving communities in heritage preservation efforts, blockchain enhances the transparency, authenticity, and inclusivity of cultural heritage projects. As the world grapples with the challenges of preserving its diverse heritages, blockchain emerges as a beacon of innovation, empowering institutions and communities to protect the treasures that define our shared human history.

7.8 DECENTRALIZED TOUR GUIDE PLATFORMS

As the travel industry evolves, the concept of decentralized tour guide platforms is gaining prominence, harnessing the power of blockchain technology. These platforms facilitate peer-to-peer tour guide services, leveraging smart contracts for transparent and automated payments. Beyond the financial aspects, decentralized tourism platforms empower local communities, providing them with opportunities for sustainable tourism and economic growth.

7.8.1 PEER-TO-PEER TOUR GUIDE SERVICES ON THE BLOCKCHAIN

Traditional tour guide services often involve intermediaries, leading to higher fees for tourists and reduced earnings for local guides. Decentralized tour guide platforms built on blockchain technology aim to disrupt this model by connecting travelers directly with local guides. This peer-to-peer approach not only enhances the authenticity of the experience but also fosters a more equitable distribution of earnings. Blockchain allows secure and transparent transactions, enabling travelers to pay for tours directly without the need for multiple intermediaries (Irannezhad and Mahadevan, 2021). The decentralized ledger records these transactions, providing verifiable records of engagements. This transparency is trusted by both travelers and guides, as the authenticity of services and payments can be independently verified.

Additionally, blockchain ensures that guides maintain and control their personal data and pricing strategies. Smart contracts, embedded within the platform, can automate the negotiation and execution of agreements between guides and travelers. This decentralized model not only streamlines the booking process but also gives guides the flexibility to set their own terms and conditions.

7.8.2 SMART CONTRACTS FOR TRANSPARENT AND AUTOMATED PAYMENTS

Decentralized tour guide systems rely on smart contracts to automate and transparently process payments. A smart contract containing tour details, duration, and payment is created when a traveler books a tour. After the tour is complete, the smart contract automatically transfers payments from the traveler to the guide. This automated payment system eliminates third-party payment processors, reducing transaction fees and ensuring guides receive their money immediately. Smart contracts reduce disputes since all parties have an immutable record of agreed-upon terms and service completion. Smart contracts can include reviews and ratings to enhance tour guide service management. Commenting on the blockchain provides future travelers with valuable information and builds a reputation mechanism that rewards guides for excellent service.

7.8.3 EMPOWERING LOCAL COMMUNITIES THROUGH DECENTRALIZED TOURISM PLATFORMS

Decentralized tour guide systems can empower local communities economically and culturally. These platforms help towns showcase their histories, traditions, and hidden gems by connecting tourists with local guides. This ensures that local guides and communities benefit financially from tourism while promoting a more authentic travel experience. Communities can leverage decentralized structures to diversify their tourism offerings beyond mainstream attractions (Warren and Dinnie, 2018). Whether it's a gastronomic tour of local markets, an exploration of traditional craftsmanship, or a visit to lesser-known historical sites, decentralized tour guide platforms allow communities to create and share their own narratives. Additionally, the financial benefits of decentralized tourism help local communities thrive. Community-based guides become responsible tourism ambassadors. Travelers become more aware of and appreciative of the local culture. Ultimately, decentralized tour guide platforms transform the tourism industry by connecting tourists and locals more directly and equitably. Blockchain for transparent transactions and smart contracts for automated payments enhance tour guide services and empower local communities economically and culturally. Decentralized tourism systems demonstrate blockchain technology's potential to provide a more sustainable and inclusive global future as the travel industry evolves.

7.9 ENSURING DATA PRIVACY IN TRAVEL

In an era where digital interactions play a pivotal role in the travel experience, ensuring the privacy and security of traveler data is of utmost importance. Blockchain technology emerges as a robust solution for protecting sensitive data through encryption, permissioned access, and compliance with data protection regulations to enhance data privacy within the travel industry.

7.9.1 SECURING TRAVELER DATA USING BLOCKCHAIN ENCRYPTION

Travel companies collect a significant amount of sensitive data, from passports to payment information. Protecting this data against cyber-attacks and unauthorized access is crucial. Blockchain's encryption safeguards traveler information throughout its lifecycle (Velmurugadass et al., 2021). Blockchain's decentralized structure uses cryptography to keep personal and transaction records secure. Traveler data on the blockchain is encrypted and linked to a unique cryptographic key. This makes deciphering encrypted data without the key challenging for a malicious actor who gains access to the blockchain.

Blockchain's distributed structure makes it resistant to single-point attacks because data cannot be stored centrally. Instead, every participant in the blockchain network maintains a copy of the encrypted records, increasing redundancy and system resilience against breaches.

7.9.2 PERMISSIONED ACCESS TO PERSONAL INFORMATION

Securing information is important, but so is providing essential data to authorized entities within the travel ecosystem. Blockchain allows individuals to control who can access their personal data through permissioned access. Smart contracts enable users to design and enforce data access and usage policies. Travelers can also provide airlines with specific flight ticket details without exposing unnecessary personal information. Smart contracts automate permission settings to restrict data access to approved parties. Permissioned access to data enhances privacy and reduces data misuse. Travelers can trust that their data is only shared with businesses that require it for services, preventing unwanted surveillance or abuse.

7.9.3 COMPLIANCE WITH DATA PROTECTION REGULATIONS IN THE TOURISM SECTOR

Data privacy concerns are growing, prompting governments and regulators to implement strict data security laws. The EU's General Data Protection Regulation (GDPR) and similar laws worldwide restrict how companies handle personal data. Blockchain, which emphasizes transparency and compliance, aligns well with data protection regulations. Smart contracts can be programmed to automatically enforce compliance with specific policies, ensuring that the processing and sharing of traveler data adhere to legal requirements (Wang et al., 2019).

Blockchain's decentralization aids regulatory compliance. Blockchain distributes data management and security across a network of participants. This eliminates regulatory violations due to centralized failure or negligence. Travel agencies and providers can demonstrate data privacy and compliance by utilizing blockchain. The transparency and traceability of blockchain transactions enable auditing, allowing organizations to prove regulatory compliance. In conclusion, data privacy is a complex issue that requires robust solutions. Blockchain technology's encryption, permissioned access framework, and data protection compliance provide a comprehensive foundation for traveler data protection. As the travel sector navigates digital interactions and regulatory landscapes, blockchain offers a more secure, transparent, and privacy-centric future for managing sensitive global travel data.

7.10 BLOCKCHAIN AND SUSTAINABLE TOURISM

Sustainable tourism is becoming increasingly vital in addressing the environmental and social impacts of the travel industry. Blockchain technology emerges as a transformative force, offering innovative solutions to track, verify, and incentivize sustainable practices. From monitoring carbon offset initiatives to promoting eco-friendly travel through transparent initiatives, blockchain is reshaping the landscape of sustainable tourism.

7.10.1 MONITORING AND VERIFYING SUSTAINABLE PRACTICES IN THE INDUSTRY

One of the biggest challenges in sustainable tourism is ensuring that companies adhere to eco-friendly practices. Complex, centralized sustainability tracking systems are prone to errors, fraud, and greenwashing. Blockchain enables transparent and decentralized monitoring and verification of sustainable practices. Smart contracts on the blockchain can be used to create a decentralized ledger of sustainable projects and certifications (Khan et al., 2021). Green accommodations, airlines, and tour operators can register their sustainability efforts on the blockchain for verification and immutability. This ledger can include waste reduction, energy efficiency, and community engagement data. The transparent and tamper-resistant nature of blockchain ensures accurate statistics, allowing travelers to make informed choices and support sustainable initiatives. Blockchain's decentralization allows a global network to identify and share sustainability certifications and best practices. Collaboration among companies and organizations worldwide promotes sustainable tourism.

7.10.2 CARBON OFFSET PROGRAMS ON THE BLOCKCHAIN

Carbon offset initiatives help reduce the environmental impact of tour-related sports. Blockchain technology provides a reliable and traceable framework for carbon credits, improving carbon offset transparency and efficiency. Traditional carbon offset applications use shopping carbon credits to offset emissions. However, the carbon credit market's lack of openness and accountability can cast doubt on its environmental impact. Blockchain overcomes those concerns by creating a decentralized, auditable carbon offset ledger. Each blockchain-tokenized carbon credit represents a precise amount of carbon dioxide offset. Smart contracts automate carbon credit accumulation and retirement to meet predefined conditions. This automation eliminates fraud and provides a clear, verifiable record of offsets' environmental impact. Blockchain tokenization of carbon credits enables more precise emissions tracking and offsetting. Travel companies can inform passengers about their trips' environmental impact and carbon offset responsibilities.

7.10.3 ENCOURAGING ECO-FRIENDLY TRAVEL THROUGH TRANSPARENT INITIATIVES

Blockchain technology helps travelers make eco-friendly choices by providing transparency about travel carriers' sustainability factors (Prados-Castillo et al., 2023). Decentralized applications (DApps) on the blockchain can give tourists real-time data on airlines, hotels, and other service providers' environmental practices.

A DApp may use blockchain data to provide tourists with a sustainability score for each route choice based on carbon footprint, waste reduction, and network engagement. This helps sustainable tour firms by allowing holidaymakers to align their ideals with their choices. Blockchain also allows for tokenized green tour awards and incentives. Travelers who choose sustainable options can receive tokens or discounts, establishing a positive feedback loop that promotes widespread environmental awareness. Blockchain's transparency ensures that these incentives are legitimate and directly linked to sustainability actions.

Blockchain also allows travelers to micro-donate to local environmental programs. Vacationers can support community-based environmental, cultural, and social responsibility projects with cryptocurrency or tokenized assets. Blockchain technology is transforming sustainable tourism by documenting, validating, and encouraging green practices. Blockchain is revolutionizing the tourism industry by recording sustainability responsibilities, improving carbon offset transparency, and encouraging green journeys with tokenized incentives. Blockchain in tourism connects with global environmental conservation initiatives and creates a more responsible and resilient tourism sector as vacationers choose sustainability.

7.11 MITIGATING FRAUD IN TRAVEL INSURANCE

Fraud within the travel insurance sector poses significant challenges for both insurers and policyholders, leading to increased costs and eroded trust. Blockchain technology emerges as a game changer in mitigating fraud, providing transparency, automation, and trust-building measures that reshape the landscape of travel insurance.

7.11.1 BLOCKCHAIN FOR TRANSPARENT AND FRAUD-RESISTANT INSURANCE CLAIMS

Manual processes and a lack of transparency make it difficult for traditional insurance systems to detect and prevent fraudulent claims. Blockchain provides a decentralized, tamper-resistant ledger that ensures claim transparency and integrity. From policy issuance to claim filing, the blockchain records every travel insurance policy transaction (Liu et al., 2019). Insurers, policyholders, and legal intermediaries can access this clear and unchangeable audit trail. If a medical claim occurs during a trip, the blockchain securely records the incident, medical care, and related charges. This prevents data manipulation after the fact, boosting claim statistics' credibility. Blockchain's decentralization reduces internal fraud by preventing individual control of data. Transparency unites insurers and policyholders, making the environment more stable and fraud-resistant.

7.11.2 SMART CONTRACTS FOR AUTOMATED CLAIMS PROCESSING

Smart contracts automate and streamline travel insurance claim processing (Vo et al., 2017). Blockchain contracts automatically execute actions when certain conditions are met. Smart contracts speed up travel insurance claim approval and disbursement. The smart contract can evaluate a policyholder's claim against preset criteria, policy terms, and blockchain records. The smart contract automatically pays the policyholder if the conditions are met, reducing processing time and manual involvement. Smart contracts can also include weather or flight data to support claims. If a policyholder cancels a trip due to bad weather, the smart contract can automatically check official meteorological data to ensure the claim matches the documented events. The automation and transparency of smart contracts expedite claim processing and reduce fraud. Policyholders benefit from faster and more efficient claims settlements, while insurers can better focus resources on legitimate claims.

7.11.3 TRUST-BUILDING MEASURES IN TRAVEL INSURANCE THROUGH BLOCKCHAIN

Trust between insurers and policyholders is key to travel insurance success. Blockchain adds trust-building features that enhance the integrity and reliability of insurance processes. First, the blockchain ledger's decentralization and transparency provide all parties with equal information. Policyholders can independently verify their policies, insurance, and claims histories, minimizing data asymmetry and encouraging a more informed and trusted relationship. Second, blockchain allows policyholders to have virtual identities and reputation scores. Each blockchain transaction and interaction boosts the policyholder's reputation by demonstrating their policy compliance and claim reporting honesty. This reputation score helps insurers assess risk and adjust policies. Third, blockchain tokenization can make premium payments and disbursements secure and traceable. Tokenized assets secure and streamline financial transactions, reducing the risk of fraud or unauthorized transfers. Blockchain also allows insurers, reinsurers, and other stakeholders to form a blockchain community. This collaboration shares fraud detection and prevention ideas, best practices, and analytics. Insurers can improve fraud mitigation by using a shared platform. Technology transforms travel insurance fraud mitigation through transparency, automation, and trust-building. Blockchain enhances the travel insurance industry by creating a fraud-resistant claims process with transparent ledgers and smart contracts, gaining insurer and policyholder trust. As blockchain adoption grows, the travel insurance sector can continue to reduce fraud, increase trust, and improve insurance strategy performance.

7.12 CONCLUSION

In conclusion, blockchain has transformed the tourism industry in many ways. This exploration of blockchain programs in tourism, from supply chain transparency to sustainable practices and fraud mitigation, demonstrates that blockchain is a catalyst for redefining how an industry operates and serves its customers. Blockchain technology in supply chain management has transformed transparency by securely recording goods and service origins on a decentralized ledger. This improves stakeholder agreement and streamlines processes, creating a greener, more resilient tourism environment. Decentralized tour guide systems empower community groups, change tour narratives, and connect travelers and guides. Smart contracts automate processes, making transactions clear, consistent, and trust-driven. Blockchain impacts sustainable tourism as well. Blockchain allows vacationers to make value-based choices by monitoring and verifying sustainable behaviors and simplifying carbon offset applications. This supports the industry's environmental goals and fosters accountability among carriers and clients. Blockchain protects traveler data by encrypting, permissioning, and adhering to information security regulations. This addresses data and privacy issues and helps tourists and service firms build trust. The development of blockchain in tourism continues. New programs and use cases will refine and broaden blockchain's impact on enterprises as technology progresses. Governments, corporations, and scholars must explore these options to unleash blockchain's full potential in transforming tourism.

The promising results and insights from current implementations emphasize the importance of research and experimental persistence. Governments, corporations, and technology developers must collaborate more to create innovation and foster blockchain solutions. As we progress, it becomes evident that blockchain can adapt to the tourism industry's changing needs. Its impact goes beyond technology, transforming tourist experiences, management, and sustainability. The tourism industry can embrace this transformative force to pursue greater efficiency, transparency, and sustainability, fostering trust and creating a future where travelers and stakeholders benefit from blockchain-driven tourism.

KEYWORDS

- **blockchain in tourism**
- **data privacy in travel**
- **decentralized tour guide platforms**
- **fraud mitigation in travel insurance**
- **smart contracts in hospitality**
- **supply chain transparency in tourism**
- **sustainable tourism**
- **travel industry revolution**

REFERENCES

Chauhan, R., & Kaul, V. (2021). Blockchain technology: A panacea for medical distribution ailments. In *Blockchain for Healthcare Systems* (pp. 43–60). CRC Press. https://doi.org/10.1201/9781003141471-4.

Chauhan, R., & Kaul, V. (2022). Blockchain technology: Unlocking the business model maze for evolving businesses and start-ups. In *Transformations Through Blockchain Technology: The New Digital Revolution* (pp. 291–309). Springer International Publishing. https://doi.org/10.1007/978-3-030-93344-9_13.

Cook, D. (2020). The freedom trap: Digital nomads and the use of disciplining practices to manage work/leisure boundaries. *Information Technology and Tourism, 22*(3), 355–390. https://doi.org/10.1007/s40558-020-00172-4.

Dadkhah, M., Rahimnia, F., & Filimonau, V. (2022). Evaluating the opportunities, challenges, and risks of applying blockchain technology in tourism: A Delphi study approach. *Journal of Hospitality and Tourism Technology, 13*(5), 922–954. https://doi.org/10.1108/JHTT-04-2021-0115.

Dutta, S. K. (2020). Smart contracts. In *The Definitive Guide to Blockchain for Accounting and Business: Understanding the Revolutionary Technology* (pp. 61–78). Emerald Publishing. https://doi.org/10.1108/978-1-78973-865-020201005.

Hari, A., & Lakshman, T. V. (2016). The Internet blockchain: A distributed, tamper-resistant transaction framework for the Internet. *Proceedings of the 15th ACM Workshop on Hot Topics in Networks (HotNets 2016)*, 204–210. https://doi.org/10.1145/3005745.3005771.

Hasan, H. R., Salah, K., Jayaraman, R., Omar, M., Yaqoob, I., Pesic, S., Taylor, T., & Boscovic, D. (2020). A blockchain-based approach for the creation of digital twins. *IEEE Access, 8,* 34113–34126. https://doi.org/10.1109/ACCESS.2020.2974810.

Ichikawa, D., Kashiyama, M., & Ueno, T. (2017). Tamper-resistant mobile health using blockchain technology. *JMIR mHealth and uHealth, 5*(7), e111. https://doi.org/10.2196/mhealth.7938.

Irannezhad, E., & Mahadevan, R. (2021). Is blockchain tourism's new hope? *Journal of Hospitality and Tourism Technology, 12*(1), 85–96. https://doi.org/10.1108/JHTT-02-2019-0039.

Jain, P., Singh, R. K., Mishra, R., & Rana, N. P. (2023). Emerging dimensions of blockchain application in tourism and hospitality sector: A systematic literature review. *Journal of Hospitality Marketing & Management, 32*(4), 454–476. https://doi.org/10.1080/19368623.2023.2184440.

John William, A. D., Rajendran, S., Pranam, P., Berry, Y., Sreedharan, A., Gul, J., & Paul, A. (2022). Blockchain technologies: Smart contracts for consumer electronics data sharing and secure payment. *Electronics, 12*(1), 208. https://doi.org/10.3390/electronics12010208.

Khan, S. N., Loukil, F., Ghedira-Guegan, C., Benkhelifa, E., & Bani-Hani, A. (2021). Blockchain smart contracts: Applications, challenges, and future trends. *Peer-to-Peer Networking and Applications, 14*(5), 2901–2925. https://doi.org/10.1007/s12083-021-01127-0.

Lee, E. (2022). Technology-driven solutions to banks' de-risking practices in Hong Kong: FinTech and blockchain-based smart contracts for financial inclusion. *Competition & Change, 51*(1, 2), 83–108. https://doi.org/10.1177/14737795211071095.

Li, J., & Kassem, M. (2021). Applications of distributed ledger technology (DLT) and blockchain-enabled smart contracts in construction. *Automation in Construction, 132,* 103955. https://doi.org/10.1016/j.autcon.2021.103955.

Lin, J., Lobo, A., & Leckie, C. (2017). The role of benefits and transparency in shaping consumers' green perceived value, self-brand connection, and brand loyalty.

Journal of Retailing and Consumer Services, 35, 133–141. https://doi.org/10.1016/j.jretconser.2016.12.011.

Liu, J. L., Wu, X. Y., Yu, W. J., Wang, Z. C., Zhao, H. L., & Cang, N. M. (2019). Research and design of travel insurance system based on blockchain. *Proceedings of the 4th International Conference on Intelligent Informatics and Biomedical Sciences (ICIIBMS 2019)*, 121–124. https://doi.org/10.1109/ICIIBMS46890.2019.8991444.

Morais, D. B., Dorsch, M. J., & Backman, S. J. (2005). Building loyal relationships between customers and providers. *Journal of Travel & Tourism Marketing, 18*(1), 49–57. https://doi.org/10.1300/J073v18n01_05.

Nguyen, T. V., Lê, L. S., Dao, B., & Nguyen-An, K. (2019). Leveraging blockchain in monitoring SLA-oriented tourism service provisioning. In *Proceedings of the 2019 International Conference on Advanced Computing and Applications (ACOMP 2019)* (pp. 42–50). IEEE. https://doi.org/10.1109/ACOMP.2019.00014.

Ozuem, W., Thomas, T., & Lancaster, G. (2016). The influence of customer loyalty on small island economies: An empirical and exploratory study. *Journal of Strategic Marketing, 24*(6), 447–469. https://doi.org/10.1080/0965254X.2015.1011205.

Prados-Castillo, J. F., Guaita Martínez, J. M., Zielińska, A., & Gorgues Comas, D. (2023). A review of blockchain technology adoption in the tourism industry from a sustainability perspective. *Journal of Theoretical and Applied Electronic Commerce Research, 18*(2), 814–830. https://doi.org/10.3390/jtaer18020042.

Rashideh, W. (2020). Blockchain technology framework: Current and future perspectives for the tourism industry. *Tourism Management, 80,* 104125. https://doi.org/10.1016/j.tourman.2020.104125.

Riesco, R., Larriva-Novo, X., & Villagra, V. A. (2020). Cybersecurity threat intelligence knowledge exchange based on blockchain: Proposal of a new incentive model based on blockchain and smart contracts to foster the cyber threat and risk intelligence exchange of information. *Telecommunication Systems, 73*(2), 259–288. https://doi.org/10.1007/s11235-019-00613-4.

Shinde, S. M., & Gurrala, V. R. (2023). Securing trustworthy evidence for the robust forensic cloud-blockchain environment for immigration management with improved ECC encryption. *Expert Systems with Applications, 229,* 120478. https://doi.org/10.1016/j.eswa.2023.120478.

Treiblmaier, H. (2022). Blockchain and tourism. In *Handbook of e-Tourism* (pp. 475–495). Springer International Publishing. https://doi.org/10.1007/978-3-030-48652-5_28.

Vasilakakis, K., & Sdrali, D. (2023). Supplier selection criteria in the Greek hotel food and beverage divisions. *Journal of Hospitality and Tourism Insights, 6*(2), 447–463. https://doi.org/10.1108/JHTI-07-2021-0169.

Veljan, A. (2021). Regulating the uncontrollable: The development of card scheme fees in payments markets in light of recent policy intervention. *Research in Law and Economics, 29,* 89–110. https://doi.org/10.1108/s0193-589520210000029006.

Velmurugadass, P., Dhanasekaran, S., Shasi Anand, S., & Vasudevan, V. (2021). Enhancing blockchain security in cloud computing with IoT environment using ECIES and cryptography hash algorithm. *Materials Today: Proceedings, 37(Part 2),* 2653–2659. https://doi.org/10.1016/j.matpr.2020.08.519.

Vo, H. T., Mehedy, L., Mohania, M., & Abebe, E. (2017). Blockchain-based data management and analytics for micro-insurance applications. In *Proceedings of the*

International Conference on Information and Knowledge Management (CIKM) (pp. 2539–2542). ACM. https://doi.org/10.1145/3132847.3133172.

Wang, S., Ouyang, L., Yuan, Y., Ni, X., Han, X., & Wang, F. Y. (2019). Blockchain-enabled smart contracts: Architecture, applications, and future trends. *IEEE Transactions on Systems, Man, and Cybernetics: Systems, 49*(11), 2266–2277. https://doi.org/10.1109/TSMC.2019.2895123.

Warren, G., & Dinnie, K. (2018). Cultural intermediaries in place branding: Who are they and how do they construct legitimacy for their work and themselves? *Tourism Management, 66,* 302–314. https://doi.org/10.1016/j.tourman.2017.12.012.

Westerkamp, M., Victor, F., & Küpper, A. (2020). Tracing manufacturing processes using blockchain-based token compositions. *Digital Communications and Networks, 6*(2), 167–176. https://doi.org/10.1016/j.dcan.2019.01.007.

Wojcik, D., MacDonald-Korth, D., & Zhao, S. X. (2017). The political–economic geography of foreign exchange trading. *Journal of Economic Geography, 17*(2), 267–286. https://doi.org/10.1093/jeg/lbw014.

Zhu, X., & Badr, Y. (2018). Identity management systems for the Internet of Things: A survey towards blockchain solutions. *Sensors, 18*(12), 4215. https://doi.org/10.3390/s18124215.

Zorlu, O., Ozsoy, A., & Sert, S. A. (2023). A role-based access control management model on blockchain for restricted facilities: An airport example. In *Proceedings of the 10th International Conference on Recent Advances in Air and Space Technologies (RAST 2023).* IEEE. https://doi.org/10.1109/RAST57548.2023.10197974.

CHAPTER 8

Technological Edge: "Artificial Intelligence and Blockchain" – Its Impact and Role in National Defense

SUJIT S. NAIR,[1] SHIKHA MISHRA,[1] and MONICA CHOUDHARY[2]

[1]*Amity Business School, Noida, U.P., India*

[2]*SP Jain School of Global Management, Australia*

*Corresponding author

"Professional knowledge and professional competence are the main attributes of leadership."

—Field Marshal Sam Manekshaw

ABSTRACT

Commercial and military applications of artificial intelligence (AI) systems have emerged for the first time in recent years. Some modern militaries have deployed AI in areas such as logistics support, cyber defense, medical evacuation robots, and integrated surveillance. The integration of blockchain technology, along with AI, into military operations represents a tremendous combination that could further transform critical aspects of modern warfare. Dynamic adaptation to threats; AI's ability to learn and adapt, with Blockchain's dynamic recognition protocol, allows defense systems to respond effectively to the rapidly changing military environment, ensuring that only authorized organizations have access to sensitive data and critical systems.

Collective cybersecurity intelligence: AI data analysis capabilities and secure data sharing through blockchain power collective cybersecurity efforts. Military organizations can respond together in real-time to

Blockchain Technology: Unlocking the Value of 3Ts in the Business World.
Richa Chauhan & Vidhi Kaul (Eds.)

DOI: 10.1201/9781998511914-8

emerging cyber threats, creating stronger defenses against sophisticated adversaries. Blockchain adds additional security, transparency, and trust to military operations with decentralized and stronger records; this review aims to analyze the impact of artificial intelligence on security operations and national defense (Brynjolfsson & McAfee, 2014).

8.1 INTRODUCTION

The term "artificial intelligence" (AI) refers to the ability of machines to perform tasks normally performed by organic intelligence (Mitchell, 2019). This means recognizing patterns, learning through repetition, making conclusions, and acting in a certain way. It helps us understand reason and think about the possible calculation of our actions. Artificial intelligence can be used to detect, respond, and predict threats to selected assets, as well as improve defense mechanisms against threats to critical infrastructure such as scientific assets, power grids, and inventory markets (Cummings, 2017). We can optimize our general abilities. The general definition when looking at artificial intelligence refers to the range of capabilities superimposed on the human mind from these elements. In military terms, the US Defense Science Board defines artificial intelligence as "the ability of laptop architecture to perform tasks that normally require the human mind." Collectively, these provide the building blocks that can be used in the definition of artificial intelligence.

The rapid development and expansion of time changed the human struggle and world politics. Artificial intelligence is now infiltrating the military and civilian sectors and is driving change. But artificial intelligence is not the main element of change. Repeated naval missions speed up options, change the pace of operations, and sooner or later may push warfare beyond human cognitive capacity. The purpose of this review is to analyze the impact of artificial intelligence on military operations and its impact on national security.

AI-powered smart drones can be used to hijack and manipulate enemy air defense systems. This increases the effectiveness of various military operations, such as air interdiction, surface attacks, long-range vectors, and naval operations, especially when an enemy wants to degrade, compete, or deny access.

The integration of AI into military processes not only increases efficiency but also introduces a new paradigm in strategic thinking and execution. The combination of risk reduction, advanced decision support, and enhanced operational capabilities positions AI as a transformative force in modern warfare. However, ethical considerations, international norms, and potential unintended consequences must be taken into account to ensure the responsible and legal use of AI in military settings such as:

1. **Enhance National Security:** Commercial and military applications of AI systems have been at the forefront in recent years. Some modern militaries have deployed AI militarily in areas such as logistics support, cyber defense, medical evacuation robots, and integrated surveillance. However, AI will accelerate the speed of operations for strategic advantages in space, cyber, air, land, and maritime domains while impacting survivability in operational defense. Military strategists considering war in the 21st century must consider new war technologies using artificial intelligence, which include:

    i. **AI on Defense:** This work does not aim to accurately capture the technical aspects of AI but rather focuses on how it is perceived by society. The use of AI in defense can be divided into three categories: support and assistance; kinetic and non-kinetic; resistance and movement.

    ii. **Operations and Logistics:** In short, the use of artificial intelligence and support in defense uses artificial intelligence to support "back office" functions such as operations and logistics.

    iii. **Inelastic Additive and Non-Kinetic:** AI includes applications related to cyber defense and cyber-crime functions. This refers to the use of adversarial and kinetic AI systems in combat systems, such as "target" identification and the use of lethal autonomous weapon systems (LAW).

    iv. **Technical Justification:** While the technical potential of AI is the subject of much research and development in many defense sectors, reliability, ethics, legality, and ethical issues can pose similar challenges. Effective deployment of AI-powered capabilities must be driven by intelligence.

v. **Decision Making:** A strong caveat for the use of AI in the context of defense concerns situations where AI-driven machines can make life-threatening decisions and/or take them out of the direct control of human operators. Take care of life with advanced AI tool profiles.

2. **Understanding Attitudes Towards AI:** Research and analysis based on the fact that there is currently no consistent modern standard for AI in defense. Designing such a volume will help identify key trends in public attitudes and understand the key determinants of those attitudes, as well as potential influences on attitudes due to information bias and misinformation.

We hope this technology will allow developers to review aspects of the reputation of their work; helping policymakers formulate appropriate strategies; AI will help those involved in implementation to communicate effectively with communities and influencers. Therefore, the main goal of this study is to develop a reliable and psychometrically valid scale capable of measuring attitudes towards the use of AI in defense.

8.2 ANALYSIS: CURRENT CAPABILITIES AND FUTURE TRAJECTORIES

Because artificial intelligence is a military reality, and not a special program, but a whole system, it can be said that the armed forces of the world see its introduction in one way. For example, weapons control systems can make independent decisions without human input (other than setting parameters before they start working), and large data sets can be analyzed to identify patterns.

Artificial intelligence systems and advanced computers and networks will be able to make fast, consistent, and autonomous decisions in an uncertain environment. Intelligent systems that are adept at drawing conclusions based on random data will learn from the battlefield, past actions, and parallel machine actions (NITI Aayog, 2018).

At the cutting edge of artificial intelligence research, self-learning algorithms, even with limited data, are rapidly developed to deal with complex and nonlinear data. In general, ongoing AI research focuses on non-defense areas, such as civilian vehicle manufacturing, the Internet, and healthcare. In general, the ability to categorize asymmetric information and use it as a basis for decision-making has several strategic applications. Early research promises that AI will be used across the spectrum from tactics to strategy (Levy, 2001).

8.3 AI ATTRIBUTES

8.3.1 AUTOMATION

The word AI refers to technology that is intelligent enough to recognize its surroundings and respond accordingly. Neural networks, robotic process automation, machine learning, and natural language processing are some of the applications of artificial intelligence. Robots can react to external input on their own thanks to artificial intelligence. Input that the software cannot always foresee because it cannot be directly manipulated by it (Cave et al., 2018). Hence, AI is a catch-all word for any computer system that has the ability to see, process, learn from, and respond to its environment. Digital assistants, chatbots, and learning algorithms are a few of the AI applications common worldwide.

8.3.2 TYPE OF AI

AI functions in four domains. Artificial intelligence is one of them; it deals with the computerization of cognitive or manual, routine or irregular tasks. AI promotes people's faster and better task performance. Better decision-making is facilitated by enhanced intelligence, and autonomous intelligence enables decision-making to be automated without the need for human involvement. An overview of some of the words used frequently in the literature on artificial intelligence will help one better grasp how they are applied (Brynjolfsson & McAfee, 2014).

8.3.3 METHOD OF ARTIFICIAL INTELLIGENCE

Two distinct theoretical approaches to artificial intelligence are discussed in this chapter "Some Specific Technical Aspects of Artificial Intelligence": general artificial intelligence, both weak and strong artificial intelligence, and narrow artificial intelligence (Mitchell, 2019). The approaches are:

1. **Research and Development:** Research initiatives are underway to develop powerful artificial intelligence and virtual machines that

can perform tasks and adapt like human beings, such as universal artificial intelligence. Robust artificial intelligence utilizes unprocessed information to construct its model. The strong artificial intelligence hypothesis holds that we will eventually be able to substitute human minds with robots that can talk and think. These robots will be able to reason, think, and carry out every action that a person can.

2. **Weak AI:** These applications use pre-defined problem domain models. Soft AI enables machines to behave intelligently in specific situations, such as computers playing chess based on pre-programmed actions.
3. **Machine Learning (ML) and Deep Learning (DL):** ML uses algorithms to analyze data, learn from it, and identify or recognize things related to the data. Machines are trained on machine learning algorithms to learn how to perform tasks. Computer vision is key to machine learning. DL is a branch of ML based on a set of algorithms that attempt to model high-level data abstraction.

8.4 GLOBAL COMPARATIVE ANALYSIS OF AI DEVELOPMENT IN DEFENSE APPLICATIONS

1. **Russia:**
   - Vladimir Putin, the president of Russia, declared in 2017 which nations would lead the world in artificial intelligence in areas such as (Latiff, 2017):
   - Russia is developing AI-powered missiles that have the ability to choose a new target while in mid-air. Russia thinks that a lot of drones won't be able to leave the battlefield in the future.
   - Russia is testing a number of semi-autonomous and autonomous combat systems, such as AI-powered cameras, machine guns, and Kalashnikov's Neural Net Battle Module, which can make choices on their own without human assistance.
   - The Russian government suggested the prohibition on LAWs be lifted in light of its vehement rejection of the law.
   - The biggest telecom company in Russia's AI sector has established a $100 million venture capital fund for companies, complete with a startup accelerator.

2. **China:**
   - China views itself as the US's closest competitor in AI development (Kai-Fu, 2018).
   - The Chinese military is trying to gain an advantage by changing the paradigm of war with its new war. The close relationship between Silicon Valley and China and the open nature of the American research community have made the most advanced AI technologies in the West available to China:
   - There are many achievements in AI in Chinese industry; for example, Baidu passed the Chinese language recognition index in 2015.
   - By 2030, China aims to achieve major breakthroughs in various AI technologies such as big data and autonomous intelligent systems. By 2017, China's roadmap aims to create a US$ 150 billion AI industry by 2030.
   - By 2025, China plans to make major breakthroughs in AI to become a leader in AI development and a key driver of economic transformation. Until then, China wants to be a leading player in research and development, using AI extensively in fields such as manufacturing, medicine, and national defense. China's core AI industry will exceed RMB 400 billion (about $59 billion), with AI-related industries valued at RMB five trillion (about $740 billion).
   - China's leadership supports an "innovation-based" strategy for civilian and military development, aiming to become the world's "first innovation center" in AI by 2030.
   - China also pursues military-civilian integration and aims to develop AI supported by the People's Liberation Army.
   - The Chinese industry has made significant progress in AI; in 2015, Baidu passed the Chinese language recognition index, for instance.
   - China hopes to make significant strides in a number of AI fields by 2030, including big data and autonomously intelligent systems. China's strategy foresees a US$ 150 billion AI industry by 2030. It was published in 2017.

3. USA:
    - It considers itself to be the nearest AI rival of the US (Brose, 2020).
    - By adopting a new battle paradigm, the Chinese military is attempting to get the upper hand. China now has access to the most cutting-edge AI technologies available in the West thanks to the close ties between the tech industry in Silicon Valley and China, as well as the transparent character of the American academic community. The United States thinks that the next wave of conflict will be shaped by the quick advancements in AI. With a 50% rise over 2018 and an anticipated $1 billion in federal spending in 2020, artificial intelligence is predicted to be one of the fastest-growing sectors of technology investment.
    - The Sea Hunter autonomous naval vessel, which is built to sail into and out of ports and operate at sea for extended periods of time without a crew, is one of the other US military AI warfare programs.
    - The administration of the United States believes that breakthroughs in artificial intelligence will define the next phase of the battle and has invested billions in preparing for it.

8.5 BLOCKCHAIN IMPLEMENTATION FOR NATIONAL SECURITY AND DEFENSE: AN IN-DEPTH TECHNICAL EXPLORATION

With its decentralized, secure, and transparent architecture, blockchain technology has great potential to transform national security and defense operations (Tapscott & Tapscott, 2016). Looking at technical complexity opens up a multifaceted approach to changing key aspects of defense strategy, including:

1. **Cryptographic Security:** Blockchain protects data with sophisticated encryption methods like public cryptography. End-to-end encryption is used in military information and communication security to ensure that sensitive data can only be accessed and

decrypted by authorized people who possess the necessary private keys.

- *Example:* Key cryptography is used to encrypt messages in military communication systems. Secure and confidential interaction is ensured by limiting access to the private key to only the intended recipient.

2. **Store Immutable Data:** The immutability of blockchain guarantees that once data is recorded, it cannot be updated or altered. This capability protects sensitive documents, including mission plans, reports on intelligence, and strategic decisions in defense applications.

    - *Example:* Mission-critical data, including strategic target coordinates, are jotted down in a notebook. Once recorded, the information is safe and unchangeable.

3. **Security Identity Management:** A more secure and decentralized method for securely storing employee identities is offered by blockchain-based identity management. As a result, there will be less chance of identity theft and illegal access to defense systems.

    - *Example:* A soldier's identification is kept in a notebook along with their records of training and security chapter work. This guarantees that confidential information and specific resources are only accessible to those who are permitted.

4. **Collaboration System:** Blockchain establishes a shared, decentralized ledger, which fosters interoperability. Data may be easily shared across ally nations and various military branches while preserving data security and integrity.

    - *Example:* Blockchain ensures that every organization has access to the most accurate and up-to-date information by enabling real-time information sharing amongst the Air Force, Navy, and Army during joint operations.

5. **Smart Contract and Automatic Processing:** Tasks are automated by smart contracts when predetermined criteria are satisfied. This may handle logistical management, procurement procedures, and

requirements for the release of secret information in the defense industry

- *Example:* A smart contract guarantees that the necessary resources are constantly available without the need for manual intervention by triggering supply purchases whenever the inventory level drops below a certain threshold.

6. **Decentralized Command and Control:** Blockchain makes it easier for the control and command structure to become decentralized, which enhances and adapts the security process. A decentralized blockchain might be able to continue having decision-making and communication capabilities in a realistic scenario.
   - *Example:* A decentralized network built on blockchain technology enables military units to connect efficiently in the event of a cyber-attack against a centralized command center, guaranteeing operational continuity.
7. **Cyber Security and Threat Analysis:** Because blockchain is distributed, important data is dispersed, reducing the danger of cyber-attacks. Through the sharing of threat intelligence, it can be utilized to improve collaborative cyber security operations for military organizations.
   - *Example:* To enable real-time information sharing regarding emerging cyber threats and vulnerabilities with all users, military institutions augment a system based on blockchain technology with threat intelligence.
8. **Automatic Vehicles and Drones:** Drones and autonomous cars can communicate and share data thanks to blockchain technology. Because smart contracts provide safe and verified communication, they can lower the risk of cyberattacks on these vital resources.
   - *Example:* Blockchain-based smart contracts ensure that only authorized parties can access or receive data from drone assets by authenticating and verifying communications between military drones.
9. **Dynamic Recognition Protocol:** Blockchain provides a dynamic and adaptive authentication protocol, ensuring that only authorized

organizations can access sensitive data or monitor critical systems in a rapidly changing military environment.

- *Example:* A blockchain-based authentication protocol in dynamic military operations.

8.6 TRANSFORMATION OF KEY PROCESSES

AI and blockchain technology combined offer a powerful combination that has the potential to revolutionize many facets of contemporary combat (Tapscott & Tapscott, 2016). With decentralized and more durable records, blockchain enhances military operations' security, transparency, and trustworthiness. Implementing blockchain technology can enhance the following:

1. **Risk and Volatility Mitigation Records:** The resilience of blockchain technology guarantees the accuracy of military data, mitigating the possibility of manipulation of information or cyberattacks. This increases the AI system's level of assurance and significantly lowers total operational risk.
2. **General Military Operations with Enhanced Confidence:** The decentralized structure of blockchain fosters confidence in the information-sharing process among armed forces groups. It can be utilized to support a coherent strategy for military operations and boost confidence in information shared in conventional and gray war situations, establishing a transparent and safe communications infrastructure.
3. **Safe Decision Making and Forecasting:** By guaranteeing the integrity of important data, blockchain can strengthen the security of AI-powered war games and simulations. Because the data utilized in this process is more durable and dependable, the secure environment aids in the production of forecasts and decisions that are more accurate.
4. **Consistent Intelligence Records from Large Databases:** Blockchain technology can be used in conjunction with AI to produce unchangeable recordings of intelligence gleaned from huge datasets. This ensures the data's accuracy and boosts the dependability

of the intelligence that can be derived from this database, offering a transparent and auditable route.

5. **Enhanced Security for Drones and Smart Operations:** By generating an unchangeable record of drone activity, blockchain technology enhances security in smart drone operations. This enhances the overall efficacy of air interdiction operations and other military operations by guaranteeing that the drone's guidance information, such as blueprints and navigation data, remains secure and undamaged.

Blockchain technology and artificial intelligence work together to improve military operations while also resolving issues with trust, integrity, and security of data. The military can respond to changing threats in a complicated and contested environment more efficiently and adaptably when various technologies are used concurrently. However, rigorous attention to ethical, legal, and international standards is necessary to guarantee the appropriate and accountable use of this technology in military contexts.

8.7 NATIONAL SECURITY AI-BLOCKCHAIN APPLICATION: IMPLEMENTATION

The use of AI tools in defense improves the processing and utilization of data, which in turn increases the speed of decision-making on the battlefield (Brynjolfsson & McAfee, 2014). Some military applications of AI are:

1. **Reasons for Military Computing:** It addresses the issues that people encounter in the military and concentrates on making wise choices in combat. Examine the conflict's trajectory and take appropriate action based on the information provided by the directive set (course of action or COA). These programs are capable of successfully exploiting weaknesses held by adversaries in a controlled environment, like a military laboratory.
2. **Intelligent Unmanned Weapon Systems and Autonomous Unmanned Vehicles:** Artificial intelligence is directly applied in robotics and the most sophisticated weapons utilized by the armed

services. These weapon systems are equipped with the intelligence needed to locate, identify, and eliminate enemy targets.

3. **AI-Powered Data Processing and Intelligent Analysis:** Satellite photography can be utilized to gather raw intelligence and interpret sensor data using deep learning algorithms. This knowledge can aid in making informed decisions during combat and training, as well as when constructing national infrastructure.
4. **Communication and Computer Communication:** Visibility is an essential component of every military mission. Combat forces are less able to make decisions when they lack quality information. Future military effectiveness will hinge on preserving network connectivity and implementing adaptable intelligence systems from several sources, which will be significantly improved by artificial intelligence.
5. **Cyber Defense and Cyber Warfare:** Big data analysis, machine learning, statistical analysis, and pattern matching can all be used to identify and stop cyberattacks. Cyber networks can be completely destroyed by malicious cyber operations in a matter of minutes. Because AI is applied so quickly, it serves as a safeguard against malevolent cyberattacks. Applications for advanced generative networks (GANs) in cyber detection are planned.
6. **Electronic Warfare (EW):** These comprehends, secures, and transmits information using the electromagnetic spectrum, which includes radio waves, infrared signals, and radar. It can also be employed to stop adversaries from utilizing electronics. AI is being used directly in learning behavior for the nation's Adaptive Radar Security measures (ARC) and Adaptive Electronics Warfare (BLADE). Real-time automatic aerial surveillance (EW) systems against evasive radars are made possible by ARC.
7. **Pattern Recognition:** A branch of study that examines data pattern recognition systems' architecture and operation. When an app scans something, it frequently matches patterns it detects in the image, such as handwriting, fingerprints, or facial recognition. With military applications, key application areas comprise image analysis, speech analysis, character recognition, human identification of devices, and personnel identification.

8. **Bioinformatics:** Data on biology is managed using computer technology. AI offers strong algorithms and methods to address important bioinformatics issues. Artificial neural networks, latent Markov models, Bayesian networks, and kernel techniques are examples of approaches that operate well in situations where there is more data than theory.
9. **Collect Data:** An AI program that can search databases for pertinent information and identify patterns in the data. Modern businesses utilize data mining, an increasingly crucial technique, to transform data into intelligence that can be put to use.
10. **Expert System:** A system with expertise is computer software that uses certain experts' knowledge to solve issues or provide guidance. It is essentially a multi-section, multi-module, sub-module system that is based on knowledge. Almost every industry, including knowledge-based applications, commerce, science, engineering, and the military, has seen an increase in efficiency due to AI.
11. **Robotic Surgery:** Surgeons may be able to perform robotic surgery, life-saving procedures, and remote monitoring thanks to this. By combining the functional exoskeleton with AI robots, futuristic superheroes (similar to Iron Man) may be developed. This will be beneficial not only for weightlifting but also for warfare, building, LOPES (an exoskeleton), ReWalk, Spurs/Musculature, and Human Universal Loader. Artificial intelligence will enable the use of prostheses, regenerative medicine, brain and retinal implants, body implants, young troops, and life extension. The armed forces' ability to fight and their static strength will both be greatly enhanced by this.
12. **Self-Modulating Modular Robot (SCMR):** AI will enable SCMR to become a reality that can revolutionize the design of many physical structures and machines.
13. **Swarm Robots:** AI has the potential to enable autonomous collaboration among populations of dispersed robots, including self-driving cars, autonomous automation, nanorobotics, particle swarm optimization, and multi-agent systems.

14. **Invisibility Cloak:** AI can provide microscopic advice cloaked in optical frequencies, thus solving the military challenges of the past. Hidden objects under infrared light.
15. **Immersive Virtual Reality:** AI can also provide an immersive artificial environment similar to what the user experiences in augmented reality. This will add realism to war games and simulations and can save training costs.
16. **Social Media-Based AI:** While large publishers can now utilize AI to create and produce articles on the same premise, more traffic, social media sites are the go-to source of news for younger people and news groups that increasingly depend on social networking sites to make broadcasts. This is supported by government agencies to advance a positive story.
17. **Logistics:** Various application research models have been used in the effective management of logistics operations. AI has great potential in supply chain planning and servicing.
18. **AI Tactical Applications:** The nature of defense may change due to the rise of AI's tactical applications. The development of AI has the potential to fundamentally alter defense, resulting in a shift from information-based warfare of the present to intelligence systems of the future, when AI will be essential to defense systems. An unmanned aerial vehicle (UAV) for minesweeping that looks for, covers, or transports mines using feature maps.

8.8 CONCLUSION

Blockchain technology and AI have a lot of potential to improve national security in the future. A synergistic framework that meets the changing difficulties of defense operations is created by combining the distributed, secure, and transparent properties of blockchain technology with AI analytics capabilities. In actuality, the union of AI and blockchain not only fortifies the basis of national security but also enables defense forces to prosper in a globe growing more intricate and interconnected by the day. This collaborative strategy foresees and tackles potential hazards, guaranteeing that nations are equipped to adjust, react, and safeguard their concerns when confronted with constantly changing worldwide security issues. At some point, blockchain technology and artificial intelligence will be used in every facet of combat. Creating mission-driven strategies for strategically significant AI technologies can be aided by a thorough long-term vision.

KEYWORDS

- **AI robots**
- **artificial intelligence**
- **blockchain**
- **body implants**
- **expert system**
- **functional exoskeleton**
- **human universal loader**
- **long-term vision**
- **remote monitoring**
- **retinal implants**
- **robotic surgery**

REFERENCES

Brose, C. (2020). *The Kill Chain: Defending America in the Future of High-Tech Warfare*. Hachette Books.

Brynjolfsson, E., & McAfee, A. (2014). *The Second Machine Age: Work, Progress, and Prosperity in a Time of Brilliant Technologies*. W. W. Norton & Company.

Cave, S., Craig, C., Dihal, K., Dillon, S., Montgomery, J., Singler, B., et al. (2018). *Portrayals and Perceptions of AI and Why They Matter*. The Royal Society.

Center for Strategic Studies. (2017). *The Future of Artificial Intelligence and Defense, and its Strategic Implications for Small and Medium-Sized Military Providers*. The Hague.

Cummings, M. L. (2017). *Artificial Intelligence and the Future of War* (Research Paper).
International Security Agency and the US-US Program.

Kai-Fu, L. (2018). *AI Superpowers: China, Silicon Valley, and the New World Order*. Houghton Mifflin Harcourt.

King, B. (2018). *Augmented: Life in the Smart Lane*. Wiley.

Latiff, R. H. (2017). *Future War: Preparing for the New Global Battlefield.* Public Affairs.

Levy, S. (2001). *Crypto: How the Code Rebels Beat the Government—Saving Privacy in the Digital Age.* Viking.

Mitchell, M. (2019). *Artificial Intelligence: A Guide for Thinking Humans.* Farrar, Straus and Giroux. NITI Aayog. (2018). *National Strategy on Artificial Intelligence: Discussion Paper.* NITI Aayog.

https://www.niti.gov.in (accessed on 14 December 2024).

Tapscott, D., & Tapscott, A. (2016). *Blockchain Revolution: How the Technology Behind Bitcoin and Other Cryptocurrencies is Changing the World.* Penguin.

CHAPTER 9

An Analytic Hierarchy Process Approach to Prioritize the Critical Success Factors of Blockchain Technology: A Study in the Healthcare Sector

RUCHI JAIN GARG,[1] POOJA TIWARI,[1] MOHD. SALMAN SHAMSI,[2] and PARUL SAXENA[1]

[1]*Sharda School of Business Studies, Sharda University, Greater Noida, Uttar Pradesh, India*

[2]*Department of Marketing, Faculty of Business Administration, University of Tabuk, Tabuk, Saudi Arabia*

*Corresponding author

ABSTRACT

In the present scenario, many challenges can be witnessed in the healthcare sector. Currently, blockchain technology has played a crucial role in addressing these challenges. The role of this technology has been crucial in the automation and mining of different health records, which helps make diagnoses more accurate. This work aims to prioritize the different critical success factors (CSFs) that will help organizations in contingency situations. Based on the literature review, 21 critical success factors in healthcare have been adopted, categorized into six dimensions, and confirmed with the help of expert opinions. Additionally, an analytic hierarchy process (AHP) was conducted to prioritize the critical success factors in the healthcare sector.

Blockchain Technology: Unlocking the Value of 3Ts in the Business World.
Richa Chauhan & Vidhi Kaul (Eds.)

DOI: 10.1201/9781998511914-9

The findings indicated that at the criteria level, among the six dimensions, the "organization" dimension has been prioritized first, followed by the "customer service" dimension, then the "development," "technology," "environment," and "cost" dimensions. At the sub-criteria level, the priority ranking of the critical success factors is as follows: "ATL" of the "organization" dimension, "PI" of the "customer service" dimension, "PS" of the "development" dimension, "PWR" of the "environment" dimension, "TRU" of the "customer service" dimension, and so forth. According to the available literature, it can be inferred that the criteria for measuring CSFs in the healthcare sector are mutually exclusive. However, in other practical scenarios, there is a possibility that these sub-criteria may depend on each other. To address this issue, it is advised that future research be conducted employing an analytic network process (ANP).

The outcomes of the study indicate that practitioners in the healthcare sector must utilize the studied critical factors based on their weights, which will facilitate organizations in attaining a competitive advantage. To the best of our knowledge, it has been inferred that none of the studies have conducted the prioritization of critical success factors for blockchain technology in the healthcare sector. The results presented in the chapter are trustworthy and objective and may be generalized within the healthcare segment.

9.1 INTRODUCTION

Blockchain technology, which was introduced with the advent of Bitcoin in 2008, has emerged as a game-changing force showcasing qualities like decentralization, transparency, and anonymity. Bitcoin itself has achieved success with 400 million transactions completed as of March 2019, serving as a tangible example of how blockchain can be effectively implemented. This achievement has sparked discussions and considerations about its applications in data-focused fields, particularly the healthcare sector. According to IBM reports, 70% of healthcare torchbearers trust that blockchain would significantly contribute to the improvement of clinical trial management, ensuring adherence to regulations along with establishing a decentralized structure for disseminating electronic health records (EHR). This positive outlook is supported by projections that indicate the worldwide blockchain technology market will surpass $500 million in the healthcare sector by 2022 (Hasselgren et al., 2020).

The healthcare sector is often perceived as a traditional sector with resistance to change and an unwillingness to adopt new practices, making it inherently difficult to measure and evaluate the global focus on issues in healthcare technology, which is recognized as a viable solution. Blockchain has the potential to significantly improve the immediacy of healthcare practices, especially in improving healthcare delivery and supporting quality care. A recent report by Market Research Future (MRFR) highlights the growing importance of blockchain technology in healthcare, estimating a significant value of over $42 million, and expects an annual growth rate of an amazing composite growth of 71.8% by 2023 (Biswas and Gupta, 2019). Its increased transparency, advanced security and privacy policies, better searchability, increased productivity, and, as a result, reduced costs. Blockchain's anticipated positive impact on healthcare is underlined by its potential to transform information systems, reduce privacy concerns, and streamline processes in the healthcare sector. Leveraging blockchain's decentralized nature, the industry aims to foster greater transparency and security while addressing prevailing challenges (Deloitte, 2018).

The allure of integrating blockchain technology into healthcare stems from its ability to transform information systems. Leveraging its inherent insights, blockchain provides the ability to address ongoing challenges in healthcare, including issues of data security, performance management, and accessibility. The promise of creating a verifiable shared library for health information is of interest to the sector. The transparent and immutable properties of blockchain hold the potential to enhance the accuracy and security of healthcare data, offering solutions to age-old problems that hinder effective information management in the healthcare sector (Deloitte, 2018). As a result, the industry is delving into the revolutionary possibilities brought about by blockchain, envisioning a future in which health information is not only secure but also shared and accessed with ease in healthcare parts of the system.

Thus, the recognition of critical success factors has become evident for the proper implementation of blockchain technology in the healthcare segment. In this direction, Bali et al. (2023) have conducted a study in this arena, and with the help of the DEMATEL approach, they have identified 21 critical success factors for the implementation of blockchain technology in the healthcare segment. The study has further categorized these 21

CSFs into six dimensions. The study has also provided direction for future research, that is, to apply other MCDM techniques such as AHP, ANP, and alike. Therefore, in this study, we have applied the Analytic Hierarchy Process (AHP) approach to six dimensions and 21 CSFs for the execution of blockchain technology in the healthcare sector. AHP is the multi-criteria decision-making technique used to prioritize the 21 CSFs to handle crucial situations that occurred during the COVID-19 pandemic efficiently. The prioritization of these critical success factors will contribute to mitigating economic loss and suffering that occurred during the pandemic in a prompt manner. To conduct this study, two major objectives are framed as:

- To prepare an AHP model through six dimensions and 21 CSFs for the implementation of blockchain technology in the healthcare sector.
- To identify the relative priorities of the dimensions and critical success factors so that the contingency situations that occur in the healthcare sector can be handled promptly.

To achieve the said objectives, this chapter is planned in the following manner. Firstly, the CSFs for the implementation of blockchain technology in healthcare were identified through a literature review and finalized through the opinions of an expert panel. Secondly, a multilevel AHP model was developed in order to prioritize the recognized dimensions and CSFs. Thirdly, the relative priorities of these CSFs were identified. Finally, the inferences of the identified priorities were deliberated with regards to the healthcare sector.

9.2 LITERATURE REVIEW

According to Biswas and Gupta (2019), blockchain can be understood as a "digital ledger," which is a cryptographically signed ledger, and it is grouped as a block. As this grouping of blocks has been done using technology, it is referred to as blockchain technology. In the early 1990s, blockchain came into existence, but it gained attention when this technology was utilized for an electronic credit system, that is, Bitcoin, in the year 2008. According to the survey conducted by Deloitte (2018), it is evident that blockchain technology is growing at a rapid rate, but the real understanding of this technology and its potential is still underexplored.

This provides an opportunity to explore the future developments and potential implementation of blockchain technology, which can bring radical changes in the social environment and also includes bringing changes and improvements in the domain of the healthcare sector.

To store digital data, this technology provides a solution in a decentralized manner, which facilitates the secure storage of the generated or meta-data across different networks and users. This technology uses cryptography, and it includes a list of records, and these records are linked through cryptography (Zhang et al., 2018). Modification of data is not possible due to the presence of this design: It can be understood that once the data is recorded in any one block, alteration of data is not possible; to alter the data, data has to be altered in all the subsequent blocks, which requires the agreement of the major networks. This technology works as a decentralized ledger system; it allows the different agents of an independent nature to integrate within the ecosystem, which ensures the transparency and timely recording of the information. This improves the processing speed and generation of revenue, providing enhanced security while reducing the risk and cost associated (Deloitte, 2018).

In contrast to the conservative way of information storage at one single point, this technology practices a peer-to-peer (P2P) network, which includes numerous copies of similar data stored in various locations and on different devices. This peer networking allows computing resources to be shared, and it will help the central coordination by the servers, which can be used by other participants. Preservation, coordination, and security of business data exchanges can be ensured by the different networks when these nodes adopt various roles (Van Rijmenam and Ryan, 2018). Satoshi Nakamoto developed the concept of blockchain technology via Bitcoin, and it is due to his novel and unprecedented efforts that we are now moving towards adopting blockchain technology. This processing approach was brought to the healthcare industry, which is one of the most crucial industries for humanity (Engelhardt, 2017).

Blockchain technology has shown its presence in almost all sectors. It is a cross-domain technology widely used by almost all the various industries and has left its impression in the healthcare sector as well. Various methods are adopted by the healthcare sector for recordkeeping and transfer of data. The main matter of concern is to maintain the privacy and accessibility of these records (Calduch et al., 2021). In the conservative approach, record maintenance was considered to be a tedious task. Data storage and retrieval are major concerns and also cumbersome in nature. For intruders, it was considered an easy task to misuse and tamper with the data. There are various resources used by organizations to overcome these challenges to protect against data tampering. With the advancement of technology, storage of data became a manageable task, but securely transferring the data was always a challenge (Azaria et al., 2016; Mamoshina et al., 2018).

The need for sharing various records, like prescriptions, cannot be overlooked in healthcare, where multi-level sharing is a routine task. This sharing of data should be done securely, as any interference or tampering with this data can be dangerous. As already pointed out, the storage of data electronically is vulnerable to security breaches (Khanra et al., 2020). According to Massachusetts General Hospital, Boston, the importance of securing health data for any nation cannot be overstated (Smith and Eloff, 1999). Blockchain technology is highly useful in the healthcare sector, as the same data is accessed and used by many users (Irving and John, 2016). This technology can be instrumental in facilitating and improving the medical care provided (Attaran, 2020). Different critical success factors have been identified by Bali et al. (2022). Although blockchain technology has been implemented in many sectors, there are limited studies conducted in the healthcare sector. Therefore, this study has adopted the model of critical success factors from Bali et al. (2022) and, using AHP, has prioritized the 21 critical success factors divided into six dimensions. Table 9.1 represents the various 21 critical success factors, which are further divided into six dimensions.

9.3 AHP METHODOLOGY

The Analytic Hierarchy Process (AHP) is a decision-making methodology formulated by Saaty (1980). Specifically, it is a structured technique used to organize and analyze complex decisions, especially when multiple criteria and alternatives are to be taken into account. In practice, this approach involves breaking down a decision problem into a hierarchical structure, then judging the relative importance of different criteria before finally evaluating how different alternatives will satisfy these criteria, hence leading to objective choice-making in decision-making. Bali et al. (2023) have identified the CSFs for the implementation of blockchain technology in the healthcare sector and suggested applying the AHP

TABLE 9.1 Dimensions and Critical Success Factors

Dimensions	CSFs	Description	References
Technology	Data transparency (DT)	Transparency must be maintained regarding personal healthcare data in the healthcare system to enable technology adoption among stakeholders and foster relationships. This is crucial for linking the healthcare system and patients. The visibility of transactions is ensured for all patients.	Siyal et al. (2019); Randall et al. (2017)
	Real-time processing (RTP)	RTP refers to adherence to deadlines while processing and accumulating data. Real-time monitoring of patients in the healthcare system is a top priority. Additionally, tracking all transactions in the domain is essential.	Rahman et al. (2018); Zhang et al. (2017)
	Reliability (REL)	The healthcare sector must ensure failure-free data transmission. Reliability involves operating without failure within specific durations and conditions. Blockchain technology (BT) improves system reliability for user availability.	Shardeo et al. (2020)
	Security (SEC)	BT ensures robust encryption and secure transmission of patient data. Security and privacy are the most important features of BT, which help in data protection and environmental security.	Randall et al. (2017); Yang et al. (2017)
Customer service	Track and traceability (TT)	This technology helps track activities performed by patients in the healthcare system. A secure model provided by BT allows continuous observation and monitoring, ensuring traceability for all data packets, with clearly defined origins and destinations.	Abu-Elezz et al. (2020); Olnes and Jansen (2018)
	Trust (TRU)	Patient data cannot be tampered with, ensuring data integrity and authentication. BT creates an environment of trust in healthcare, providing secure data storage and transfer, ultimately building users' confidence.	Liang et al. (2017)
	Process improvement (PI)	BT provides a secure system for storing clinical trial data. It helps address emerging challenges in the healthcare sector by securely recording data and reducing errors, thereby improving patient satisfaction.	Yang et al. (2017)
	Immutable (IMT)	Ensuring immutability (i.e., no one is authorized to modify records) is a top priority in healthcare. BT's inherent immutability feature fulfills this requirement.	Yli-Huumo et al. (2016)

TABLE 9.1 *(Continued)*

Dimensions	CSFs	Description	References
Cost	Implementation cost (IC)	Adopting blockchain technology involves significant implementation costs, including expenses related to entities and infrastructure.	Shardeo et al. (2020)
	Reduced manpower cost (RMP)	Healthcare often requires high manpower costs. BT can help reduce recruitment expenditures while improving productivity.	Olnes and Jansen (2018)
	Training and skill up gradation cost (TSGC)	Upgrading individual skills and competencies necessitates training on novel technologies in the healthcare sector.	Farooque et al. (2020)
	Effective claim management (ECM)	Claiming insurance is vital in healthcare. Privacy, security, and fraud prevention through increased transparency are essential for effective claim management.	Kuo et al. (2017)
Organization	Top management support (SEC)	Top management support is critical for organizational transitions. As BT is novel in healthcare, endorsement from leadership is essential.	Khatoon (2020); Shardeo et al. (2020)
	Management stability (MS)	Stability in management is an indicator of organizational success. However, priorities and preferences may vary across different management structures.	Farooque et al. (2020)
	Appropriate team leadership (ATL)	Positive leadership fosters a congenial working environment. A project leader should be self-assured and proactive in managing their team.	Raj and Sah (2019)
	Creating detailed road-map (CDR)	Developing a proper roadmap is essential for project efficiency and effectiveness. It ensures work prioritization and enhances the likelihood of project success.	Shardeo et al. (2020)
Environment	Paperwork reduction (PWR)	Digitally sharing data reduces paperwork and fosters an environmentally friendly system. Paperwork is prone to tampering and delays, while digital systems improve efficiency.	Radanovic and Likic (2018); Singh (2019); Siyal et al. (2019)
	Green initiatives (GI)	With increasing ailments, trials, and drug development, data storage is critical. BT helps maintain records while promoting greener and healthier environments.	Ahmad et al. (2019); Hölbl et al. (2018); Maslove et al. (2018)

TABLE 9.1 *(Continued)*

Dimensions	CSFs	Description	References
Development	Enhanced collaboration (EC)	BT facilitates collaboration and data sharing in healthcare, fostering mutual benefits for all stakeholders.	Mamoshina et al. (2018); Ray et al. (2020); Zhang et al. (2018)
	Process simplification (PS)	The decentralized nature of BT reduces work complexities in healthcare. It ensures integrity, data security, and effective tracking.	Abu-Elezz et al. (2020); Randall et al. (2017)
	Government support (GS)	Governments offer incentives, subsidies, and infrastructure support to implement blockchain technology.	Balasubramanian et al. (2021); Batubara et al. (2018); Raj and Sah (2019)

Source: Adapted from: Bali et al. (2021).

technique to prioritize those CSFs. So, as per the directions for future research, the authors have applied the AHP method to identify the relative significance of the critical success factors to handle contingency situations like COVID-19 type situations promptly so that economic and health emergencies can be handled effectively.

The process of the AHP methodology is as follows (Garg et al., 2012). First, figure out the decision problem that needs to be made and break down the decision problem into a hierarchical structure with the goal at the top. Break down the decision into a hierarchy of criteria, sub-criteria, and alternatives. Next, contrast every element at criterion, sub-criteria, and alternatives at the same level. These pairwise comparisons are to be done by using Saaty's (1994) nine-point scale. These pairwise comparisons will be done with the help of experts and decision-makers. Within a hierarchical structure, comparisons are to be done between the items that are associated with the same node, with a particular emphasis placed on the elements that are located beneath a common node. To elucidate, if "n" elements are present beneath a node, the number of comparisons that are carried out within that node is equal to n multiplied by$(n\text{-}1)/2$.

Let us take into consideration a collection of elements as $A_1, A_2, A_3, \ldots, A_n$ within the node being represented by the letter "D" and the elements being given mathematical weights as $x_1, x_2, x_3, \ldots, x_n$. As a result of this, the values of the relative weights of the pairwise comparisons are depicted in the matrix P. The pairwise comparisons between the various elements A_1, $A_2, A_3, \ldots, A_n$ are laid out in this matrix.

$$P = \begin{matrix} A_1 \\ A_2 \\ \cdot \\ \cdot \\ \cdot \\ A_n \end{matrix} \begin{bmatrix} x_1/x_2 & x_1/x_2 & \dots & x_1/x_n \\ x_2/x_1 & x_2/x_2 & \dots & x_2/x_n \\ \cdot & \cdot & \cdot & \cdot \\ \cdot & \cdot & \cdot & \cdot \\ \cdot & \cdot & \cdot & \cdot \\ x_n/x_1 & x_n/x_2 & \dots & x_n/x_n \end{bmatrix}$$

$$P = \begin{matrix} A_1 \\ A_2 \\ \cdot \\ \cdot \\ \cdot \\ A_n \end{matrix} \begin{bmatrix} b_{11} & b_{12} & \dots & b_{1n} \\ b_{12} & b_{22} & \dots & b_{2n} \\ \cdot & \cdot & \cdot & \cdot \\ \cdot & \cdot & \cdot & \cdot \\ \cdot & \cdot & \cdot & \cdot \\ b_{n1} & b_{n2} & \dots & b_{nn} \end{bmatrix}$$

where; $b_{ij} = x_i / x_j$ $(i, j = 1, 2, \ldots, n)$. This expression signifies the computed comparative significance between the pair of elements A_i and A_j. When $i = j$, then $b_{ij} = 1$, and if $b_{ij} > 0$, $b_{ij} = 1/ b_{ji}$.

Further, the priority weights of the elements will be determined by using the maximum eigenvectors and eigenvalues. As per the study conducted by Saaty (1994), the formula to calculate the largest eigenvalue is:

$$\lambda_{max} = \sum_{j=1}^{n} a_{ij} \frac{X_j}{X_i}$$

where; X represents the eigenvector; and λ_{max} represents the largest eigenvalue of matrix P.

The formula to compute eigenvector M is:

$$(P\text{-}\lambda_{max} I)M = 0$$

Additionally, a consistency test is to be performed in order to identify if any deviation exists in the values from a consistent matrix. Hence, the consistency index (CI) and consistency ratio (CR) are calculated using the following formulae:

$$CI = \frac{\lambda_{max} - m}{m - 1}$$

$$CR = \frac{CI}{RI}$$

Here; *m* represents the order of the matrix; and *RI* represents the random index. The values of *RI* are depicted in Table 9.2. The maximum acceptable limit for the consistency test for both *CI* and *CR* is 0.1; otherwise, the pairwise comparison is found to be inconsistent and is hence discarded (Saaty, 1994).

TABLE 9.2 Random Index (RI)

N	2	3	4	5	6	7	8	9
RI	0.00	0.58	0.90	1.12	1.24	1.32	1.41	1.45

Source: Saaty (1994).

At last, the local and global weights of the elements are to be calculated with regards to the objectives of the research, and the relative priorities of the elements will be tabulated and can be depicted in the form of a graph.

9.3.1 MODEL DEVELOPMENT

On the basis of the steps illustrated in the previous section, a multilevel hierarchical model is developed, shown in Figure 9.1.

The first level of the model denotes the study's objectives, the next level, which is known as the criteria level, is represented by the six dimensions considered for the implementation of blockchain technology in healthcare, and the last level, that is, the sub-criteria level, is represented by the 21 critical success factors adopted for the implementation of blockchain in healthcare.

9.4 ANALYSIS AND RESULTS

The analysis and results of the AHP model are discussed in this section. It is as follows:

1. **Pairwise Comparison of Dimensions:** The six dimensions for the implementation of blockchain technology in healthcare are "cost," "development," "technology," "customer service," "organization," and "environment." Pairwise comparisons of these dimensions are performed, keeping in focus the goal of the model,

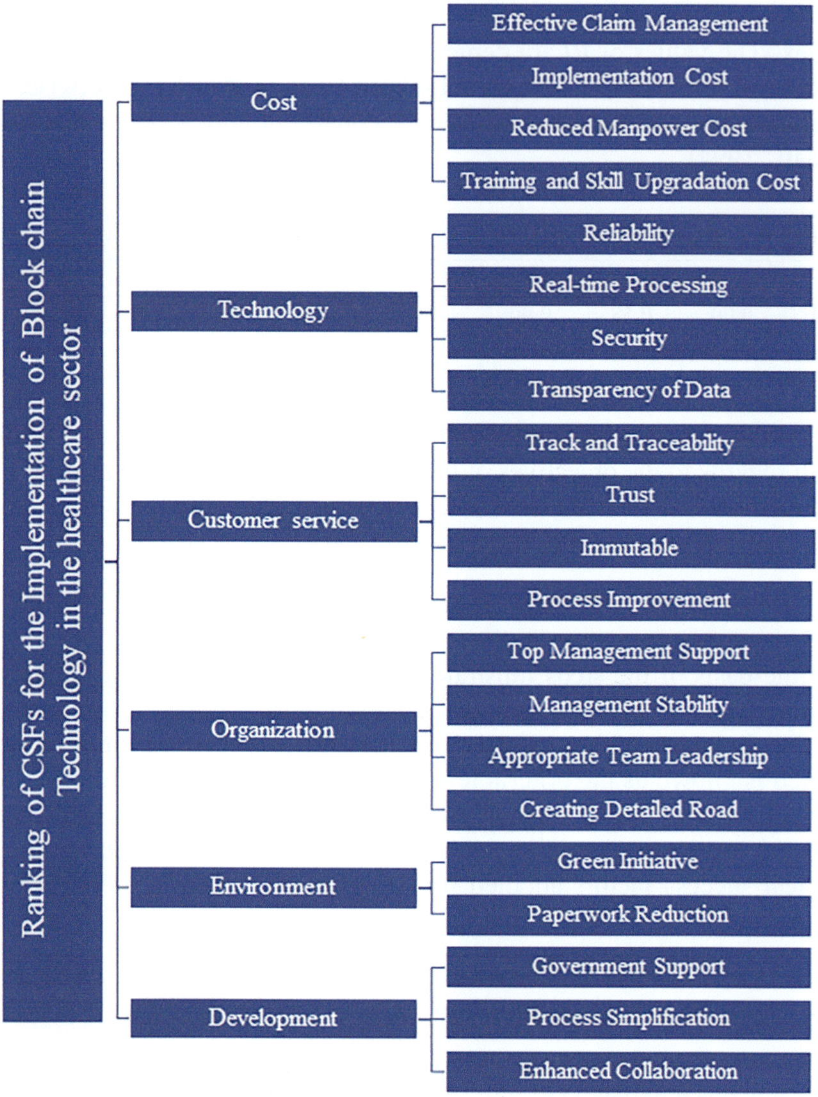

FIGURE 9.1 An analytic hierarchy process model.

that is, ranking of CSFs for the implementation of blockchain technology in the healthcare sector. The pairwise comparison matrix is shown in Table 9.3.

TABLE 9.3 Pairwise Comparisons of Dimensions

Goal	Cost	Development	Organization	Technology	Environment	Customer Service
Cost	1.00	0.14	0.25	1.00	0.20	0.17
Development	7.00	1.00	0.50	2.00	4.00	0.50
Organization	4.00	2.00	1.00	4.00	5.00	1.00
Technology	1.00	0.50	0.25	1.00	2.00	0.50
Environment	5.00	0.25	0.20	0.50	1.00	0.50
Customer service	6.00	2.00	1.00	2.00	2.00	1.00

2. **Pairwise Comparison of Critical Success Factors:** After commencing pairwise comparisons of dimensions at the criteria level, the next step is to conduct the pairwise comparisons of 21 critical success factors with respect to six dimensions. Under the "cost" dimension, there are four CSFs, and their pairwise comparisons are shown in Table 9.4; for the "development" dimension, there are three CSFs, and their pairwise comparisons are shown in Table 9.5; for the "technology" dimension, there are four CSFs, and their pairwise comparisons are shown in Table 9.6; for the "customer service" dimension, there are four CSFs, and their pairwise comparisons are shown in Table 9.7; for the "organization" dimension, there are four CSFs, and their pairwise comparisons are shown in Table 9.8; and for the "environment" dimension, there are two CSFs, and their pairwise comparisons are shown in Table 9.9.

TABLE 9.4 Pairwise Comparisons of Critical Success Factors with Respect to "Cost" Dimension

Cost Dimension	IC	RMP	ECM	TSGC
IC	1.00	6.00	0.50	8.00
RMP	0.17	1.00	0.14	4.00
ECM	2.00	7.00	1.00	8.00
TSGC	0.13	0.25	0.13	1.00

TABLE 9.5 Pairwise Comparisons of Critical Success Factors with Respect to "Development" Dimension

Development Dimension	EC	PS	GS
EC	1.00	0.50	4.00
PS	2.00	1.00	4.00
GS	0.25	0.25	1.00

TABLE 9.6 Pairwise Comparisons of Critical Success Factors with Respect to "Technology" Dimension

Technology Dimension	SEC	REL	RTP	TOD
SEC	1.00	3.00	5.00	2.00
REL	0.33	1.00	3.00	2.00
RTP	0.20	0.33	1.00	0.20
TOD	0.50	0.50	5.00	1.00

TABLE 9.7 Pairwise Comparisons of Critical Success Factors with Respect to "Customer Service" Dimension

Customer Service Dimension	TRU	TT	PI	IMT
TRU	1.00	3.00	0.50	3.00
TT	0.33	1.00	0.25	3.00
PI	2.00	4.00	1.00	5.00
IMT	0.33	0.33	0.20	1.00

TABLE 9.8 Pairwise Comparisons of Critical Success Factors with Respect to "Organization" Dimension

Organization Dimension	TMS	MS	ATL	CDR
TMS	1.00	3.00	0.20	0.50
MS	0.33	1.00	0.20	0.50
ATL	5.00	5.00	1.00	2.00
CDR	2.00	2.00	0.50	1.00

TABLE 9.9 Pairwise Comparisons of Critical Success Factors with Respect to "Environment" Dimension

Environment Dimension	PWR	GI
PWR	1.00	0.14
GI	7.00	1.00

9.4.1 CALCULATION OF LOCAL AND GLOBAL WEIGHTS

The pairwise comparisons of the elements at criteria and sub-criteria levels are followed by calculating the local and global weights of all the elements. The local as well as global weights of all the six dimensions and 21 CSFs are depicted in Table 9.10.

TABLE 9.10 Local and Global Weights of the Elements

Dimensions	Relative Weights	Critical Success Factors	Local Weights	Global Weights
		IC	0.347158	0.017289
Cost	0.049802	RMP	0.098023	0.004882
		ECM	0.511358	0.025467
		TSGC	0.043461	0.002164
		EC	0.34595	0.070697
Development	0.204357	PS	0.543753	0.11112
		GS	0.110297	0.02254
		TMS	0.155748	0.047356
Organization	0.304059	MS	0.090293	0.027454
		ATL	0.520215	0.158176
		CDR	0.233744	0.071072
		SEC	0.463563	0.044586
Technology	0.096181	REL	0.242433	0.023317
		RTP	0.069304	0.006666
		TOD	0.2247	0.021612
		TRU	0.284784	0.071478
		TT	0.147279	0.036966
Customer service	0.25099	PI	0.488735	0.122668
		IMT	0.079202	0.019879
		PWR	0.875	0.082784
Environment	0.09461	GI	0.125	0.011826

9.4.2 RANKING OF THE CRITICAL SUCCESS FACTORS

In this last step, we have to identify the final priorities of all the 21 CSFs. The rankings of these CSFs are identified by the values of global weights shown in Table 9.10. The pictorial representation of these priorities is illustrated in the form of the graph shown in Figure 9.2.

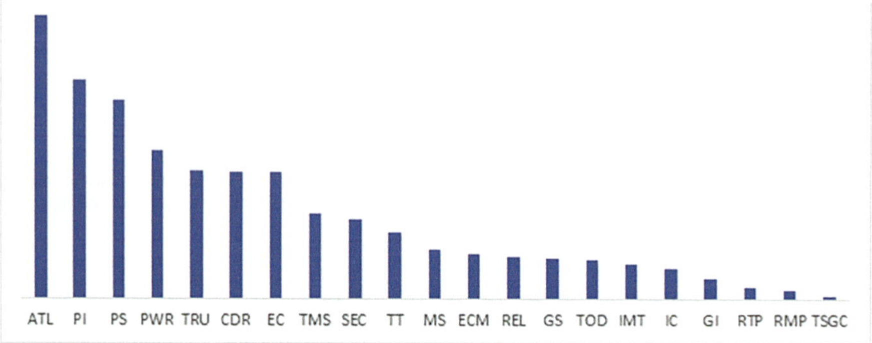

FIGURE 9.2 Ranking of critical success factors.

9.5 DISCUSSION AND IMPLICATIONS

This study has prioritized the various critical success factors using AHP, and the findings of this study will help the healthcare industry to understand the relevant prioritization of various CSFs. From the findings, it can be concluded that at the criteria level, out of six dimensions, the "organization" dimension has been prioritized first, the "customer service" dimension comes second in the list, followed by the "development," "technology," "environment," and "cost" dimensions. At the sub-criteria level, the priority level of the CSFs is as follows: "ATL" of the "organization" dimension, "PI" of the "customer service" dimension, "PS" of the "development" dimension, "PWR" of the "environment" dimension, "TRU" of the "customer service" dimension, and so on. The findings of the study will be crucial for different professionals in healthcare, insurance, policy, and governance, and also for various technology enablers to comprehend the priority of different factors in the execution of blockchain technology in healthcare. It can be inferred from the findings that to implement any novel technology, the role of an appropriate leader is crucial. Appropriate leaders will be instrumental in identifying and involving the people in bringing change to the organization. Additionally, to implement the novel technology, process improvement is essential, and it ensures that customer services can be improved when we have worked on our processes. Also, simplifying the process is imperative, and as we have moved into the digital era, paperwork reduction has become a necessity. In the end, in the healthcare sector, the most vital aspect is to maintain the trust of the patient, and storing and sharing medical data is crucial. Therefore, these factors are essential for the healthcare industry to implement blockchain technology.

KEYWORDS

- **analytic hierarchy process**
- **blockchain**
- **cost dimension**
- **critical success factors**
- **customer service dimension**
- **electronic health records**
- **healthcare industry**
- **sub-criteria of organization**
- **technology dimension**

REFERENCES

Abu-Elezz, I., Hassan, A., Nazeemudeen, A., Househ, M., & Abd-Alrazaq, A. (2020). The benefits and threats of blockchain technology in healthcare: A scoping review. *International Journal of Medical Informatics, 142*, 104246. https://doi.org/10.1016/j.ijmedinf.2020.104246.

Ahmad, S. S., Khan, S., & Kamal, M. A. (2019). What is blockchain technology and its significance in the current healthcare system? A brief insight. *Current Pharmaceutical Design, 25*(12), 1402–1408. https://doi.org/10.2174/1381612825666190423160455.

Attaran, M. (2020). Blockchain technology in healthcare: Challenges and opportunities. *International Journal of Healthcare Management, 15*(1), 1–14. https://doi.org/10.1080/20479700.2020.1734204.

Azaria, A., Ekblaw, A., Vieira, T., & Lippman, A. (2016). MedRec: Using blockchain for medical data access and permission management. *In 2nd International Conference on Open and Big Data (OBD)*, 25–30. https://doi.org/10.1109/OBD.2016.16.

Balasubramanian, R., Shukla, V., Sethi, J. S., Islam, N., & Saloum, R. (2021). A readiness assessment framework for blockchain adoption: A healthcare case study.

Technological Forecasting and Social Change, 165, 120536. https://doi.org/10.1016/j.techfore.2021.120536.

Bali, S., Bali, V., Mohanty, R. P., & Gaur, D. (2021). Analysis of critical success factors for blockchain technology implementation in the healthcare sector. *Benchmarking: An International Journal, 28*. https://doi.org/10.1108/BIJ-07-2021-0433.

Batubara, F. R., Ubacht, J., & Janssen, M. (2018). Challenges of blockchain technology adoption for e-government: A systematic literature review. *In Proceedings of the 19th Annual International Conference on Digital Government Research: Governance in the Data Age* (pp. 1–9).

Calduch, E. N., Muscat, N. A., Krishnamurthy, R. S., & Ortiz, D. N. (2021). Technological progress in electronic health record system optimization: Systematic review of systematic literature reviews. *International Journal of Medical Informatics, 152*, 104507. https://doi.org/10.1016/j.ijmedinf.2021.104507.

Deloitte. (2018). *Breaking Blockchain Open—Deloitte's 2018 Global Blockchain Survey*. Deloitte Insights, 48. https://doi.org/10.1002/ejoc.201200111.

Engelhardt, M. A. (2017). Hitching healthcare to the chain: An introduction to blockchain technology in the healthcare sector. *Technology Innovation Management Review, 7*(10), 1–14. https://doi.org/10.22215/timreview/1103.

Farooque, M., Jain, V., Zhang, A., & Li, Z. (2020). Fuzzy DEMATEL analysis of barriers to blockchain-based life cycle assessment in China. *Computers and Industrial Engineering, 147*, 32–46. https://doi.org/10.1016/j.cie.2020.106598.

Hasselgren, A., Kralevska, K., Gligoroski, D., Pedersen, S. A., & Faxvaag, A. (2020). Blockchain in healthcare and health sciences–A scoping review. *International Journal of Medical Informatics, 134*, 104040. https://doi.org/10.1016/j.ijmedinf.2019.104040.

Irving, G., & John, H. (2016). How blockchain-timestamped protocols could improve the trustworthiness of medical science. *F1000 Research, 3*, 1–7. https://doi.org/10.12688/f1000research.8762.1.

Khanra, S., Dhir, A., Islam, A. N., & Mäntymäki, M. (2020). Big data analytics in healthcare: A systematic literature review. *Enterprise Information Systems, 14*(7), 878–912. https://doi.org/10.1080/17517575.2019.1666098.

Khatoon, A. (2020). A blockchain-based smart contract system for healthcare management. *Electronics, 9*(1), 94. https://doi.org/10.3390/electronics9010094.

Kuo, T. T., Kim, H. E., & Ohno-Machado, L. (2017). Blockchain distributed ledger technologies for biomedical and health care applications. *Journal of the American Medical Informatics Association, 24*(6), 1211–1220. https://doi.org/10.1093/jamia/ocx068.

Liang, X., Zhao, J., Shetty, S., Liu, J., & Li, D. (2017). Integrating blockchain for data sharing and collaboration in mobile healthcare applications. *In IEEE 28th Annual International Symposium on Personal, Indoor and Mobile Radio Communications (PIMRC)*, 1–5. https://doi.org/10.1109/PIMRC.2017.8292689.

Mamoshina, P., Ojomoko, L., Yanovich, Y., Ostrovski, A., Botezatu, A., Prikhodko, P., Izumchenko, E., Aliper, A., Romantsov, K., Zhebrak, A., Ogu, L., & Zhavoronkov, A. (2018). Converging blockchain and next-generation artificial intelligence technologies to decentralize and accelerate biomedical research and healthcare. *Oncotarget, 9*, 5665–5690. https://doi.org/10.18632/oncotarget.24422.

Maslove, D. M., Klein, J., Brohman, K., & Martin, P. (2018). Using blockchain technology to manage clinical trials data: A proof-of-concept study. *JMIR Medical Informatics, 6*(4), 11949. https://doi.org/10.2196/11949.

Mettler, M. (2016). Blockchain technology in healthcare: The revolution starts here. *18th International Conference on E-Health Networking, Applications, and Services (Healthcom)*, 1–3. https://doi.org/10.1109/HealthCom.2016.7749514.

Olnes, S., Ubacht, J., & Janssen, M. (2017). Blockchain in government: Benefits and implications of distributed ledger technology for information sharing. *Government Information Quarterly, 34*(3), 355–364. https://doi.org/10.1016/j.giq.2017.09.007.

Radanovic, I., & Likic, R. (2018). Opportunities for the use of blockchain technology in medicine. *Applied Health Economics and Health Policy, 16*(5), 583–590. https://doi.org/10.1007/s40258-018-0400-1.

Rahmadika, S., & Rhee, K. H. (2018). Blockchain technology for providing an architecture model of decentralized personal health information. *International Journal of Engineering Business Management, 10*, 1847979018790589. https://doi.org/10.1177/1847979018790589.

Rahman, M. A., Hassanain, E., Rashid, M. M., Barnes, S. J., & Hossain, M. S. (2018). Spatial blockchain-based secure mass screening framework for children with dyslexia. *IEEE Access, 6*, 61876–61885. https://doi.org/10.1109/ACCESS.2018.2874591.

Raj, A., & Sah, B. (2019). Analyzing critical success factors for implementing drones in the logistics sector using a grey-DEMATEL-based approach. *Computers and Industrial Engineering, 138*, 106118. https://doi.org/10.1016/j.cie.2019.106118.

Randall, D., Goel, P., & Abujamra, R. (2017). Blockchain applications and use cases in health information technology. *Journal of Health and Medical Informatics, 8*(3), 8–11. https://doi.org/10.4172/healthmed.1000311.

Rane, S. B., & Narvel, Y. A. M. (2021). Re-designing the business organization using disruptive innovations based on blockchain-IoT integrated architecture for improving agility in future Industry 4.0. *Benchmarking: An International Journal, 28*(5), 1883–1908. https://doi.org/10.1108/BIJ-01-2021-0074.

Ray, P. P., Kumar, N., & Dash, D. (2020). BLWN: Blockchain-based lightweight simplified payment verification in IoT-assisted e-healthcare. *IEEE Systems Journal, 15*, 134–145. https://doi.org/10.1109/JSYST.2020.2963739.

Shardeo, V., Patil, A., & Madaan, J. (2020). Critical success factors for blockchain technology adoption in freight transportation using fuzzy ANP–Modified TISM approach. *International Journal of Information Technology and Decision Making, 19*(6), 1549–1580. https://doi.org/10.1142/S0219622020500274.

Singh, P. (2019). Lean in healthcare organization: An opportunity for environmental sustainability. *Benchmarking: An International Journal, 26*(1), 205–220. https://doi.org/10.1108/BIJ-05-2018-0172.

Siyal, A. A., Junejo, A. Z., Zawish, M., Ahmed, K., Khalil, A., & Soursou, G. (2019). Applications of blockchain technology in medicine and healthcare: Challenges and future perspectives. *Cryptography, 3*(1), 3. https://doi.org/10.3390/cryptography3010003.

Smith, E., & Eloff, J. H. (1999). Security in healthcare information systems—Current trends. *International Journal of Medical Informatics, 54*(1), 39–54. https://doi.org/10.1016/S1386-5056(99)00042-7.

Van Rijmenam, M., & Ryan, P. (2018). *Blockchain: Transforming your Business and Our World*. Routledge.

Yang, Z., Zheng, K., Yang, K., & Leung, V. C. (2017). A blockchain-based reputation system for data credibility assessment in vehicular networks. *In 28th IEEE Annual International Symposium on Personal, Indoor, and Mobile Radio Communications (PIMRC)*, 1–5. https://doi.org/10.1109/PIMRC.2017.8292695.

Yli-Huumo, J., Ko, D., Choi, S., Park, S., & Smolander, K. (2016). Where is current research on blockchain technology? — A systematic review. *PLOS ONE, 11*(10), 1–27. https://doi.org/10.1371/journal.pone.0163477.

Zhang, A., & Lin, X. (2018). Towards secure and privacy-preserving data sharing in E-health systems via consortium blockchain. *Journal of Medical Systems, 42*(8), 1–18. https://doi.org/10.1007/s10916-018-1000-7.

CHAPTER 10

Building Digital Advertising Through Blockchain Technology

ANJU GUPTA and ALKA KUMAWAT

Khandelwal Vaish Girls Institute of Technology, Jaipur, Rajasthan, India

*Corresponding author

ABSTRACT

Blockchain technology, initially conceptualized for cryptocurrencies, has evolved into a versatile tool applicable across industries, including digital marketing and advertising. This chapter explores blockchain potential to transform the marketing landscape through its decentralized, transparent, and secure framework. By leveraging features such as immutable records, cryptographic security, and smart contracts, blockchain addresses issues like ad fraud, data privacy, and inefficient lead generation. The chapter outlines how blockchain can enhance advertising experiences by providing direct compensation to consumers, securing data transactions, and enabling transparent supply chain tracking. Key applications in affiliate marketing, user data protection, and influencer marketing are examined. The discussion extends to challenges in scalability, cost, and cybersecurity, highlighting the dual-edged impact of this innovative technology. This chapter underscores blockchain role in fostering efficient and ethical practices in marketing, setting a foundation for its broader adoption in the digital economy.

Blockchain Technology: Unlocking the Value of 3Ts in the Business World.
Richa Chauhan & Vidhi Kaul (Eds.)

DOI: 10.1201/9781998511914-10

10.1 INTRODUCTION

Blockchain is an emerging concept that solves many business issues. It is not a new concept, as people have been organizing transactions and keeping records of goods traded and received since the beginning of human history. However, it was not until 2008 that the digitalization and transfer of records and information were made possible. Various intermediaries and middlemen were able to provide services for the efficient operation of transactions, such as records storage, validation, and verification. The democratization of data acquisition, storage, and processing, along with the development of algorithms for the collective creation of digital distributed ledgers, have enabled the development of chattels and abilities that go far beyond the capabilities of traditional chapter-backed ledgers. Currently, blockchain is mainly used for the purchase, sale, and trading of cryptocurrencies, but it can also be used in marketing to create additional protected relationships between brands and customers, allowing them to create more targeted advertisements without having to share too much information with many companies.

As we continue to explore blockchain technologies, many different opportunities will open up for a brand. Many industries will look to improve efficiencies in their operational models. In the case of advertising, the ways blockchain can potentially add value are as follows:

- Blockchain technology has the potential to revolutionize the advertising experience by providing users with direct compensation for viewing ads.
- It also offers secure data and transactions that are kept completely confidential, as well as transparent and reliable marketing that allows customers to accurately identify where products come from, where they are prepared, and what materials are used.
- Companies will also be able to collect more accurate leads due to the rigorous data they will receive from users.
- Finally, blockchain technology can help brands and websites to work together more cost-effectively, eliminating the need for middleman bloat in advertising costs.

In this chapter, we'll cover the basics, processes, and potential applications of blockchain in marketing and advertising, as well as the prevailing hurdles and mandatory steps and agreements that companies need to implement. However, we'd also like to point out the extent of the use of blockchain across other industries from which advertising needs to take a closer look. Blockchain has become an essential tool for marketing. Its technology, such as decentralized ledger, allows marketing and promotion teams to better handle data, analyze customer communications via ad campaigns, and build meaningful customer relationships.

With any new innovation that comes along, there are always challenges regarding its safety and feasibility. However, blockchain technology has proven to have a lot more advantages than disadvantages – especially in the marketing sector. It may take some time before blockchain marketing really takes off, but it is important to understand the direction the industry is taking so that one can begin to adopt some new tools and best practices to make marketing strategies more effective (Figure 10.1).

FIGURE 10.1 Mechanism of blockchain.

- **Definition:** Digital marketing is defined as "An adaptive, technology-enabled process by which firms collaborate with customers and partners to jointly create, communicate, deliver, and sustain value for all stakeholders" (Kannan and Li, 2017, 2023).

As per IBM, a blockchain is a distributed, shared ledger that records transactions and tracks assets within a network. An asset can be a physical asset such as property, a house, a car, or an intangible asset such as digital currency, an intellectual property right, and so on. The main purpose of a blockchain is to store data and track its activity in a distributed environment.

A blockchain is a database or a decentralized, distributed digital ledger that is stored in a peer-to-peer network with no central authority controlling it. A block is a record of each verified transaction (a data exchange) on a blockchain. Every block contains transaction information, a timestamp, and the previous block's cryptographic signature. A blockchain requires special permission to record entries, but in most cases, the ledger itself is accessible to the public. One of the most important features of a verified block is that it can't be changed, making it one of the immutable aspects of a blockchain. The term "blockchain" is a common one. There are already many blockchains in existence, and many more are in the pipeline. The most popular and broadly used one is the one that keeps track of every bit of transactions using bitcoin. The bitcoin blockchain protocol is essentially a giant spreadsheet whose primary purpose is to track and exchange the value of bitcoin.

Ethereum (ETH) is another blockchain. It was designed to support a wider range of applications. A lot of the apps currently being developed are based on Ethereum. It is the basis for a lot of them. Ether is the currency in which value is stored. Ethereum facilitates scripting language and API support, both of which are powerful tools. Smart contracts are one of the most important features of Ethereum. Smart contracts create and execute self-executing agreements between multiple parties.

10.1.1 APPLICATION OF BLOCKCHAIN

At first, Bitcoin was the only cryptocurrency that came into existence, but then other cryptocurrencies followed suit. Some of them rose to fame, while others fell behind. Bitcoin rapidly expanded its scope and real value to a wide range of untraditional applications, including the healthcare sector, enterprise software creation, finance, banking, and insurance, among others. In today's era, blockchain technology is changing the current technology paradigms covering every industry. According to official statistical websites, the global blockchain market is expected to expand by $20.7 billion in 2024. The main applications of blockchain are discussed in subsections.

10.1.1.1 BANKING AND PAYMENT SERVICES

Blockchain applications are becoming increasingly popular in the banking and payment system landscape. Cryptocurrencies such as Bitcoin-based remittances can be used to regulate payment systems with no geopolitical restrictions. For example, ABRA is a case in point. A bank transfer can take up to three days to settle, which is not only inconvenient for the consumer, but it is also logistically challenging for the bank. Nowadays, a simple bank transfer bypasses the complicated intermediaries from the bank to the custodial service, allowing it to reach the recipient quickly. This is where blockchain comes into play in banking. Blockchain acts as a decentralized ledger that keeps track of transactions in a transparent and public way, eliminating the need for custodial services and allowing transactions to be settled in a public blockchain. These blockchain applications are one of the main ways in which blockchain can make transactions faster and simpler.

The application of blockchain technology in banking services is as follows:

1. **Securities:** Banks will need to keep a record of who owns what in order to trade debt, stocks, or commodities. In order to do so, they'll need to liaise with a number of exchanges and brokerages, as well as clearing houses and custody banks. The combination of these parties and the outdated chapter ownership system slows down the process and increases the risk of inaccuracies and fraud. Blockchain technology is revolutionizing the banking industry by creating a decentralized ledger of digital and distinct assets. A distributed ledger makes it easier to transfer assets through tokens representing the assets "off-chain." The advantages of blockchain in banking come from the creation of tokenized security, which has the potential to eliminate middlemen and reduce asset exchange fees.

2. **Loans:** Credit reports are the primary basis for banks to underwrite loans. However, with the introduction of blockchain technology into consumer banking, the possibility of P2P lending has emerged as one of the top investment opportunities in FinTech. When a consumer applies to get a loan, banks assess the risk they would face if they don't pay it back. They do this based on the consumer's credit score, equity position, and debt-to-income ratio. This information is collected through credit reports, which is a centralized system that is often unfriendly to the consumer. One of the benefits of blockchain in banking is the emergence of a substitute lending platform that provides an efficient, cost-effective, and safe way to offer personal loans to the consumer. In

the case of a blockchain-based banking system, payment history is stored in a decentralized registry, which makes it easier for consumers to apply for personal loans.

3. **Customer KYC:** The answer to how blockchain works is also the answer to how long customer KYC takes in the banking division. In most cases, it takes up to three months to go through all of the KYC processes, such as photo, address, and biometric verification. Not only does this take a lot of time, but it also costs a lot of money for banks to carry out these processes. Blockchain technology is being used in retail banking to make the KYC process easier. Now, you can see how blockchain is being used in banking by storing the customer's information on the blockchain, which allows banks to access information relevant to KYC. This event reduces the personnel cost by 10%, which is equivalent to $160 million per year.

10.1.1.1.1 Benefits of blockchain in banking

In various ways, blockchain revolutionizes the banking sector. One of the main advantages of blockchain is its ability to ensure a secure and transparent record of transactions. In a traditional banking system, transactions are usually recorded in a centralized database. With blockchain, transactions are recorded on a partially decentralized network where only the most trusted participants can host validating nodes, making it almost impossible for one point of failure to disrupt the entire system and reducing the risk of fraud or cyber-attacks. Blockchain also has the potential to improve efficiency and lower costs. In a traditional banking system, there are many intermediaries that can add a lot of costs and delay transactions. However, blockchain technology can reduce these intermediaries, making transactions faster and cheaper. This can lead to substantial cost savings for banks as well as their customers.

10.1.1.2 CYBER SECURITY

Blockchain relies on cryptography to validate and secure data. This prevents unauthorized modifications and hacks. It ensures the security of the system by preventing any third party from making unauthorized modifications. An asymmetric cryptographic key and digital signature are often the foundation of blockchain technology. This helps to verify the ownership and integrity of data. In some instances, these elements can even eliminate passwords, which are often used by cybercriminals to gain unauthorized access to accounts and networks.

No one can change a record on the blockchain ledger. If there is an error in a record, it can be fixed by making another transaction, and both transactions will be readable. IoT (Internet of Things) protection, with the increasing use of IoT devices in various industries, means that cybercriminals often target IoT devices due to their vulnerability. Blockchain provides additional protection for IoT devices. A DDoS attack is a type of cyber-attack that targets a server by flooding it with requests. The target is usually an IP address or a small number of IP addresses. A blockchain-based DNS system neutralizes that single focal point. In data privacy, although blockchain is transparent, and everyone can see immutable transactions, only trusted networks can view or participate in the transactions. This can be done with minimal governance. In addition, blockchain does not have the traditional targets that cybercriminals look for, making it more difficult for them to gain unauthorized access through privacy rules. A smart contract (also called a smart contract token) is a block of code that stores information on the blockchain. Smart contracts implement a transaction when certain conditions are fulfilled. This kind of automation simplifies the process of making payments. The blockchain remains secure because it has components that are verified for authentication, data integrity, access controls, and business logic verification.

10.1.1.2.1 The key cyber risks affecting blockchains

Physical device theft is a risk that falls within the scope of traditional cybersecurity, which means that best practices can mitigate these risks to information security. Theft of keys—no matter how secure a blockchain is, if a cybercriminal is able to steal keys, it can cause a lot of damage. For instance, in 2016, hundreds of millions of dollars' worth of Bitcoin was stolen from the Hong Kong exchange Bitfinex when it was accessed from users' separate wallets. This amount equates to approximately 0.75% of all Bitcoin currently in circulation. As a result, the value of Bitcoin dropped by 23%. Phishing attacks are one of the primary ways cybercriminals gain access to a wallet key owner's access credentials. Phishing involves sending emails that appear to be from an authority, asking for access credentials through fake links. If the owner of the key is fooled by a phishing attempt, the cybercriminal will take action that harms them and the blockchain. Blockchain provides an anonymous, secure, and convenient way to store money, but there is usually no recourse when it comes to stolen keys.

A code exploit is a method used by hackers to exploit vulnerabilities. For example, in 2016, an attacker used a code exploit to steal a large sum of money—$50 million—from the Ethereum platform, which is the currency for blockchain transactions. Hackers are well-versed in intercepting data as it moves through the data transfer process. For blockchain transactions, hackers can steal data on the way to the internet service provider (ISP), extracting sensitive information or even money. Bitcoin mining, where computers use their computing power to perform complex cryptographic calculations, is also a common attack vector on public blockchain networks, known as a 51% attack. A 51% attack is a hostile takeover of a blockchain by an individual or group that has 51% or more of the network's mining power, which gives them control over the ledger access and the capability to operate it. Attacks like these serve as a reminder that blockchain isn't infallible and that cybersecurity best practices must be followed.

Physical device theft: It's well known that hot wallets, such as mobile wallets, desktop wallets, and browser wallets, are not as secure as cold wallets because they're stored offline. Even if a blockchain was completely secure (which it isn't), it wouldn't be safe because access credentials would still be out there. If a hacker can get their hands on a physical device that's connected to the internet, they'll have the ability to make large fraudulent transactions.

10.1.1.2.2 Application of blockchain in cyber security

In the field of cybersecurity, the Central Intelligence Agency's (CIA) triad model is used as a benchmark to evaluate an organization's security model. The CIA triad is made up of: Secrecy, Integrity, Availability. With blockchain, we can ensure that all of these policies are met:

1. **Confidentiality:** Ensure only interested parties and parties having authority can access the pertinent data. Ensure that blockchain records are completely encrypted so that unauthorized parties can't access them while they're circulating on an untrusted network. Ensure that access controls are directly at the application level to prevent attacks from within the network. Blockchain can provide advanced security by using public vital infrastructure to verify parties and encrypt their communications. However, backup storage space of private keys within secondary storage presents a high risk for private key theft. To address this, key management processes, such as the Information Technology Forum (ITF) or the Original Research Framework (RFC), should be implemented. Ensure encryption algorithms based on an integer factorization problem.
2. **Integrity:** Organizations can leverage blockchain technology's built-in properties, such as being immutable and traceable, to ensure the security of their information. Consistency model protocols can also be implemented to create mechanisms for preventing and regulating the sharing of a ledger in the case of 51% cyber-attacks. With each novel version of a blockchain, a record of the previous version is maintained, providing a traceable record of the system's history. Additionally, smart contracts can also be implemented to enforce rules among parties that prohibit miners from mining a data block.
3. **Availability:** DDoS (distributed denial-of-service) is a type of cyber-attack that aims to disrupt the availability of a technology service. In recent years, DDoS attacks have become the most popular type of cyber-attack. DDoS attacks are expensive because the attacker is trying to overload the network with a large number of transactions. Blockchain systems, on the other hand, have no single failure point. This eliminates the risk of an IP-based DDoS attack that disrupts normal business processes. Data is always accessible through several nodes, so a full copy of a ledger is always available. The distributed operation and multi-node characteristics of a blockchain platform and system make it highly resilient.

10.1.1.2.3 Pros of using blockchain in cyber security

The key features of a blockchain system include: user privacy, data transparency and traceability, secure data storage and processing, and no single point of failure. User confidentiality is achieved through the use of public key cryptography, while data transparency is achieved through digitally signed transactions within the blockchain network. Traceability is accomplished by maintaining a history of all transactions, which can be traced at any time. Secure data storage and processing are ensured by the immutability of the data and the records of any changes to the data. No single node fault can cause the entire network to fail, even if there is a distributed denial-of-service (DDoS) attack on the network. Smart contracts enable the execution of agreements between parties while data is being transferred.

10.1.1.2.4 Cons of using blockchain in cyber security

Blockchain relies heavily on private keys to encrypt data. If the private key is lost, it can't be recovered, and the encrypted data will be lost forever. One of the biggest challenges that blockchain networks face is scalability and adaptability. The block volume is fixed, and there is a limit to the number of transactions per second. Therefore, it's important to check the scalability of the blockchain network. To integrate blockchain into your company's system, you'll need to completely replace it. This may be challenging. High operating costs are another major challenge that blockchain networks face. The computing power and storage capacity of blockchain require a lot of resources, which means higher operating costs compared to other blockchain applications. Blockchain concepts aren't controlled globally yet. Therefore, developing regulations and frameworks is essential for maintaining governance in blockchain apps. To learn blockchain technology, you need to know a lot of different development languages and tools. This is why there aren't many blockchain development assistants available at present.

10.1.1.2.5 Blockchain examples

Blockchain is being used for a variety of security purposes, such as traditional banking, IoT, and healthcare. For traditional banking, blockchain has been used to improve the security of fund transfers and to store customer information. For IoT, blockchain is being used to ensure the security of IoT devices. For healthcare, blockchain and AI are being used to create an ecosystem where data can be discovered and analyzed. Blockchain has also been used to secure government documents by partnering with IBM.

In China, the government and military use blockchain to secure sensitive government and military information, intelligence information, and other sensitive data. In Malta, Founders Bank aims to become the world's prime decentralized bank maintained by the people and not by a central authority. The bank will use encryption and DDLs to store and protect users' cryptocurrencies. In Colorado, the Senate passed a bill to explore the use of blockchain for record-keeping and to reduce the number of attacks on records. In New York, the traditional banking company Morgan created a blockchain-based platform called Quorum. The platform uses smart contracts and cryptography to protect private transactions. In Mountain View, California, Health Linkages wants to use blockchain to secure patient accounts and make sure that only a few people can access the records to help doctors make better decisions (Figure 10.2).

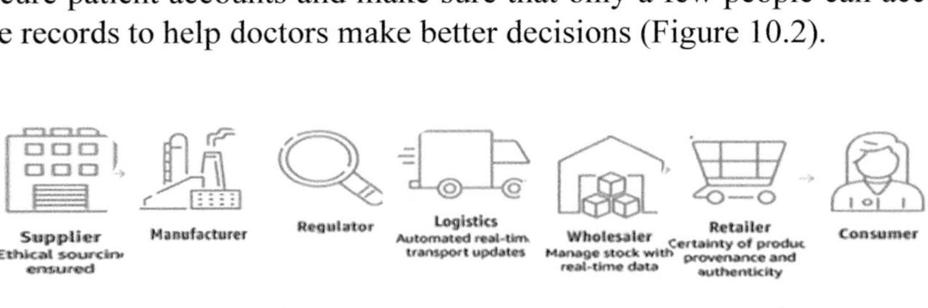

FIGURE 10.2 Supply chain.

10.1.1.3 SUPPLY CHAIN

Blockchain can improve supply chain management transparency, accountability, and feedback mechanisms throughout the supply chain. Supply Chain Management makes it possible for any product to be tracked in the supply chain. Every product movement and condition can be monitored within the blockchain with the help of IoT sensors. Block verify, Provenance is one of the best blockchain-based supply chain management systems. Blockchain technology is changing supply chain management through transparency, safety, and efficiency. Here are 10 leading uses of blockchain technology in supply chain management.

Blockchain can improve transparency and trust throughout the supply chain by enabling real-time visibility into goods and products from manufacturing to distribution and the final consumer. It can verify product authenticity and decrease counterfeiting by creating immutable and tamper-proof records of the origin and movement of a product. Blockchain can be used to track product quality as it goes through the supply chain by identifying and removing defective products, reducing waste, and improving customer satisfaction. In finance, blockchain can help ensure secure and transparent supply chain transactions. A smart contract is a type of contract that is written directly into a line of code. It automates and simplifies supply chain processes, reducing costs and increasing efficiency.

1. **Blockchain in Supply Chain:** Inventory Management:]() Blockchain technology can be used to monitor inventory levels and optimize inventory management processes, which can reduce inventory costs and improve supply chain effectiveness.
2. **Supply Chain Technology:** Reducing Administrative Work: Blockchain can digitize and automate supply chain processes, which can save time and resources by reducing administrative costs.
3. **Food Safety:** Food safety can be monitored using blockchain technology, which can help ensure food products are safe to eat by providing a traceable and transparent record of the product's journey through the food chain.
4. **Sustainable Suppliers:** Blockchain can help ensure that raw materials are sourced in an ethical and sustainable manner by providing transparency and traceability.

10.1.1.4 ONLINE DATA STORAGE

Centralized data, like OneDrive and Google Drive, is susceptible to PoD. On the other hand, blockchain allows for distributed data storage in an efficient and secure manner. Storj, for example, is an encrypted cloud storage solution. Let's now explore the benefits of blockchain on the cloud:

1. **Cloud Computing in Line with Hyperledger Blockchains:** Hyperledger is termed a decentralized open-source blockchain application that is being developed by the Linux Foundation. It makes unauthorized access and data alteration very difficult

because of the distributed copies of the exact same blockchain records. A good audit trail can be created because the record-keeping is decentralized with peer-to-peer networks. With the help of cloud computing, record sharing can take place in the cloud. Different parties can keep exact copies either in full or in part for audit authentication. In this arena, the users of Hyperledger are not very careful about staying anonymous, just like crypto-currency users. Instead, Hyperledger applications have a small group of authorized users whose information is kept by a central agency. This authority not only knows but also controls the authorized users, even though their efforts maintain the Hyperledger blockchain.

2. **Decentralization:** Cloud computing is shifting to the edge because connected IoT devices don't need to rely on central servers for data processing or decision-making. For example, someone has a biometrics recognition node that contains encrypted biometric information of each authorized person. The node will probably only have to check with the servers when a match isn't found. The node will run autonomously unless there's a need to check for updates. When multiple copies of the biometric database of the authorized personnel are moved to the edge—cloud storage that's geographically diverse—it becomes faster and easier to get official copies.

3. **More Efficient Tracking of Ownership:** When we combine the power of cloud computing with the power of blockchain, we can create many opportunities. These opportunities can be used to improve real estate title management, stock ownership management, legal agreement and contract management, and other important matters. By moving to a cloud that manages a Hyperledger system, you can dramatically improve the clearing process. Currently, a stock transfer is made through a DTC (direct to trade) or a DST (direct to trust) company. It takes around three to six business days to process a stock transfer. But with a Hyperledger blockchain and cloud storage, you can execute a stock transfer in seconds or even microseconds. Hyperledger projects are already being created to improve this process.

4. **Improved Data Security:** End-to-end encryption helps protect data when it is in transfer mode and storage. Blockchain helps protect transaction data as well as ordering data. You can have a triple layer of security with the solid peer-to-peer distribution of precise copies of the blockchain across a cloud compute network. Distributed data storage is much better than a centralized data storage model and has a much lower risk of data breaches. In distributed data storage, data files are compressed into small bits and encrypted. These encrypted bits are stored across multiple nodes in different locations.

5. **Enhanced Private Key Security:** Private keys are the key to encrypting and decrypting data in the cloud, but they are also the most vulnerable to theft and unauthorized access. To better protect them, you can implement a cloud computing process that requires a two-step authentication process to authorize transactions. To further increase security and make it harder for unauthorized access, you can set the verification responses in the two-step process for single use only. If you need even more security, you can set up cloud computing to require facial recognition/biometric recognition as a mandatory authentication process for private keys.

6. **Permanent Audit Trail:** One can create a permanent transaction record with blockchains. The proof of history (PoH) function is another innovative feature of blockchain technology. This function works with a verified delay function. It provides the cloud computing network with a time record where there is no need for trust from the users. Blockchain provides a reliable and permanent audit trail for each transaction in the correct sequence. The PoH function adds a timestamp for the permanent transactional data.

7. **Fault Tolerance:** It is possible to replicate the blockchain data across a server network as long as the servers are connected closely via collaborating clouds. This reduces the risk of failure due to a single server or node. In the event of a cloud node failure, the network will continue to operate without interruption. This provides additional fault tolerance in cloud computing with blockchain.

8. **Geo-Independence:** Nowadays, many businesses are deploying cloud applications and solutions around the world. They now have the option to store their data in multiple physical locations. Since there is enough cloud storage available across countries for a cloud computing system, the risk of data loss or disruption due to natural disasters and other causes is very low. The processing burden is simply moved to another wholly operational network node. Blockchain is basically a distributed data infrastructure. When combined with cloud-based networks, it has the potential to significantly broaden geographic scope.

9. **Quicker Disaster Recovery:** Due to the distributed nature of blockchain technology, it is possible to have multiple authorized users. This can be done by publishing the blockchain publicly or by distributing it to users. Even if one of the network nodes goes down, it will have no effect on the other copies. All other nodes will continue to operate, and even if one or two of them go down, the blockchain will continue to update. Because of the transactional nature of the blockchain entries, if a node goes down, it can compare the current state of the blockchain database almost immediately after it comes back online, especially if it has a protected timestamp.

10. **Scalability:** Blockchain networks on large-scale blockchain apps can handle many transactions at once. As a result, you need robust services to process the data. Only then can you execute many transactions quickly and enable scalable blockchain services. The cloud can then provide the computing resources needed to run the blockchain on demand. By combining blockchain with the cloud, you can create a fully integrated system that is highly scalable.

11. **Micro-Transactions:** As cloud computing increases in popularity, it also becomes more cost-effective; its scalability and ability to handle smaller transaction sizes increase; reward expenditures on the blockchain to nodes can be very low, considering that the cost of processing a transaction is even lower than the cost of the reward. For example, let's look at Bitcoin: Mining or creating Bitcoin isn't sustainable if the computing power needed to mine the Bitcoin outweighs the utility cost to mine the Bitcoin. Nowadays, cloud computing is shifting from a need to run complete virtual

servers for the processing of applications to a need to duplicate the operating system for its basic functions. The processing cost of cloud computing is steadily decreasing, making it more affordable to perform micro-transactions.

Blockchain technology can be used to establish a dispersed network of Internet of Things devices, eliminating the prerequisite for a centralized location to manage them. Insurance Blockchain is a new approach to trust management in the global insurance marketplace. It works by creating trust between participants based on mutual suspicion. An example of a blockchain-based insurance management system is "Aeternity," which automates most or all aspects of insurance (parametric), embeds policy judgment in smart contracts, allows for the execution of policies based on predetermined loss events, settles and clears transactions without manual intervention, and streamlines the management of ILS, flight delay, cancellation, and crop (insurance).

Blockchain will enable prime insurers, reinsurers, brokers, and controllers to securely exchange data in real time, systematize risk modeling and audit and compliance drafts, and bind towers of risks and contracts on one single time-stamped smart contract. Blockchain will also affect claims handling, and it can create a secure industry-wide, trustworthy, and tamper-resistant record of claims. Blockchain will reduce claims fraud by eliminating data silos and providing customers with greater control over their data, including access rights. How blockchain will impact insurance distribution: Blockchain will enable multiple parties to coordinate actions on a single online marketplace at a low cost. Blockchain can give consumers direct access to several carriers on one platform. Blockchain lets consumers manage multiple policies on a single platform. How blockchain will affect P2P insurance: Blockchain will allow consumers to make premiums or claims payments quickly, easily, and inexpensively.

10.1.1.5 GOVERNMENT

Blockchain technology can help to reduce bureaucracy, red tape, and increase competence and transparency in government processes. For example, the Dubai government has now implemented this technology. Blockchain can help to prevent government corruption, as it can replace middlemen in many electronic government services. It has a unique role to play in the fight against government corruption. Blockchains offer a unique combination of record-keeping that is tamper-proof. They can also allow states to take a decentralized approach, where appropriate. This approach promotes real-time visibility, auditability, and smart contract functionality. Smart contract technology is the foundation of decentralized finance, as it is an algorithm that executes automatically once certain prerequisites are met. Blockchain technology also makes it easier for anyone to access public information. The information shared through the blockchain follows four fundamental principles: it is open to everyone by default, it is easily accessible, it is understandable and interoperable, and it is transparent. By implementing blockchain technology in the public sector, it can help to improve efficiency and enhance citizen engagement. It also helps to streamline public affairs management.

10.1.1.6 CROWDFUNDING

Crowdfunding is one of the most widely used methods of raising funds for start-ups and projects. Blockchain crowdfunding platforms have overcome the problem of high C2P fees by building trust through the use of smart contracts and online reputation systems. New projects can issue tokens of their own, which can be converted into products, services, or cash.

10.1.1.7 ENTERTAINMENT AND MULTIMEDIA

In the entertainment field, blockchain has also emerged as a new area where the interference of third parties is excessive. The execution of blockchain in this field will remove intermediaries from the system. The implementation of blockchain has already begun in online music, for example, Mycelia and Ujo Music.

10.1.1.8 REAL ESTATE

One of the most significant innovations made by blockchain is in the field of real estate. Currently, ownership and transfer issues are among the main challenges facing the real estate system. The application of blockchain in this arena can regulate the entire real estate system by containing common ledgers.

10.2 WORKING OF BLOCKCHAIN TECHNOLOGY IN DIGITAL MARKETING

The following steps will be followed:

1. **Step 1:** Customers Compensated for Viewing Ads: Under the current system, customers are bombarded with ads on almost every channel they can access. Companies push the customer to buy, and the user gets annoyed by the huge number of ads they see. Not only does this distract the company from its target audience, but the user may miss out on the real ads that are meant for them. Blockchain could revolutionize the advertising experience. In fact, it could even shift some of the financial burden from companies that are paying for ads to provide some form of compensation to the user when they see their ads. According to Damien Martin, a marketing executive at London's Shufti Pro, blockchain allows users to opt in to view advertisements for digital currencies or tokens. This could revolutionize situations like ad fatigue, as consumers will be able to request to view ads directly instead of having them pop up every time they scroll through a web page (Figure 10.3).

FIGURE 10.3 Working of blockchain technology.

2. **Step 2:** Safe and Secure Data That Won't Be Sold: Some companies sell customer data for profit, including names and email addresses, which bombard the user's mailbox with unsolicited emails. In this case, blockchain will work well, as it is completely anonymous, allowing the user to access a secure place to store

their data. Stacks, a similar tool, has built a network of decentralized apps, which ensures that data does not have to be stored within these apps to be used. Instead, it holds the data safely in its tool and uses it to log into other apps without sharing user data.

3. **Step 3:** Transparent and Reliable Marketing: The challenge of proving the content of marketing campaigns and advertisements has become difficult. No matter what the company claims about its products, no one can prove it except blockchain. With the help of blockchain technology, brands can be more transparent than ever before, making it easier for consumers to trust businesses. Blockchain businesses can showcase different transactions by displaying the supply chain process. This way, customers can see accurately where products are sourced, where they're processed, and what resources they're sourced from. This helps customers make more ethical decisions, making companies more competitive.
4. **Step 4:** Businesses will Get More Accurate Leads: Companies spend a lot of money collecting customer data, but some campaigns are a huge waste of money because they also include untargeted consumer data, which is useless for the company. Most companies collect customer data via cookies. Blockchain will allow companies to collect much more accurate leads because they will get the exact information from the customers themselves by incentivizing them with discounts, tokens, or some other form of compensation. Since the customer will be making the choice in front of them, the lead data collected by companies will be much more precise, allowing them to market more effectively.
5. **Step 5:** More Affordable Advertising: Blockchain technology automatically verifies the ad space for brands by removing the middleman cost. This will lead to much more precise data and metrics being extracted from their ads. Brands will be able to calculate the exact number of people who saw their ads. This will make ad tracking much easier. However, because all transactions are visible to the public on the blockchain, brands will be able to see which transactions were actually made as a direct result of their advertisements and where those advertisements were placed.

10.3 BLOCKCHAIN AND MARKETING

Blockchain is becoming increasingly popular in the financial marketing industry as well as in the consumer market. Why is blockchain quickly becoming so popular in the marketing world? (Figure 10.4).

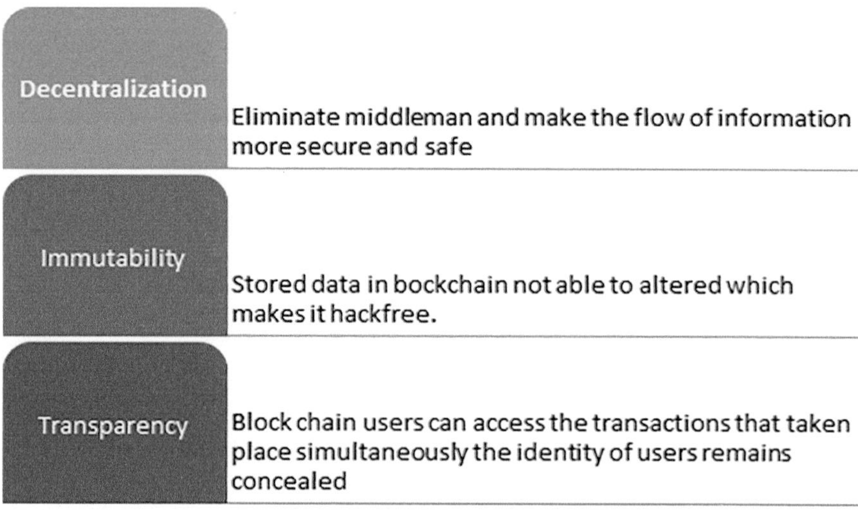

FIGURE 10.4 Blockchain and marketing.

10.4 BLOCKCHAIN STRATEGIES

Blockchain is already used in banking, BFSI, real estate, media, retail, healthcare, and so on. There is a lot of buzz around this technology, so let's explore blockchain strategies and best practices that are shaking up the marketing scenario, which include (Figure 10.5):

1. **Influencer Marketing:** Blockchain can solve the problems of return on investment (ROI), transparency, false followers, and engagement by adding smart contracts to the ecosystem. Smart contracts will pay you automatically after you complete certain tasks or after you deliver the desired results. It can also be used to verify the performance and legitimacy of the influencer.
2. **Affiliate Marketing:** It refers to the practice of an online retailer paying a commission to a third-party website in exchange for traffic or sales generated by referrals. The primary issue that

FIGURE 10.5 Blockchain strategies.

retailers face is that despite investing in affiliate marketing, brands encounter issues such as: trying to deal with shady affiliates, spending money on websites that aren't doing anything for the brand, and spending a lot of money on affiliate networks. Blockchain has the potential to help reduce this waste of money. By using cryptocurrencies, the process of paying for affiliate marketing can be simplified. Typically, payments are held on hold by affiliate networks up to a minimum payment threshold. However, blockchain-driven products can operate outside of these limits.

3. **User Data Protection:** With the increasing frequency of data breaches, businesses need to focus on safeguarding customer data. The General Data Protection Regulation (GDPR) came into force in 2018, making it mandatory for brands to collect and use customer data. Brands have been looking for ways to improve customer data security, and that's where blockchain comes in. Users can anonymize their data or sell it to a brand of their choice.

4. **Ad Fraud Prevention:** One of the biggest sources of revenue loss for businesses is advertising fraud. This can be broken down into categories such as: fraudulent traffic, domain spam, and malvertising. Blockchain technology eliminates the middleman, that

is, the network, and keeps the advertisers, publishers, and users informed. This will promote transparency.

5. **Blockchain-based Social Platforms:** Over the last decade, social media has reached billions of users. Social media has opened up new ways for people to connect with their families, friends, companies, and even famous people. However, this has come at the cost of their privacy. Because advertising is the main revenue stream for social media platforms, they monitor every interaction to enhance their targeting capabilities. This awareness, combined with recent data leaks, has led to increased caution when using social media.
6. **Social Commerce:** Social commerce isn't a new concept, but it's one that's been around since ancient times. People have always liked to shop together, and it's easy for them to trust a product/service they've already used. Social media, on the other hand, has become incredibly useful and a great source of traffic for businesses that want to increase their customer base.

In the world of e-commerce, blockchain has the potential to improve transparency in online transactions and protect fast and low-cost cross-border transactions. Excelli is one of the most prominent examples of how blockchain integration can take social e-commerce to a new level. It is a social online shopping platform that combines the power of social media with the power of online shopping. The Excelli Super App integrates a dedicated online shopping platform, allowing users to conveniently shop for their products and products from sellers around the world. The app also includes a social component, allowing users to create content, chat, recommend products, earn rewards, and more. By incorporating blockchain technology, Excelli can provide high-quality security for its users and the data stored on its platform. Blockchain also allows merchants and consumers to trade directly with each other without any middlemen. In addition, blockchain smart contracts enable automated and code-based processes that reduce costs and increase the efficiency and speed of online transactions.

10.5 CONCLUSION

Since its inception in 2009, blockchain technology has gone through many changes. Initially, it was thought to be just a cryptocurrency, but it has exploded into many new and unexpected domains. The applications of blockchain technology for marketing would bring endless changes in the way a customer interacts with a business. Blockchain brings together the value of several technologies and, most importantly, a shared ledger of transactions made by the parties involved in the chain network. The primary goal of blockchain technology is primarily to authenticate assets with the flexibility to solicit in other business operations. The bigger picture is that blockchain technology has the potential to usher in the next phase of human evolution by transforming the social economy and the political system.

Experts say blockchain will stop ad fraud and protect consumers' data. They also say more money will flow to blogs and news sites for allowing ad placements on their sites. Advertisers aren't pushing lawmakers to further regulate existing digital marketing platforms, let alone force them to adopt a blockchain model. But as the technology advances, they may adopt it for cost-effectiveness reasons. Overall, it's hard to say how blockchain will affect digital marketing trends in the short term. But keep an eye on this ever-evolving industry. Blockchain technology is providing us with far more benefits than drawbacks, especially in marketing. It will take time to understand where blockchain marketing is going, but it's important to understand where the industry is headed so the market can start adopting new tools and best practices in their marketing strategies.

KEYWORDS

- **blockchain**
- **blockchain integration**
- **business operations**
- **cross-border transactions**
- **digital advertising**
- **digital marketing**
- **E-commerce**
- **online transactions**
- **secured data**
- **social commerce**
- **social media**

REFERENCES

Avital, M. (2018). *Blockchain Technology Market Size*. Copenhagen Business School. Retrieved from: https://www.grandviewresearch.com/industry-analysis/blockchain-technology-market (accessed on 14 December 2024).

Blockchain Council. (n.d.). *Impact of Blockchain in Marketing and Advertising in 2020*. Retrieved from: https://www.blockchain-council.org/blockchain/impact-of-blockchain-in-marketing-and-advertising-in-2020/ (accessed on 14 December 2024).

Emarsys. (2021). *Blockchain Changing Marketing*. Retrieved from: https://emarsys.com/learn/blog/blockchain-changing-marketing/ (accessed on 14 December 2024).

Kannan, P., & Li, H. (2017). Digital marketing: A framework, review, and research agenda. *International Journal of Research in Marketing, 34*(1), 22–45. https://doi.org/10.1016/j.ijresmar.2016.11.006.

Kumar, S. O. (2021). *Impact of Blockchain in Marketing and Advertising in 2020*. Blockchain Council. Retrieved from: https://www.blockchain-council.org/blockchain/impact-of-blockchain-in-marketing-and-advertising-in-2020/ (accessed on 14 December 2024).

Marketing Muses. (2020). *What Blockchain Could Mean for Marketing*? Retrieved from: https://www.marketingmuses.com/what-blockchain-could-mean-for-marketing/ (accessed on 14 December 2024).

Raju, L. (2021). *An Analysis of Blockchain Technology that Can Benefit Marketing*. Retrieved from: https://www.researchgate.net/publication/349335797_An_Analysis_On_Blockchain_Technology_That_Can_Benefit_Marketing (accessed on 14 December 2024).

ResearchGate. (2021). *How Blockchain Technology Can Benefit Marketing: Six Pending Research Areas*. Retrieved from: https://www.researchgate.net/publication/339354050_How_Blockchain_Technology_Can_Benefit_Marketing_Six_Pending_Research_Areas (accessed on 14 December 2024).

CHAPTER 11

Emerging Role of Blockchain Technology in the Banking Sector

PRIYANKA YADAV,[1,*] ANSHUL JAIN,[1] KHYATI KOCHHAR,[1] and NANDITA MITTAL[2]

[1]*WISDOM, Banasthali Vidyapith, Tonk, Rajasthan, India*

[2]*Barnet and Southgate College, Wood Street Campus London*

*Corresponding author

ABSTRACT

Blockchain is a decentralized, irrevocable database, making it easier for networks to monitor assets and record transactions. The usage of blockchain technology is very potent compared to other contemporary world technologies, as it allows for faster and more expedient transactions. Financial technology is now a very attractive idea in the blockchain. Innovations like artificial intelligence, automation of processes, and blockchain technology might drastically change the structure and guiding principles of the banking sector. The primary tasks of banks include transforming assets, storing liquidity, and offering financial services by using economies of scale. Banks utilizing a centralized intermediary make it easier to keep assets for usage and find funding sources with several channels for stakeholders. Since the assets of the stakeholders are at the core of all financial transaction activities, they are all managed as a group in a centralized ledger system for payments, services, and storage. The advent of blockchain software has been hailed as a new era that will alter the scope and setup of the banking industry and the conduct of business transactions.

Blockchain Technology: Unlocking the Value of 3Ts in the Business World.
Richa Chauhan & Vidhi Kaul (Eds.)

DOI: 10.1201/9781998511914-11

In the modern world, the banking system faces many issues, such as a lack of data security, increased fraud, the need for more manpower, the length of time required, and the high volume of paperwork. The banking sector, which is heavily dependent on technology, can be completely transformed by blockchain technology (BT). Therefore, a system has been proposed that will eliminate the third party on one hand and help bank users conduct financial transactions more securely and quickly on the other. In the banking sector, blockchain is used for several purposes, such as the transfer of payment, settlement and clearance of transactions, trade finance, approval of customers' KYC (Know Your Customer), compliance, and financial reporting, and so on. Presently, consumers have been emphasizing online services. Thus, future digital bank branches must interact with consumers in an environment built on technologies that give them easy access to the data they need. A beneficial new mode of commerce, safer and superior to real cash, is created by blockchain using encryption technology. The blockchain operates as a digital ledger that primarily contains data about financial transactions. It is a communal digital ledger accessible to everyone on the computer network. Decentralization, transparency, and immutability are three crucial attributes of blockchain technology that give network participants security and confidence. The World Economic Forum viewed blockchain as the lifeblood of the financial industries. By scrutinizing the possibilities of BT in the realms of banking and finance and highlighting its distinctive properties like decentralization, immutability, and transparency, it draws in organizations, especially in the financial industry. The various benefits of BT (like increased trust, reduced cost, ensured security, increased efficiency and speed, and many more) have been corroborated by the banking sector to perform their functions properly. However, certain issues (including scalability, safety, confidentiality, and user-related challenges) have been identified that restrain the use of BT in the banking industry. Since August 2021, it has increased by 10%, indicating a steady rise in investment. According to the records of CNBCTV18, the top five banks that are investing the most in blockchain technology include Morgan Stanley, Goldman Sachs, BNY Mellon, Commonwealth Bank of Australia (CBA), and Citigroup. Therefore, it has been concluded that the general public's use of cryptocurrency in blockchain technology is expanding tremendously. Furthermore, the analysis suggests that the government and banking sector should focus on eliminating the barriers, thus increasing the integration of blockchain with the financial sector.

11.1 INTRODUCTION

> *"Financial institutions must be able to deliver an easy-to-navigate, seamless digital platform that goes far beyond a miniaturized online banking offering."*
>
> —Jim Marous

With the creation of standalone PCs, banks have started moving towards information technology, powered by computer systems and the introduction of core banking. The primary objective of adopting ICT is to allow "anywhere-anytime banking" for financial organizations (Jena, 2022). Technologies that enable services such as e-banking, online money transfers, linking bank branches, and ATMs have also profoundly transformed the financial framework (Jena, 2022). However, there has also been a significant enhancement in the development of advanced information and communication technology. Breakthroughs like BT, artificial intelligence, process automation, and others might drastically change the structure and guiding principles of the banking sector. Financial technology (FinTech) is currently showing high interest in the idea of blockchain. It uses various ICT (information and communication technology) techniques, including consensus methods, distributed data storage, and encryption algorithms. It has moreover been recognized as a transformative breakthrough of the Internet age (Guo and Liang, 2016). Employing blockchain, telebanking, online, and mobile banking, as well as other innovative digital emerging technologies over time, has helped to provide new experiences (Osmani et al., 2021). Blockchain technology is majorly influencing the fourth industrial revolution. The development of blockchain technology will radically change how banks offer their services (Jena, 2022). Gibraltar is one of the first nations globally to provide a tailored license for financial technologies utilizing distributed ledger technology (Osmani et al., 2021). Several advanced nations, like the US, the UK, and Japan, have closely followed the growth of blockchain and looked into its potential uses in a variety of industries (Guo and Liang, 2016; Osmani et al., 2021).

11.1.1 ROLE OF BLOCKCHAIN AND BANKING

Banks' primary tasks include transforming assets, storing liquidity, and offering financial services using economies of scale (Jena, 2022). Banks utilizing a unified agent make it easier to keep assets for usage and find funding sources with several channels for stakeholders. Since the assets of the stakeholders are at the core of all financial events, they are all administered through a set of centralized transaction ledgers that facilitate storage, payment, and services. Therefore, the arrival of blockchain technology has been hailed as a new era that will alter the scope of the financial and banking sectors as well as the manner in which commercial deals are carried out (Osmani et al., 2021). In the financial sector, BT is a critical element of the new fundamental shift in financial innovation that fosters growth (Osmani et al., 2021). It is among the latest technologies currently available on the market, grabbing the interest of businesses, startups, and the media. BT has the potential to bring forth substantial changes in various industries, resulting in the possibility of more fair, secure, transparent, and efficient procedures. As a result of the digitization of records, massive amounts of data are produced daily, making it crucial for any organization to manage security concerns effectively while reducing costs (Shah and Jani, 2018). Here is where blockchain comes into play, promising decentralized ownership, data integrity, and data encryption (Shah and Jani, 2018). While creating secure, trustworthy records of transactions and agreements, blockchain technology can streamline administrative processes in the financial and banking services sector (Treleaven et al., 2019).

In contrast to conventional systems, blockchain is flexible enough to take the lead in deployment in a volatile market environment. The paramount supremacy of a blockchain is that it assures that every party has a trail recorded in an open ledger. It is a ledger frequently shared across numerous users, forming a duplicate record that is shared among these consumers and can only be accessed by those with the required access rights. Consensus, authenticity, data integrity, and finality are some of the different elements it works with, ensuring that all of these facets combine to form a fair whole (Shah and Jani, 2018). With a compound annual growth rate (CAGR) of 62.1%, the blockchain market for finance and bank services increased on a global scale, rising from $1.89 billion in 2022 to $3.07 billion in 2023 (GlobeNewswire, 2023).

"The action of conducting or executing a trade or exchange to completion or settlement" is how a "transaction" is typically defined. These days, a person "transacts" with numerous entities daily in various online formats. But regardless of their scale or complexity, all transactions share the fundamental requirement that the parties involved must have faith in one another or abide by a framework that makes this faith possible (Niti Aayog, 2020). The most well-known of these systems, banks, are best known for helping people build confidence while exchanging money. Almost solely, enforcement agencies and specific governmental entities exist to implement rules or laws that foster secure conditions for stakeholders to conduct business (Niti Aayog, 2020). According to the Corruption Perceptions Index reported by Transparency International, the estimated amount of misconduct in public "trust systems" is particularly low, with a rank of 85 out of 180 nations (Transparency.org, 2021). With the assistance of blockchain technology, any organization and government can quickly build trust among their citizens. The principal motivation is that it is substantially less expensive, and the processing of transactions happens much quicker. Banks have been in search of approaches to employ BT for trade financing, clearing, and settlement. Banks will be able to complete the payments quickly and accurately while also spending less on transaction processing. The blockchain is based upon certain principles, and they are:

- On a blockchain, every party is entitled to all information and its history. Information and specifics are under multiple parties' control. By doing away with all middlemen, each party has the ability to autonomously examine the documentation pertaining to their counterparts in the transaction (Yoo, 2017).
- Interactions between peers take place without the aid of a central node. Each individual node is tasked with both storing and transmitting information to all other nodes (Yoo, 2017).
- All individuals who possess the privilege of accessing the system have the capability to observe every transaction and the accompanying value. Every user or node on a blockchain is uniquely identified by an alphanumeric address that has more than 30 characters. Participants have the choice of either maintaining their anonymity or revealing their identity to other participants. Exchanges occur between blockchain addresses (Yoo, 2017).
- After incorporating the transactions into the repository and revising the financial statements, alterations to the records become unfeasible due to their interdependence with all preceding transaction records (Yoo, 2017).

11.1.2 BLOCKCHAIN IN OTHER SECTORS

In a variety of industries, such as manufacturing, healthcare, finance, supply chains, international trade, taxation, and corporate operations, blockchain technology (BT) has been extensively researched and examined. With the help of blockchain, financial, operational, managerial, commercial activities, agriculture, energy and construction, medical, and so on, have been improving (Chauhan and Kaul, 2021). In the financial sector, it acts as a ledger system for Bitcoin cryptocurrency that maintains data integrity (Al-Megren et al., 2018). It has minimized the risk of money laundering and has been utilized to bring cutting-edge goods and services to the securities and insurance sectors (Kimani et al., 2020; Lewis et al., 2017). Certain chapters are analyzed, where blockchain has been gaining popularity (Table 11.1).

TABLE 11.1 Blockchain in Other Sectors

Sector	Conclusion	Year
Energy sector	By 2020, the digital economy was projected to have grown to account for 25.5% of the world's GDP. The production and delivery of energy are being digitalized, decentralized, and carbonized in the energy sector. To facilitate this shift, distributed energy markets must develop, necessitating innovative approaches to data and energy exchange. In order to integrate small renewable energy producers and balance the supply and demand of energy, smart grids are essential. Distributed energy markets can benefit from peer-to-peer trading made possible by distributed ledger technology like blockchain, which allows for safe, reliable transactions without the need for middlemen (Brilliantova and Thurner, 2019).	2019
Agriculture sector	The agriculture industry is investigating and implementing blockchain technology (BT) to enhance supply chain management and operations. The application of blockchain in agriculture to improve the traceability, transparency, trustworthiness, and auditability of agricultural data is the subject of present-day studies, projects, and efforts. Blockchain technology in agriculture makes it simple for producers and consumers to communicate, engage, and guarantee the delivery of wholesome food. On the other hand, there are obstacles and gaps in the research about the applicability and dependability of BT in the agricultural industry. Complexity, running expenses, and the requirement for rules to ensure its safe usage are a few of these (Alobid et al., 2022).	2022

TABLE 11.1 *(Continued)*

Sector	Conclusion	Year
Health sector	The chapter explores how BT could transform the healthcare industry by enhancing data privacy, security, and sharing. The chapter delves into the fundamentals of BT, its uses, and how it improves upon real-time and machine learning methods for disease surveillance. Due to its scalability, security, and interoperability, blockchain is a good option for surveillance systems. The study demonstrates how blockchain technology may improve patient data privacy and global health security, which will aid medical research. It also covers how blockchain technology might assist in the early detection and reporting of hazards to health authorities so that preventive actions can be taken. Blockchain technology is acknowledged by the Global Health Security Agenda as having the potential to upend established methods for disease surveillance and medical research (Chattu et al., 2019).	2019
Blockchain oracle	Bitcoin scripts and smart contracts are enabling the application of distributed ledger technology, like blockchain, in industries other than cryptocurrency, such as supply chain, healthcare, and finance. To retrieve data from the real world, these applications, however, rely on oracles, and there is currently a dearth of academic study on blockchain oracles. In this work, a bibliometric analysis identifies organizations and writers that are actively interacting with the oracle literature, addressing topics, paths, and integrated research projects. The research indicates that although oracles on DeFi networks handle a substantial quantity of money, their roles and activities are frequently overlooked in the studies on blockchains; only a small fraction address oracles, and even less look into related topics. While there is still a dearth of international cooperation in blockchain oracle research, different authors and organizations are moving in the same general directions (Caldarelli, 2022).	2022

11.2 LITERATURE REVIEW

The field of finance and the banking sector is a vibrant market with fierce competition for goods and services. As a result, banks are steadfastly committed to expanding and modernizing to compete with their rivals. Now, the competition is for a digital banking system that enables the delivery of services via digital signals. This is made possible due to the evolution of new products and services brought about by digital transformation in the electronic payment system (Kumari and Devi, 2022). Modern banks aspire to be efficient, technologically advanced facilities without bottlenecks, where customers may self-serve their needs (Kumari and Devi, 2022). A "peer-to-peer" digital currency is transferred from one person to another without the use of banks, financial institutions, or intermediaries (Nakamoto, 2020; Patki and Sople, 2020). Consumers will be the primary focus of online services. Future digital bank branches must interact with consumers in an environment built on technologies that give them easy access to the data they need.

A beneficial new mode of commerce, safer and superior to physical cash, is created by blockchain using encryption technology (Collomb, 2016). Nowadays, digital transaction ledgers are kept up to date using blockchain technology (Collomb, 2016; Osmani et al., 2021; Singh and Singh, 2016; Varma, 2019). Since the 1980s, there have been essential ideas and principles underlying blockchain technology. In 1991, an electronic record containing a contractual chain of information was employed to enable digital signatures on chapters. The idea was combined and utilized with e-currency or electronic cash in the early 2000s (Mishra and Kaushik, 2021). Blocks are compilations of transactions that occurred over 10 minutes and are logically connected to form a blockchain used to validate transaction information. The blockchain is a public ledger accessible to everyone present on the network and records every Bitcoin operation that has taken place up to this point in a linear sequence (Dutra et al., 2018; Singh and Singh, 2016; Yoo, 2017).

Blockchain is one of the ideas that have gained popularity every day since the 2008 global financial crisis. With the advent of augmented reality and artificial intelligence, it became a catalyst. It operates as a digital ledger that primarily contains data about financial transactions. It is a communal digital ledger accessible to everyone on the computer network. Decentralization, transparency, and immutability are three crucial attributes of blockchain technology that provide network participants with security and confidence. The World Economic Forum viewed blockchain as the lifeblood of the financial industries (World Economic Forum, 2016). By examining the possibilities of blockchain technology in the banking and financial sectors and highlighting its distinctive properties like decentralization, immutability, and transparency, it attracts organizations, especially in the financial industry (Mishra and Kaushik, 2021). The banking and finance industries will be profoundly disrupted by blockchain in the future. Due to blockchain's secure banking system, time, effort, and expenses associated with financial services are reduced. These blockchain-related advancements assist in the emergence of a cashless society (Jena, 2022; Yoo, 2017). The approval of implementing these technologies by bank management and personnel is crucial for the sustainability of blockchain adoption in financial systems (Jena, 2022).

11.2.1 BLOCKCHAIN'S CHALLENGES IN FINANCIAL SERVICES AND BANKING

Although blockchain has a lot of potential, there are many challenges that might hinder its widespread acceptance. Since it is a decentralized collaborative technology, every network user has access to the record of transactions and can contribute new information to the database. The open nature of the system and its decentralized coordination has negative consequences and limits the use of blockchain systems (Chang et al., 2020; Ramachandran et al., 2022). Blockchain systems worldwide have yet to have a single global standard available. As a result, it needs to be integrated into current interoperability processes. Due to this, multiple parties using the same blockchain system will have difficulty operating efficiently, leading to problems with interoperability (Bauer and Hein, 2006; Harwood-Jones, 2016; Patki and Sople, 2020). Scalability is one of the primary concerns in executing blockchain systems in the banking industry. As the volume of transactions rises, the size of the blockchain also expands (Zheng et al., 2019). Marr (2018) noted that the implementation of blockchain transactions necessitates a specific duration due to their complexity, encryption, and decentralized structure. Moreover, Werbach (2018) asserts that blockchain-based systems are insecure. Blockchain technology implementation has a security problem. Since 2009, there have been several thefts of the Ethereum and Bitcoin platforms employing blockchain as the base technology, totaling a loss of more than 600 million Yuan (Werbach, 2018). According to Zheng et al. (2019), blockchain is susceptible to attacks by dishonest cooperating miners, and various other attacks have shown that it is not very safe. Due to the inherent characteristics of blockchain technology, it is imperative to acknowledge the susceptibility of all public blockchains to either 34% or 51% of detrimental actions. Moreover, it is worth noting that the occurrence of a 51% attack manifests when malevolent entities gain control over the predominant portion of a blockchain's computational power (Chang et al., 2020). As a result, they control the blockchain completely and dominate the network. The blockchain may provide several addresses for individuals in place of their identities, making it relatively safe for users and preventing data leakage.

Nevertheless, the blockchain cannot stop the leak of financial data as every detail about transactions and balances is made publicly accessible (Kosba et al., 2016; Meiklejohn et al., 2016). Blockchain technology increases certain entities' privacy risks even though it can enable participants to create permanent and irreversible records (Till et al., 2017). In contrast, it is challenging to achieve confidentiality in public blockchain technology since the data is, by default, available to everyone on the network (Chang et al., 2020). Public blockchains enhance innovation, efficiency, and competitiveness. However, because they don't require users to verify, they also present problems for legislation against tax evasion, money laundering, and financing of terrorism (Table 11.2).

TABLE 11.2 Challenges Identified in the Literature Review

Source	Year	Challenges
Osmani et al. (2021)	2020	Scalability, security risks, interoperability, regulatory risks, reversibility
Chang et al. (2020)	2020	Scalability, security, privacy leakage, energy consumption, privacy, cybercrime, regulations and law
Patki and Sople (2020)	2019	Interoperability, security, legal framework
Islam et al. (2021)	2021	Scalability issue, forking issue, eclipse attacks, application bugs, short address, timestamp dependence, integration issue, regulatory issue
Zachariadis et al. (2019)	2019	Scalability, openness, interoperability and standard, transparency and privacy, security
Alam et al. (2021)	2021	Interoperability/compatibility, scalability, reliability, support infrastructure, high cost of operation, regulatory uncertainty, security and privacy concerns, resilience to security risks

11.2.2 A THEORETICAL PERSPECTIVE ON THE FACTORS INFLUENCING THE WIDESPREAD USE OF BLOCKCHAIN TECHNOLOGY IN THE FINANCIAL SERVICES INDUSTRY

The banking industry assumes a pivotal function in driving economic growth and is a vital component of modern society. Many banks are utilizing blockchain technology to achieve their primary objectives of embracing sustainable practices and fostering economic development. Various theories are employed in research to elucidate the extent of information technology acceptance among users. Some of these include the Theory of Planned Behavior (Ajzen, 1991), the Technology Acceptance Model (Davis, 1989), and the Unified Theory of Acceptance and Use of Technology (UTAUT; Venkatesh et al., 2003). Recently, Mansoor et al. (2023) utilized the UTAUT model to explore the utilization of blockchain technology in the provision of remittance services within the banking sector. Scholars from various backgrounds are increasingly focusing their attention on the UTAUT, particularly in the situation of implementing innovative technologies (Salem and Ali, 2019). The UTAUT model is the sole model that has amalgamated all eight models that are utilized for examining willingness, perception, and acceptance. Venkatesh et al. (2003) united the constraints that were prevalent in the eight models and formulated a comprehensive theory. This comprehensive theory encompasses four constructs, namely, effort expectancy (EE), performance expectancy (PE), social influence (SI), and facilitating conditions (FC) that indirectly influence the intention of user behavior. Various theoretical frameworks have been utilized to investigate the implementation of blockchain-based systems in different domains, when concentrating exclusively on the adoption of blockchain technology. However, many recent studies have incorporated new models by combining additional variables with the existing UTAUT variables. For instance, Wei et al. (2021) investigated the adoption of mobile devices among the younger cohorts by employing the UTAUT model and advocated for additional scholarly investigations on the impact of trust on users' inclination to engage with blockchain technology.

The first indicator of behavioral intention from the UTAUT model, PE, refers to the extent to which an individual perceives that the utilization of technology will enhance their professional capabilities (Venkatesh et al., 2003). If an individual demonstrates an increased level of anticipation towards the capabilities of a particular technology, it will lead to a boost in the user's cognitive capacity to transition into utilizing said technology (Sung et al., 2015). This elucidates how the prospects of performance impact the user's behavioral inclination. The institution's efficiency and performance can be improved by blockchain technology, which generates seven transactions in a second (Wang et al., 2018). The second indicator, EE, is the level of a person's perception regarding the convenience that technology will provide in carrying out a task (Alalwan et al., 2018). In simpler terms, when technology is deemed to be easy to grasp and use, it demands less exertion compared to engaging with intricate systems. The installation of blockchain systems does not necessitate extensive knowledge or learning for any given process, thus resulting in enhanced efficiency with minimal effort (Sharma et al., 2023). The third indicator, SI, is defined as the extent to which a person perceives that significant others hold the belief that he or she should adopt the novel system (Venkatesh et al., 2003). Social influence can be seen as a broader concept of subjective norms, an extension that can be traced back to the TPB model (Ajzen, 1991). When banks employ blockchain technology, they experience a heightened sense of rationality and sociability due to its decentralized nature. Additionally, the presence of numerous competitors amplifies the institution's sense of competitiveness. Many studies evidence the substantial relation between SI and behavioral intention (Kim and Hong, 2019; Sharma et al., 2023). The last measure, identified as facilitating conditions, involves elements like technical assistance, hardware and software, and system expertise. These factors contribute to understanding individuals' intentions to adopt a new technology, such as blockchain technology (Faqih and Jaradat, 2021). The influence of FC on the usage pattern of banks is direct, as the facilities provided by banks are frequently utilized by users (Venkatesh et al., 2012). Within the context of blockchain implementation in banks, facilitating conditions highlight the importance of having a technical infrastructure that enables acceptance and utilization of the technology.

11.3 PROBLEMS IN CURRENT BANKING SYSTEM

Banks have long served as intermediaries for various economic and financial activities, including lending, trading, processing payments, and transaction resolution. Due to the ongoing need, the industry is expanding at a consistent rate and has to advance more quickly (Naik et al., 2020). In the modern world, the banking system is facing many issues like a lack of data security, increased fraud, the need for more manpower, the length of time required, and the high volume of paperwork. All these issues are faced in most banking transactions from customer account opening to processing international payments (Shah and Jani, 2018). Moreover, there are restrictions on banking transactions depending on the different kinds of transaction modes. The limit for National Electronic Fund Transfer (NEFT) is Rs. 2 lakhs. In addition, the actual issue arises when manual labor is involved, as human mistakes are always possible.

11.4 PROPOSED SYSTEM

Digitalization has significantly shaped the financial system, drastically altering the banking system. Given the financial sector's dependence on blockchain, it has the ability to completely reshape it (Khadka, 2020). Therefore, a system has been proposed that will eliminate the third party on the one hand and help bank users conduct financial transactions more securely and quickly on the other. Blockchain technology, which can be used in maximum domains of banking, is essentially a system for tracking and storing transactions (Varma, 2019). The use of blockchain technology is exceptionally potent compared to other contemporary world technologies, as it allows for faster and more expedient transactions (Naik et al., 2020). By transforming the underlying technology that supports them, banks' credit information and payment processing systems can be enhanced and modified. Blockchain applications also foster "multi-center, weakly intermediated" situations, boosting the financial sector's effectiveness. Undoubtedly, blockchain can become the fundamental technology that drives the finance industry (Guo and Liang, 2016; Figure 11.1).

Current Pain Points	*How Blockchain Can Help*
Time-consuming process: Selection of members based on financial soundness and industry expertise, evaluation of borrower's financial background and then negotiation of term and conditions is a tedious and time consuming process for the Lead Arranger.	**Faster syndicate formation**: Automated selection criteria for syndicate formation in programmable smart contracts.
Intermediary Fees: Agents and intermediaries have to be appointed at high fees to manage and administer the process	**Technology integration**: Automated due diligence and analysis of information for loan underwriting through Blockchain, reducing TAT.
Manual Processing: The technology systems are obsolete and processes are manual and paper intensive, taking a long time as well as increasing the cost of operations.	**Digitization of documents**: Agreements, contracts, terms and condition documents, etc. are digitized on the BlockChain and validations and checks are automated.
Duplication of effort: The lack of technology integration for due diligence and underwriting causes referencing of different applications and sources during the process. Document duplication also leads to risk of fraud.	**Document immutability:** Immutability feature of the Blockchain eliminates need for multiple copies of the same documents being held.
Delayed settlement cycles: Delayed settlement cycles for payments lock up capital and increase default risk.	**Reduced settlement periods:** Blockchain can facilitate near real-time loan funding and payment settlements with activities executed via smart contracts.

FIGURE 11.1 Comparison of conventional banking systems and blockchain-assisted banking systems.

Source: Reprinted from Shah and Jani (2018). https://creativecommons.org/licenses/by/4.0/

11.4.1 DIFFERENT USES OF BLOCKCHAIN FOR BANKS

1. **Transfer of Payment:** Currently, added fees and delayed payments cause billions of dollars to be wasted regularly. On average, the conventional banking system takes 1–5 business days and charges between $40 and $50 for any international payment (Khadka, 2020). Therefore, to better handle these issues, banks are currently using blockchains. Blockchain technology allows direct foreign payments economically and efficiently (Guo and Liang, 2016). Moreover, by adopting blockchain, banks can also handle conventional bank deals more quickly and minimize the need for third-party verification.
2. **Settlement and Clearance:** To execute a standard bank payment, it might take up to three days. Because of this, both customers and institutions have to face many challenges. Compared to more established methods like SWIFT (Society for Worldwide Interbank Financial Telecommunications), the decentralized blockchain technology can let banks settle deals directly and maintain proper track of them. It implies that the transactions can be resolved in the open database rather than depending on custodial services.
3. **Trade Finance:** This is provided by financial intermediaries is a type of credit support that offers the settlement of trade transactions. An important role is played by the banks in financing the worldwide flow of products. Approximately 80–90% of worldwide commerce, according to the World Trade Organization (WTO), is funded by trade finance (Khadka, 2020). Loans and credits are the most used forms of trade finance. In the international flow of finance, giving a loan is a complex process. Thus, blockchain technology helps in cutting down the cost and time of deal execution. Smart contracts may be utilized to carry out the transaction when every party to it is a member of the blockchain network, allowing information to be exchanged on a discreetly distributed ledger (Guo and Liang, 2016).
4. **Customer KYC (Know Your Customer):** KYC is regarded as another crucial application of blockchain in finance. The KYC procedure is typically completed by institutions in approximately 26 days (Petrov, 2019). The bank performs KYC on customers to

record their information and ensure its validity before initiating any financial relationship (Hassani et al., 2018). Currently, customers have to provide this information to each bank separately in which they want to open an account. However, blockchain technology can save this detail in the form of blocks, and blocks can be shared between banks whenever the information is required (Gupta and Gupta, 2018; Khadka, 2020). It improves operational efficiency and eliminates repetitious tasks.

5. **Compliance and Financial Reporting:** All financial institutions must prepare a regular report of tax returns, audits, and other financial reports. This process will consume much precious time of staff. The BT has the potential to substantially reduce time and cost through automatic reporting (Khadka, 2020). Thus, all the chapter work is eradicated (Petrov, 2019). Moreover, the work of auditors and regulators will be made more accessible by Blockchain's capacity to record and monitor transactions.

11.4.2 CHARACTERISTICS OF BLOCKCHAIN TECHNOLOGY

1. **Decentralized System:** Unlike the traditional system of transactions, which is centralized and requires validation by a third party, blockchain is a decentralized system in which every party can look up past transactions (Wang et al., 2018). With restricted boundary conditions, decentralized networks have successfully managed blockchain and its associated activities (Pereira et al., 2019).
2. **Secrecy of Operators:** Every person on a blockchain possesses an exclusive alphanumerical identification that they can choose to make private or public, and the transactions in the blockchain occur between these addresses (Tapscott and Tapscott, 2017). Users may now connect directly with the blockchain network by utilizing the generated address, eliminating the requirement for a centralized third party to store their private information (Wang et al., 2018).
3. **Consensus Mechanism:** A consensus mechanism is any method used in a decentralized computer network to achieve consensus, confidence, and security (Huang et al., 2019). Since the entire

network lacks a single trustworthy agent, a consensus mechanism is added, the goal of which is to reach a consensus regarding the validity of each record. Thus, any error can be easily detected (Chang et al., 2020).

4. **Execution:** A computer network executes the transaction when certain conditions are met. Users can employ algorithms and guidelines in order to initiate transactions among nodes. A smart contract is one in which the blockchain executes the command, even if specific requirements are fulfilled (Tapscott and Tapscott, 2017).

11.4.3 BENEFITS OF BLOCKCHAIN FOR THE BANKING SECTOR

1. **Increased Trust:** The main benefit of blockchain is tracking and verifying transactions. It relieves individuals and organizations from dependency on any third party (New Age Banking Summit Europe, 2018). Moreover, the risk that the opposing party won't be able to fulfill its responsibilities will be significantly reduced when transactions are settled almost immediately, which could save banks a lot of money (London Speaker Bureau; Wang et al., 2018). Therefore, when the risk is minimized, users will generate trust in blockchain technology.
2. **Reduced Cost:** For the smooth running of the system, the finance industry pays many operating expenses. Some of these expenses include the time and money required to undertake substantial infrastructure improvements, the cost of electricity from ATMs, the gas and water consumed by employees, and waste production (Cocco et al., 2017). However, by connecting to blockchain technology, banks can eliminate all these expenses. Blockchain technology can speed up payments at lower costs than banks by providing a decentralized payment record (Sharma, 2022).
3. **Ensure Security:** Blockchain technology is secure because it cannot be changed once the data is entered into a block. It can be viewed by anyone who uses the technology (New Age Banking Summit Europe, 2018). Blockchain is more resistant to attacks because of the network's decentralized structure, which guarantees no single point of failure (T-Hub, 2018). Moreover, it is shared across many users, making hacking difficult and ensuring

transparency. Security is also ensured by the availability of two security keys – public and private keys. Every user has access to a public key, while the parties to a particular exchange share a private key (Sharma, 2022).

4. **Increased Efficiency:** Conventional chapter-intensive processes often need a third party, consume a lot of time, and are affected by human error. Thus, blockchain technology is used to automate these processes, which can carry out transactions more speedily and efficiently. The blockchain eliminates the need for chapter by storing documents and transaction information together (IBM Blockchain). Clearing and settling may occur much faster because various ledgers do not need to be balanced. When the settlement is quick, the entire system's efficiency will improve.

5. **Increased Speed:** Transacting internationally can take a few days to a few weeks due to the traditional banking methods and regional banking norms that impose various funding requirements, which cause severe delays in finishing a single transaction (Hassani et al., 2018). Through the certainty it offers by allowing people and businesses to trade directly, the BT can assist institutions in reducing the time to settlement. Eliminating third parties cuts the transaction time in half and the payment time to just seconds (Tapscott and Tapscott, 2017). Additionally, it allows 24/7 processing of transactions (New Age Banking Summit Europe, 2018; Table 11.3).

TABLE 11.3 Perceived Benefits of Blockchain

Perceived Benefit	References
Increased trust	Christopher, M. (2016); Garg et al. (2021); London Speaker Bureau (n.d.); New Age Banking Summit Europe (2018); Shah and Jani (2018); Wang et al. (2018)
Reduced cost	Cocco et al. (2017); Hassani et al. (2018); Sharma (2022)
Ensure security	Garg et al. (2021); Khalfan et al. (2022); New Age Banking Summit Europe (2018); Sharma (2022); T-Hub (2018); Wang et al. (2018)
Increased efficiency	Garg et al. (2021); IBM Blockchain (n.d.); T-Hub (2018)
Increased speed	Baiod et al. (2021); Hassani et al. (2018); Khalfan et al. (2022); Tapscott and Tapscott (2017)

11.4.4 TOP BANKING INSTITUTIONS INVESTING IN BLOCKCHAIN-BASED TECHNOLOGY

The general public's use of cryptocurrencies and blockchain technology is expanding quickly. Even though the crypto market is quite volatile, institutional investors are pouring enormous sums of money into this. The financial sector also attempts to get a share of the over $1 trillion cryptocurrency market. According to Blockdata (2022), of the majority of the top 100 banks, precisely 61 of them have made substantial investments in technology that is based on blockchain. Since August 2021, this has increased by 10%, indicating a steady rise in investment. As per the records of CNBCTV18 (2022), the top five banks investing the most in blockchain technology include Morgan Stanley, Goldman Sachs, BNY Mellon, Commonwealth Bank of Australia (CBA), and Citigroup. The investment by these banks between 2021 and 2022 is as follows:

- Morgan Stanley was the leading bank engaging in cryptocurrency and blockchain-related businesses between 2021 and 2022. Over the course of this time, the international investment bank invested close to $1.1 billion in the cryptocurrency industry. In its most recent financing rounds, the business added to its investment in New York Digital Investment Group and made a new investment in a company that provides blockchain technology and services (Blockdata, 2022; Morgan Stanley, 2018).
- Goldman Sachs invested approximately $698 million into the cryptocurrency market. Five blockchain and cryptocurrency-based businesses shared this sum, including Coin Metrics, CertiK, Blackdaemon, and Elwood. In total, 11 cryptocurrency companies are part of Goldman Sachs' portfolio (Blockdata, 2022; Morgan Stanley, 2018).
- Bank of New York Mellon Corporation (BNY Mellon) has invested roughly $690 million in three financing phases. The institution is also among the top financiers of Fireblocks, which offers crypto custody services to BNY Mellon (Blockdata, 2022; Morgan Stanley, 2018).
- An estimated $421 million was invested by the Commonwealth Bank of Australia in four distinct blockchain and cryptocurrency companies. This included funding in Lygon, a startup in blockchain finance, and Gemini, a cryptocurrency exchange (Blockdata, 2022; Morgan Stanley, 2018).

- Citigroup has been expanding its holdings in the cryptocurrency industry, investing in nine distinct businesses in 2021 and six more businesses as of May 2022. It has raised $215 million in crypto and blockchain technology. Citigroup invested in blockchain and cryptocurrency companies like Talos, Blockdaemon, and many more (Blockdata, 2022; Morgan Stanley, 2018; Figure 11.2).

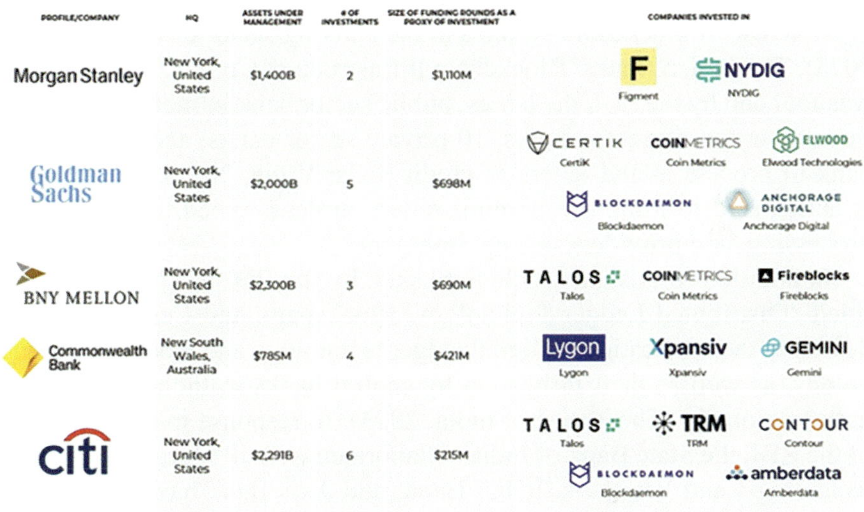

PROFILE/COMPANY	HQ	ASSETS UNDER MANAGEMENT	# OF INVESTMENTS	SIZE OF FUNDING ROUNDS AS A PROXY OF INVESTMENT	COMPANIES INVESTED IN
Morgan Stanley	New York, United States	$1,400B	2	$1,110M	Figment, NYDIG
Goldman Sachs	New York, United States	$2,000B	5	$698M	CertiK, Coin Metrics, Elwood Technologies, Blockdaemon, Anchorage Digital
BNY MELLON	New York, United States	$2,300B	3	$690M	Talos, Coin Metrics, Fireblocks
Commonwealth Bank	New South Wales, Australia	$785M	4	$421M	Lygon, Xpansiv, Gemini
citi	New York, United States	$2,291B	6	$215M	Talos, TRM, Contour, Blockdaemon, Amberdata

FIGURE 11.2 Top banks investing in cryptocurrency and blockchain technology (2021–2022).

Source: Blockdata (2022).

11.5 BLOCKCHAIN IN THE INDIAN BANKING SECTOR

Potentially, blockchain technology can completely change the global banking industry by offering a variety of options for how individuals handle money and values. Each block in a blockchain acts as a storehouse for information, and these interconnected structural elements offer a logical chain that acts as the main conduit for the principal transaction (Shah and Jani, 2018). In relation to blockchain technology, the Indian banking sector is actively testing and analyzing some business scenarios, such as syndicated loans, customer loyalty programs, and vendor financing (Deloitte, 2017). Since blockchain is considered an extremely secure network, it is being used in government banking institutions in India. It dramatically boosts transaction volume and the reliability of networks, enabling central banks to handle RTGS transactions more quickly and securely (Anand, 2023). The Reserve Bank of India, State Bank of India, Yes Bank, and Axis Bank are among the well-known Indian banks that are considering integrating blockchain technology as it provides faster, clearer, and safer means to better serve their clients (Garg et al., 2021). The technology research division of RBI has also conducted a workshop for all bankers, financial regulators, and other stakeholders to examine the possible use of blockchain systems in the banking sector of India (Anand, 2023). The Indian Banks' Blockchain Infrastructure Co. Pvt. Ltd. (IBBIC) was founded by the SBI, the largest public sector bank in India, along with three other public sector banks, 10 private sector banks, and one foreign bank to process inland letters of credit (LCs; Vahia, 2021). This change is anticipated to minimize administrative burdens, speed up transaction processing, and provide a safe environment.

In addition to this, an article published by the RBI in February 2020, titled "Distributed Ledger Technology, Blockchain, and Central Banks," describes the application of digital ledger technology and blockchain technology, as well as their utilization by central banks in India and various nations globally (The Times of India, 2023). In response to this initiative of the RBI, the State Bank of India collaborated with JP Morgan to use this technology, and Yes Bank, ICICI Bank, and Axis Bank have also joined JP Morgan's Interbank Information Network (The Times of India, 2023).

11.6 RESULTS, DISCUSSION, AND CONCLUSION

The traditional banking system has faced many challenges, such as lack of data security, increased fraud, the need for more manpower, the length of time required, and the high volume of paperwork. To overcome these challenges, a system reliant on blockchain technology has been suggested to eliminate the third party and help bank users conduct financial transactions more securely and quickly. The advent of blockchain technology has been hailed as the beginning of a fresh era that will change the landscape of the banking and financial sector as well as how business transactions are conducted. Thus, the study aims to acknowledge the role of blockchain technology in the banking domain and also study the various benefits and challenges offered by this technology. The decentralized blockchain system provides numerous advantages to the banking sector, such as reduced risk, higher security of data, increased customer trust, increased transaction speed, and overall efficiency. It has been analyzed throughout the study that blockchain has a lot of potential, but there are many challenges that might hinder its widespread acceptance. These challenges may include a lack of a centralized mechanism, problems with interoperability, scalability, data hacking, high operating costs, and many more. Moreover, it has been analyzed that the use of cryptocurrencies and blockchain technology is expanding among banks in India. The RBI has been conducting workshops for all bankers, financial regulators, and other stakeholders to examine the possible use of blockchain technology in the banking sector. In addition to this, other banks such as the State Bank of India, Yes Bank, and Axis Bank are among the well-known Indian banks that are considering integrating blockchain technology as it provides faster, clearer, and safer means to better serve their clients. However, it can be said that there is still a need to spread awareness about blockchain technology in India. For this, various workshops and training programs need to be organized. Intensive research and pilot projects must be conducted to commercialize blockchain technologies on a wide scale. By enhancing the speed, security, transparency, and efficiency of financial transactions, blockchain will bring a paradigm shift to the Indian banking industry in the near future.

KEYWORDS

- **banking sector**
- **blockchain**
- **business scenarios**
- **cryptocurrency**
- **digital banking**
- **financial transactions**
- **global banking industry**
- **information storage**
- **secure network**
- **transaction volume**
- **vendor financing**

REFERENCES

Ajzen, I. (1991). The theory of planned behavior. *Organizational Behavior and Human Decision Processes, 33*(1), 52–68. https://doi.org/10.47985/dcidj.475.

Alalwan, A. A., Dwivedi, Y. K., Rana, N. P., & Algharabat, R. (2018). Examining factors influencing Jordanian customers' intentions and adoption of internet banking: Extending UTAUT2 with risk. *Journal of Retailing and Consumer Services, 40*(July 2017), 125–138. https://doi.org/10.1016/j.jretconser.2017.08.026.

Alam, S., Shuaib, M., Khan, W. Z., Garg, S., Kaddoum, G., Hossain, M. S., & Zikria, Y. Bin. (2021). Blockchain-based initiatives: Current state and challenges. *Computer Networks, 198*(August), 108395. https://doi.org/10.1016/j.comnet.2021.108395.

Al-Megren, S., Alsalamah, S., Altoaimy, L., Alsalamah, H., Soltanisehat, L., Almutairi, E., & Pentland, A. S. (2018). Blockchain use cases in digital sectors: A review of the literature. In *Proceedings–IEEE 2018 International Congress on Cybermatics: 2018 IEEE Conferences on Internet of Things, Green Computing and Communications, Cyber, Physical and Social Computing, Smart Data, Blockchain, Computer and Information Technology, IThings/Green* (pp. 1417–1424). https://doi.org/10.1109/Cybermatics_2018.2018.00242.

Alobid, M., Abujudeh, S., & Szűcs, I. (2022). The role of blockchain in revolutionizing the agricultural sector. *Sustainability (Switzerland), 14*(7), 1–15. https://doi.org/10.3390/su14074313.

Anand, B. (2023). *Applications of Blockchain Technology in the BFSI Sector in India.* Retrieved from: https://www.knowledgehut.com/blog/blockchain/applications-of-blockchain-technology-in-india (accessed on 14 December 2024).

Baiod, W., Light, J., & Mahanti, A. (2021). Blockchain technology and its applications across multiple domains: A survey. *Journal of International Technology and Information Management, 29*(4), 78–119. https://doi.org/10.58729/1941-6679.1482.

Bauer, K., & Hein, S. E. (2006). The effect of heterogeneous risk on the early adoption of Internet banking technologies. *Journal of Banking and Finance, 30*(6), 1713–1725. https://doi.org/10.1016/j.jbankfin.2005.09.004.

Brilliantova, V., & Thurner, T. W. (2019). Blockchain and the future of energy. *Technology in Society, 57*, 38–45. https://doi.org/10.1016/j.techsoc.2018.11.001.

Bunea, D., Karakitsos, P., Merriman, N., & Studener, W. (2016). Occasional paper series: Coverage rules for central banks. *169*.

Caldarelli, G. (2022). Overview of blockchain oracle research. *Future Internet, 14*(6). https://doi.org/10.3390/fi14060175.

Chang, V., Baudier, P., Zhang, H., Xu, Q., Zhang, J., & Arami, M. (2020). How blockchain can impact financial services–The overview, challenges, and recommendations from expert interviewees. *Technological Forecasting and Social Change, 158*(May), 120166. https://doi.org/10.1016/j.techfore.2020.120166.

Chattu, V. K., Nanda, A., Chattu, S. K., Kadri, S. M., & Knight, A. W. (2019). The emerging role of blockchain technology applications in routine disease surveillance systems to strengthen global health security. *Big Data and Cognitive Computing, 3*(2), 1–10. https://doi.org/10.3390/bdcc3020025.

Chauhan, R., & Kaul, V. (2021). Blockchain technology: A panacea for medical distribution ailments. *Blockchain for Healthcare Systems*, 43–60. https://doi.org/10.1201/9781003141471-4.

CNBC TV18. (2022). *5 Top Banks that have Invested in Crypto and Blockchain*. Retrieved from: https://www.cnbctv18.com/photos/cryptocurrency/crypto-and-blockchain-invested-in-by-5-top-banks-14203232.htm (accessed on 14 December 2024).

Cocco, L., Pinna, A., & Marchesi, M. (2017). Banking on blockchain: Cost savings thanks to blockchain technology. *Future Internet, 9*(3), 1–20. https://doi.org/10.3390/fi9030025.

Collomb, A. (2016). Blockchain / distributed ledger technology (DLT): What impact on the financial sector? *Digiworld Economic Journal, 103*, 93. Retrieved from: www.comstrat.org.

Counterpoint Global Insights. (2018). *May*.

Davis, F. D. (1989). Perceived usefulness, perceived ease of use, and user acceptance of information technology. *MIS Quarterly: Management Information Systems, 13*(3), 319–339. https://doi.org/10.2307/249008.

Deloitte. (2017). Blockchain in banking: While the interest is huge, challenges remain for large-scale adoption. *Deloitte*. Retrieved from: https://www2.deloitte.com/content/dam/Deloitte/in/Documents/strategy/in-strategy-innovation-blockchain-in-banking-noexp.pdf (accessed on 14 December 2024).

Dutra, A., Tumasjan, A., & Welpe, I. M. (2018). Blockchain is changing how media and entertainment companies compete. *MIT Sloan Management Review*. Retrieved from: https://sloanreview.mit.edu/article/blockchain-is-changing-how-media-and-entertainment-companies-compete/ (accessed on 14 December 2024).

Faqih, K. M. S., & Jaradat, M. I. R. M. (2021). Integrating TTF and UTAUT2 theories to investigate the adoption of augmented reality technology in education: Perspective from a developing country. *Technology in Society, 67*(October 2021). https://doi.org/10.1016/j.techsoc.2021.101787.

Ganatra, M. (2023). The use of blockchain in financial services. *The Times of India*. https://timesofindia.indiatimes.com/blogs/voices/the-use-of-blockchain-in-financial-services/ (accessed on 14 December 2024).

Garg, P., Gupta, B., Chauhan, A. K., Sivarajah, U., Gupta, S., & Modgil, S. (2021). Measuring the perceived benefits of implementing blockchain technology in the banking sector. *Technological Forecasting and Social Change, 163*(October), 120407. https://doi.org/10.1016/j.techfore.2020.120407.

GlobeNewswire. (2023). *Blockchain in Banking and Financial Services Global Market*. Retrieved from: https://www.globenewswire.com/news-release/2023/04/03/2640115/0/en/Blockchain-In-Banking-And-Financial-Services-Global-Market-Report-2023.html (accessed on 14 December 2024).

Guo, Y., & Liang, C. (2016). Blockchain application and outlook in the banking industry. *Financial Innovation, 2*(1). https://doi.org/10.1186/s40854-016-0034-9.

Gupta, A., & Gupta, S. (2018). Blockchain technology basics. *Delhi Business Review, 19*(2), 1–5. Retrieved from: https://www.spheregen.com/blockchain-technology-basics/ (accessed on 14 December 2024).

Harwood-Jones, M. (2016). Blockchain and T2S: A potential disruptor. *Standard Chartered, 20*. Retrieved from: https://www.sc.com/BeyondBorders/wp-content/

uploads/2016/06/2016-06-16-BeyondBorders-Report-SCB_Nema_Block-Chain-Paper-Final.pdf (accessed on 14 December 2024).

Hassani, H., Huang, X., & Silva, E. (2018). Banking with blockchain-ed big data. *Journal of Management Analytics, 5*(4), 256–275. https://doi.org/10.1080/23270012.2018.1528900.

Huang, J., Kong, L., Chen, G., Wu, M. Y., Liu, X., & Zeng, P. (2019). Towards secure industrial IoT: Blockchain system with credit-based consensus mechanism. *IEEE Transactions on Industrial Informatics, 15*(6), 3680–3689. https://doi.org/10.1109/TII.2019.2903342.

IBM. (n.d.). *Benefits of Blockchain*. Retrieved from: https://www.ibm.com/topics/benefits-of-blockchain (accessed on 14 December 2024).

Islam, M. R., Rahman, M. M., Mahmud, M., Rahman, M. A., Mohamad, M. H. S., & Embong, A. H. (2021). A review of blockchain security issues and challenges. *2021 IEEE 12th Control and System Graduate Research Colloquium, ICSGRC 2021–Proceedings, August*, 227–232. https://doi.org/10.1109/ICSGRC53186.2021.9515276.

Jena, R. K. (2022). Examining the factors affecting the adoption of blockchain technology in the banking sector: An extended UTAUT model. *International Journal of Financial Studies, 10*(4). https://doi.org/10.3390/ijfs10040090.

Khadka, R. (2020). The impact of blockchain technology in banking: How can blockchain revolutionize the banking industry? *October*, 37.

Khalfan, M., Kemyani, A., Al, J., Abdul, R., Tariq, R., Kindi, A., Al, I. Y., Chandan, M., & Tiwari, K. (2022). Blockchain applications in accounting and finance: Qualitative evidence from the banking sector. *Quest Journals Journal of Research in Business and Management, 10*(4), 2347–3002. Retrieved from: www.questjournals.org (accessed on 14 December 2024).

Kim, N., & Hong, A. (2019). Conceptualizing blockchain technology in the utilization of social welfare services for the disabled. *Sustainable Development Research (ISSN 2690–9898 e-ISSN 2690–9901), 1*(1), p35. https://doi.org/10.30560/sdr.v1n1p35.

Kimani, D., Adams, K., Attah-Boakye, R., Ullah, S., Frecknall-Hughes, J., & Kim, J. (2020). Blockchain, business, and the fourth industrial revolution: Whence, whither, wherefore, and how? *Technological Forecasting and Social Change, 161*(May), 120254. https://doi.org/10.1016/j.techfore.2020.120254.

Kosba, A., Miller, A., Shi, E., Wen, Z., & Papamanthou, C. (2016). Hawk: The blockchain model of cryptography and privacy-preserving smart contracts. In *Proceedings–2016 IEEE Symposium on Security and Privacy, SP 2016* (pp. 839–858). https://doi.org/10.1109/SP.2016.55.

Kumari, A., & Devi, N. C. (2022). The impact of Fintech over financial services. *Technology Innovation Management Review, 12*(1).

Lewis, R., McPartland, J. W., & Ranjan, R. (2017). Blockchain and financial market innovation. *Economic Perspectives, 41*(7). Retrieved from: https://www.chicagofed.org/publications/economic-perspectives/2017/7#ftn5 (accessed on 14 December 2024).

London Speaker Bureau. (n.d.). *Blockchain–7 Benefits for the Financial Industry*. Retrieved from: https://londonspeakerbureau.com/blockchain-7-benefits-financial-industry/ (accessed on 14 December 2024).

Mansoor, M., Abbasi, A. Z., Abbasi, G. A., Ahmad, S., & Hwang, Y. (2023). Exploring the determinants affecting the usage of blockchain-based remittance services: An empirical study on the banking sector. *Behavior and Information Technology*, 1–19. https://doi.org/10.1080/0144929X.2023.2241564.

Marr, B. (2018). The 5 big problems with blockchain everyone should be aware of. *Forbes*. Retrieved from: https://www.forbes.com/sites/bernardmarr/2018/02/19/the-5-big-problems-with-blockchain-everyone-should-be-aware-of/?sh=6cd5f62f1670 (accessed on 14 December 2024).

Martin Christopher, C. (2016). The bridging model: Exploring the roles of trust and enforcement in banking, bitcoin, and the blockchain. *Nevada Law Journal, 17*, 139–180.

Meiklejohn, S., Pomarole, M., Jordan, G., Levchenko, K., McCoy, D., Voelker, G. M., & Savage, S. (2016). A fistful of Bitcoins: Characterizing payments among men with no names. *Communications of the ACM, 59*(4), 86–93. https://doi.org/10.1145/2896384.

Mishra, L., & Kaushik, V. (2021). Application of blockchain in dealing with sustainability issues and challenges of financial sector. *Journal of Sustainable Finance and Investment, 0*(0), 1–16. https://doi.org/10.1080/20430795.2021.1940805.

Naik, V., Pejawar, R., Singh, R., Aher, A., & Kanchan, S. (2020). Expeditious banking using Blockchain Technology. *International Conference on Computational Intelligence for Smart Power System and Sustainable Energy, CISPSSE 2020*. https://doi.org/10.1109/CISPSSE49931.2020.9212253.

Nakamoto, S. (2020). Bitcoin: A peer-to-peer electronic cash system. *Transforming Government: People, Process and Policy, 15*(4), 580–596. https://doi.org/10.1108/TG-06-2020-0114.

National Institution for Transforming India (NITI Aayog). (2020). *Blockchain: The India Strategy Towards Enabling Ease of Business, Ease of Living, and Ease of Governance.*

New Age Banking Summit Europe. (2018). *4 Key Advantages of Using Blockchain in Banking* (6th ed.). Retrieved from: http://newagebankingsummit.com/europe/4-key-advantages-of-using-blockchain-in-banking/ (accessed on 14 December 2024).

Osmani, M., El-Haddadeh, R., Hindi, N., Janssen, M., & Weerakkody, V. (2021). Blockchain for next-generation services in banking and finance: Cost, benefit, risk and opportunity analysis. *Journal of Enterprise Information Management, 34*(3), 884–899. https://doi.org/10.1108/JEIM-02-2020-0044.

Patki, A., & Sople, V. (2020). Indian banking sector: Blockchain implementation, challenges and way forward. *Journal of Banking and Financial Technology, 4*(1), 65–73. https://doi.org/10.1007/s42786-020-00019-w.

Pereira, J., Tavalaei, M. M., & Ozalp, H. (2019). Blockchain-based platforms: Decentralized infrastructures and their boundary conditions. *Technological Forecasting and Social Change, 146*, 94–102. https://doi.org/10.1016/j.techfore.2019.04.030.

Petrov. (2019). The impact of blockchain and distributed ledger technology on financial services. *International Scientific Journal, 4*(2), 88–91. https://stumejournals.com/journals/i4/2019/2/88.full.pdf (accessed on 14 December 2024).

Ramachandran, N., Shiju, N. K., Suresh, S., & Jain, A. (2022). Blockchain: Distinguishing facts from myths. In *Blockchain Technology: Exploring Opportunities, Challenges, and Applications* (Issue February). https://doi.org/10.1201/9781003138082-1.

Salem, S., & Ali, N. (2019). A proposed adoption model for blockchain technology using the Unified Theory of Acceptance and use of Technology (UTAUT). *Open International Journal of Informatics, 7*(2), 75–84.

Shah, T., & Jani, S. (2018). Finanzwirtschaftliche Anwendungen der Blockchain-Technologie [Applications of blockchain technology in finance]. *Springer Nature Link, 72*(2), 125–147. https://doi.org/10.1007/s41471-020-00090-5.

Sharma, A., Sharma, A., Singh, R. K., & Bhatia, T. (2023). Blockchain adoption in agri-food supply chain management: An empirical study of the main drivers using extended UTAUT. *Business Process Management Journal, 29*(3), 737–756. https://doi.org/10.1108/BPMJ-10-2022-0543.

Sharma, J. (2022). Top 8 ways banks benefit from blockchain tech. *FinTech Weekly*. https://www.fintechweekly.com/magazine/articles/Top-8-Ways-Banks-Benefit-From-Blockchain-Technology (accessed on 14 December 2024).

Singh, S., & Singh, N. (2016). Blockchain: Future of financial and cyber security. *Proceedings of the 2016 2nd International Conference on Contemporary Computing and Informatics, IC3I 2016*, 463–467. https://doi.org/10.1109/IC3I.2016.7918009.

Sung, H. N., Jeong, D. Y., Jeong, Y. S., & Shin, J. I. (2015). The relationship among self-efficacy, social influence, performance expectancy, effort expectancy, and behavioral intention in mobile learning service. *International Journal of U- and e-Service, Science and Technology, 8*(9), 197–206. https://doi.org/10.14257/ijunesst.2015.8.9.21.

Tapscott, A., & Tapscott, D. (2017). How blockchain is changing insurance. *Itnow, 60*(4), 16–17. https://doi.org/10.1093/ITNOW/BWY090.

Team Blockdata. (2022). *Top Banks Investing in Crypto and Blockchain May 2022 Update*. Retrieved from: https://www.blockdata.tech/blog/general/top-banks-investing-in-crypto-and-blockchain-may-2022-update (accessed on 14 December 2024).

T-Hub. (2018). *5 Ways Banks Can Benefit from Blockchain Technology*. Retrieved from: https://old.t-hub.co/blog/5-ways-banks-can-benefit-from-blockchain-technology/ (accessed on 14 December 2024).

Till, B. M., Peters, A. W., Afshar, S., & Meara, J. G. (2017). From blockchain technology to global health equity: Can cryptocurrencies finance universal health coverage? *BMJ Global Health, 2*(4). https://doi.org/10.1136/bmjgh-2017-000570.

Transparency International. (2021). *2021 Corruption Perceptions Index*. Retrieved from: https://www.transparency.org/en/cpi/2021 (accessed on 14 December 2024).

Treleaven, P., Gendal Brown, R., & Yang, D. (2019). Blockchain technology in the finance industry. *Transactions of the Korean Institute of Electrical Engineers, 68*(12), 1601–1606. https://doi.org/10.5370/KIEE.2019.68.11.1601.

Vahia, S. (2021). SBI, HDFC, ICICI, and 12 other banks are joining forces to use blockchain to power letters of credit. *Business Insider India*. https://www.businessinsider.in/cryptocurrency/news/sbi-hdfc-icici-and-12-others-banks-are-joining-forces-to-use-blockchain-to-power-letters-of-credit-a-move-that-could-be-a-boon-for-msmes/articleshow/83570874.cms (accessed on 14 December 2024).

Varma, J. R. (2019). Blockchain in finance. *Vikalpa, 44*(1), 1–11. https://doi.org/10.1177/0256090919839897.

Venkatesh, V., Morris, M. G., Davis, G. B., & Davis, F. D. (2003). *Quarterly, 27*(3), 425–478.

Venkatesh, V., Thong, J. Y. L., & Xu, X. (2012). Consumer acceptance and use of information technology: Extending the unified theory of acceptance and use of technology. *MIS Quarterly: Management Information Systems, 36*(1), 157–178. https://doi.org/10.2307/41410412.

Wang, H., Zheng, Z., Xie, S., Dai, H. N., & Chen, X. (2018). Blockchain challenges and opportunities: A survey. *International Journal of Web and Grid Services, 14*(4), 352. https://doi.org/10.1504/ijwgs.2018.10016848.

Wei, M. F., Luh, Y. H., Huang, Y. H., & Chang, Y. C. (2021). Young generation's mobile payment adoption behavior: Analysis based on an extended UTAUT model. *Journal of Theoretical and Applied Electronic Commerce Research, 16*(4), 1–20. https://doi.org/10.3390/jtaer16040037.

Werbach, K. (2018). Trust, but verify: Why the blockchain needs the law. *Energies, 6*(1), 1–8. https://doi.org/10.1177/1120700020921110.

Yoo, S. (2017). Blockchain-based financial case analysis and its implications. *Asia Pacific Journal of Innovation and Entrepreneurship, 11*(3), 312–321. https://doi.org/10.1108/apjie-12-2017-036.

Zachariadis, M., Hileman, G., & Scott, S. V. (2019). Governance and control in distributed ledgers: Understanding the challenges facing blockchain technology in financial services. *Information and Organization, 29*(2), 105–117. https://doi.org/10.1016/j.infoandorg.2019.03.001.

Zheng, Z., Xie, S., Dai, H. N., Chen, X., & Wang, H. (2019). Blockchain challenges and opportunities: A survey. *European Competition and Technology Assessment Review, 28*(2), 83–89. https://doi.org/10.54648/ecta2019011.

CHAPTER 12

A Comprehensive Review of Blockchain in the Fashion Industry: With Special Reference to Traceability, Transparency, and Product Authenticity in the Supply Chain

ADITI DHAMA,[1,*] NANDITA TRIPATHI,[2] RAJEEV SEMWAL,[2] and EBRU KEMER[3]

[1]*Amity School of Fashion Technology, Amity University, Greater Noida, Uttar Pradesh, India*

[2]*Amity Institute of Psychology and Allied Sciences, Amity University, Greater Noida, Uttar Pradesh, India*

[3]*Niğde Ömer Halisdemir University Vocational School of Social Sciences, Department of Hotel, Restaurant, and Catering Services, Dr. Öğretim Üyesi Mustafa Kemal Üniversitesi, Turkey*

*Corresponding author

ABSTRACT

This chapter examines the potential of blockchain technology in the apparel industry, with an emphasis on product authenticity, traceability, and transparency in the supply chain. Sustainability, consumer confidence, and business operations are negatively impacted by the fashion industry's

Blockchain Technology: Unlocking the Value of 3Ts in the Business World.
Richa Chauhan & Vidhi Kaul (Eds.)

DOI: 10.1201/9781998511914-12

ethical sourcing, counterfeiting, and supply chain complexities. Unique characteristics of blockchain technology, such as immutability, transparency, and smart contracts, are capable of resolving these issues. By utilizing blockchain technology, fashion firms can improve supply chain traceability, verify product authenticity, promote sustainable procurement, and enhance operational efficiency. The chapter presents case studies of blockchain implementations in different sectors of the fashion supply chain, demonstrating their efficacy in addressing supply chain challenges. However, the adoption of blockchain technology in the apparel industry is hindered by implementation complexity, scalability, data security, and adoption. The findings of this paper provide practitioners and policymakers with valuable insights, recommending collaboration, pilot projects, consumer education, and addressing implementation challenges. In addition, the implications for future research include scalability optimization, privacy-preserving mechanisms, impact evaluation, and interdisciplinary approaches. By adopting blockchain technology, the fashion industry can foster transparency, trust, and sustainability while promoting positive change within the sector.

12.1 INTRODUCTION

In the ever-evolving landscape of the global fashion industry, the pursuit of innovation and sustainability is no longer a choice but a necessity. The fashion industry, which has long captivated the world with its creativity, trends, and designs, is now at a crossroads, facing a series of challenges that demand transformative solutions. Among these challenges, the issues of traceability, transparency, and product authenticity within the supply chain have emerged as critical concerns, igniting the quest for a comprehensive and technologically driven response.

Fashion is not only about aesthetics and style; it is deeply entwined with supply chains that span the globe, sourcing materials, assembling garments, and distributing products across a vast network of manufacturers, suppliers, and retailers. Yet, this intricate web of global supply chains often conceals more than it reveals, shrouding the origins and journey of products in mystery. Global apparel and footwear retail sales surpassed \$1.9 trillion in 2019, and the fashion sector is anticipated to reach \$3 trillion by 2030. The consequences of this opaqueness are profound, as they undermine the industry's efforts towards sustainability, ethical practices, and consumer trust. Enter blockchain technology, a disruptive force that has penetrated various industries, promising unparalleled solutions to issues of transparency and traceability. Blockchain's potential to redefine the fashion supply chain is a subject of growing interest, as it aligns perfectly with the industry's increasing need for accountability and authenticity.

This chapter aims to provide a comprehensive review of blockchain's role in the fashion industry, focusing on traceability, transparency, and product authenticity within the supply chain. Through a multidisciplinary lens, we will explore how blockchain technology has the potential to revolutionize the way fashion products are produced, distributed, and consumed. By tracing the evolution of blockchain adoption in the fashion industry and assessing its practical implications, this chapter will equip industry professionals, researchers, and enthusiasts with a deeper understanding of the transformative power of this technology.

12.1.1 OVERVIEW OF THE FASHION INDUSTRY AND ITS SUPPLY CHAIN CHALLENGES

The emergence of blockchain technologies has revolutionized many industries over the past several years (Agrawal and Pal, 2019). It is a decentralized and distributed digital ledger that allows for record keeping. This technology holds great potential for the fashion and textile industry to improve in terms of transparency and traceability in various supply chain operations (Olsen and Borit, 2013a).

The fashion industry has a significant impact on the global economy, and its supply chain is comprised of numerous stakeholders and processes. Nevertheless, the industry encounters a number of obstacles that affect its efficacy, sustainability, and ethical practices.

12.1.1.1 LACK OF SUPPLY CHAIN VISIBILITY

The lack of visibility into supplier practices and working conditions is a significant obstacle. Frequently, fashion firms have difficulty tracing the origin of materials and components, making it difficult to ensure ethical sourcing and labor standards compliance.

12.1.1.2 COMPLEX AND FRAGMENTED SUPPLY CHAIN

The fashion supply chain is highly fragmented, involving multiple tiers of suppliers and subcontractors. Coordination of activities, transparency, and collaboration among these entities present obstacles. The fashion industry is characterized by fluctuating consumer demands and trends. Accurately forecasting and aligning production with fluctuating demand becomes difficult for brands, resulting in inefficient inventory management and potential waste.

12.1.1.3 SUSTAINABILITY AND ETHICAL CONCERNS

The fashion industry has a significant impact on the environment, including excessive resource consumption, pollution, and waste. In addition, the supply chain raises social and labor rights concerns regarding working conditions and equitable remuneration.

12.1.1.4 COUNTERFEITING AND INTELLECTUAL PROPERTY INFRINGEMENT

The fashion industry is plagued by counterfeiting, which not only causes legitimate brands to lose revenue but also damages their reputation. In the digital age, protecting intellectual property rights becomes increasingly difficult.

These supply chain difficulties have diverse effects on the industry. Financially, inefficiencies and lack of visibility contribute to increased costs, whereas counterfeiting results in a loss of revenue. The fashion industry contributes to resource depletion and pollution from an ecological standpoint. Concerns regarding labor rights and worker exploitation exist on a social level. In addition, supply chain issues erode consumer confidence and harm brand reputation. The resurgence of the COVID-19 pandemic has underscored the criticality of preserving ecological equilibrium and leading a healthy lifestyle. By the fourth quarter of 2023, the retail environment is anticipated to return to its pre-pandemic state (BoF-McKinsey and Company, 2021).

In response to these obstacles, the fashion industry has witnessed the introduction of initiatives and regulations. Certification programs promote ethical and sustainable practices, whereas traceability and transparency initiatives aim to increase supply chain visibility. Compliance with government regulations concerning labor rights and environmental standards remains a challenge for brands and suppliers.

Addressing the fashion industry's supply chain challenges is crucial for promoting sustainability, ethical practices, and consumer trust. Innovative solutions, such as blockchain technology, are being investigated to increase supply chain transparency, traceability, and efficiency. Collaboration and industry-wide enhancements are crucial to overcoming these obstacles and fostering positive change in the fashion industry.

12.1.2 RELEVANCE IN TODAY'S ERA

Examining and synthesizing existing research and literature concerning the implementation of blockchain technology in the fashion sector is the objective of this chapter. The chapter conducts an extensive examination and evaluation of the subject through the synthesis of numerous sources. The term "supply chain management" (SCM) is multifaceted and can be interpreted in various ways, including as a flux of products and materials, a branch of management, a philosophy of management, or a management process (Arora and Mittal, 2011). The number of stakeholders involved in supply chains varies according to their characteristics. However, in general, they may consist of producers, product assemblers, manufacturers, wholesalers, retailer merchants, and transporters (La Londe and Masters, 1994). In order to ensure uninterrupted operations, manufacturing organizations embrace a supply chain management philosophy and implement corresponding operational procedures (Mentzer et al., 2001).

12.1.2.1 IDENTIFYING RESEARCH VOIDS

Through the review process, this chapter will identify research voids and areas in need of further study. This can influence future research efforts and advance academic understanding of blockchain technology in the fashion industry.

12.1.2.2 IMPLICATIONS FOR STAKEHOLDERS

The chapter will provide practical insights and recommendations for stakeholders in the fashion industry, including fashion brands, manufacturers, retailers, and policymakers. By understanding the benefits, challenges, and potential hazards of blockchain adoption, stakeholders can make informed decisions regarding the implementation of blockchain solutions within their supply chains. Fashion, as a manifestation of deliberate notions and concepts, can be defined by an infinite number of tangential viewpoints and contentious concepts (Entwistle, 2000). According to Major and Steele (2019), the fashion industry is an element of a broader cultural and social phenomenon called the "fashion system," which encompasses the artistic and technical aspects of fashion as well as its production and consumption (Islam, 2021).

12.1.2.3 ENHANCING SUSTAINABILITY AND TRANSPARENCY

This review chapter seeks to contribute to the promotion of sustainable and ethical practices in the fashion industry by examining how blockchain technology can resolve issues of traceability, transparency, and product authenticity. Blockchain has the potential to increase supply chain transparency and consumer confidence in fashion products' authenticity and sustainability.

12.2 INTRODUCTION TO BLOCKCHAIN TECHNOLOGY AND ITS KEY FEATURES

A decentralized and distributed ledger system, blockchain technology permits multiple parties to uphold a secure and transparent ledger of digital transactions or data. By generating a tamper-resistant chain of blocks, it promotes trust, transparency, and immutability. Each block in the chain comprises a distinct set of transactions or data (Figure 12.1).

12.2.1 KEY CHARACTERISTICS OF BLOCKCHAIN TECHNOLOGY

Blockchain functions in a decentralized fashion, in contrast to conventional centralized systems where data and transactions are under the control

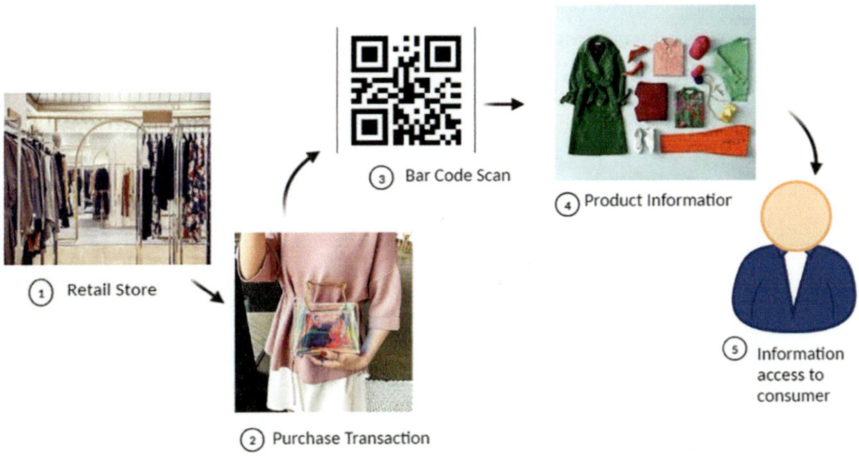

FIGURE 12.1 Application of blockchain technology to consumer.

Source: Author's own compilation.

of a central authority. It obviates the necessity for a solitary governing body by virtue of being composed of a network of nodes (computers) that collectively verify and store the blockchain's data. Blockchain technology incorporates essential functionalities such as smart contracts and consensus, which render this technology intuitive and applicable across various sectors (Kushwaha and Joshi, 2021).

1. **Distributed Ledger:** Blockchain utilizes a distributed ledger, meaning that a copy of the entire blockchain is recorded on each network node. As all participants have access to the same information, this redundancy enhances security, resiliency, and transparency.

   Blockchain provides transparency by enabling all participants to observe and authenticate blockchain-stored transactions. Once a transaction has been recorded, all participants can view it in real-time, creating a transparent and auditable system. To safeguard privacy, the actual identities of participants may be concealed.

   Blockchain uses sophisticated cryptographic techniques to ensure the security of data and transactions. Each block in the chain contains a cryptographic hash, which is a digital identifier generated from the block's data. Any modification to the data within a block would cause its hash to change, alerting the network of attempted manipulation.

Once a block has been added to the blockchain, it becomes virtually impossible to alter or delete the information it contains. The use of cryptographic hashing, consensus mechanisms, and the distributed nature of the blockchain guarantee the immutability of the recorded data, thereby enhancing trustworthiness.

2. **Consensus Mechanisms:** These are employed within the blockchain to facilitate agreement among participants of the network concerning the validity of transactions and the sequential addition of said transactions to the blockchain. Common consensus mechanisms include Practical Byzantine Fault Tolerance (PBFT), Proof of Work (PoW), and Proof of Stake (PoS).
3. **Smart Contracts:** These are blockchain-based agreements that operate autonomously to carry out predetermined actions upon the fulfillment of specific conditions. By facilitating the automation and programmability of processes and transactions, they eradicate the necessity for intermediaries and enhance efficiency.

In addition to providing transparency, blockchain also offers privacy features. Participants can maintain their privacy through pseudonymous identities or cryptographic techniques such as zero-knowledge proofs, depending on the blockchain's design, assuring confidentiality while maintaining transparency.

These defining characteristics make blockchain technology suitable for applications beyond cryptocurrencies, such as supply chain management, financial services, healthcare, and voting systems, among others. Utilizing the benefits of decentralization, transparency, security, and immutability, blockchain technology has the potential to revolutionize industries and enhance trust, efficiency, and accountability in a variety of domains.

12.2.2 BLOCKCHAIN TECHNOLOGY AND CONSENSUS MECHANISMS

Distributed ledger technology (DLT) is a broader term that incorporates various forms of decentralized and distributed ledger systems, with blockchain being one of the most well-known implementations. DLT enables multiple parties to observe a shared ledger in a synchronized and consistent manner. Key DLT features include:

1. **Replication and Redundancy:** DLT ensures that a copy of the ledger is replicated and stored across numerous network locations. This replication assures data redundancy and strengthens the system's resistance to malfunctions or attacks.
2. **Consensus Mechanisms:** DLT uses consensus mechanisms to reach consensus among network participants regarding the validity and sequence of transactions. PoW, PoS, PBFT, and other consensus techniques maintain the distributed ledger's trustworthiness and accuracy.

12.2.3 SMART CONTRACTS AND THEIR ROLE IN BLOCKCHAIN

Blockchain-programmed smart contracts execute automatically. They automate and enforce contract requirements without intermediaries. Smart contracts have several key features: Smart contracts can automatically execute many procedures and activities under specified conditions. When the prerequisites are met, they operate automatically without human intervention. Smart contracts can self-enforce agreement terms. No middlemen are needed because the transaction follows rules and logic. Programmable smart contracts can be developed in blockchain-specific languages. Complex business logic can be built due to its programmability.

Smart contracts are typically deployed and carried out on a blockchain. They utilize the blockchain's security, transparency, and immutability to ensure the integrity and dependability of contract execution.

The purpose of smart contracts in blockchain is to facilitate the secure and decentralized automation, verification, and execution of agreements and transactions. By enforcing predefined rules and conditions, they eliminate the need for intermediaries, improve efficiency, and reduce the potential for disputes or deception.

12.2.4 KEY FEATURES OF BLOCKCHAIN RELEVANT TO THE FASHION INDUSTRY

However, following a period of stagnant development during the pandemic, the fashion industry is currently undergoing a recovery phase; researchers anticipate that growth will return to pre-pandemic levels by the conclusion of 2023 (BoF-McKinsey and Company, 2021). The fashion industry can significantly benefit from blockchain technology's main features (Figure 12.2).

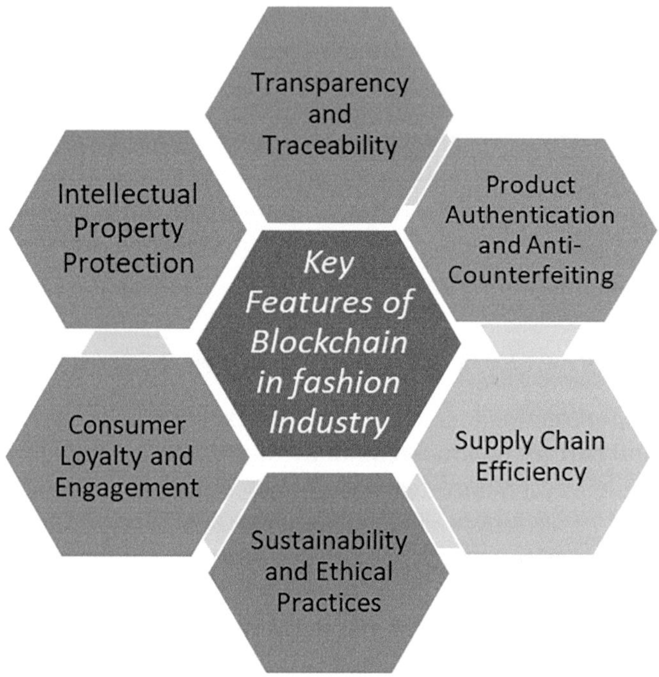

FIGURE 12.2 Key features of blockchain in the fashion industry.
Source: Author's own compilation.

Here is a comprehensive summary of how these characteristics pertain to the fashion industry.

Blockchain provides transparency and traceability by permitting all participants in the fashion supply chain to view and verify transactions recorded on the blockchain. This transparency allows stakeholders to monitor and trace the flow of products and materials throughout the supply chain. In the fashion industry, where supply chain transparency is essential for assuring ethical sourcing, sustainable practices, and conformance with labor standards, blockchain can improve supply chain transparency and traceability by providing an immutable and auditable ledger of transactions. In order to gain a fundamental understanding of the fashion industry's supply chains, it is necessary to distinguish and contrast the luxury fashion industry with the mass-produced apparel retail industry, which includes (Yang et al., 2017):

1. **Product Authentication and Anti-Counterfeiting:** In the fashion industry, counterfeiting is a significant problem that causes revenue loss and brand reputation harm. By providing a secure, tamper-resistant system for product authentication, blockchain technology can play a crucial role in combating counterfeiting. By recording the unique identifiers of authentic products on a distributed ledger, stakeholders can verify the authenticity of fashion items at each stage of the supply chain, ensuring that consumers receive authentic products and reducing the circulation of counterfeit goods. In order to provide the general public with the most recent patterns, fast-fashion brands produce vast quantities of fashion items at an accelerated rate (Tsay et al., 1999). In general, these brands outsource their supply chains (Yang et al., 2017).
2. **Supply Chain Efficiency and Collaboration:** Blockchain technology enables increased supply chain efficiency and collaboration. By utilizing a decentralized and distributed ledger, blockchain eliminates the need for intermediaries and enables stakeholders to interact and exchange information directly, thereby reducing the delays, errors, and costs associated with traditional supply chain processes. Smart contracts can automate and streamline processes such as procurement, payments, and inventory management, thereby augmenting the efficacy of the supply chain.
3. **Sustainability and Ethical Practices:** In the fashion industry, sustainability and ethical practices are gaining importance. By providing a transparent record of sustainable procurement, equitable trade, and compliance with labor and environmental standards, blockchain can support these efforts. By recording information about the origin of raw materials, manufacturing processes, and certifications on the blockchain, fashion brands can demonstrate their commitment to sustainability and ethical practices, nurturing consumer trust and promoting industry-wide improvement. Rapid-fashion brands are in search of manufacturers capable of delivering products in the shortest possible time in order to accommodate evolving patterns. For continuous

production, these strategies necessitate low-cost human labor, abundant manufacturing equipment, and simple procurement of raw materials (Igwe and Kanyembo, 2020).

4. **Intellectual Property Protection:** Infringement of intellectual property is a major concern in the fashion industry, where designs, trademarks, and branding are essential assets. The immutability and cryptographic security features of blockchain can assist in the protection of intellectual property rights by creating an immutable record of ownership and proof of creation. By registering their trademarks, designs, and copyrights on the blockchain, fashion brands can create a secure and tamper-proof system for safeguarding their intellectual property.
5. **Consumer Loyalty and Engagement:** Blockchain technology can improve consumer engagement and loyalty in the fashion industry. By providing transparent and verifiable information regarding product origins, materials, and supply chain practices, blockchain enables consumers to make informed, value-aligned purchasing decisions. In addition, blockchain-based loyalty programs can provide personalized incentives, rewards, and discounts, thereby nurturing consumer engagement and brand loyalty.

The key characteristics of blockchain, such as transparency, traceability, product authentication, supply chain efficiency, sustainability, intellectual property protection, and consumer engagement, offer the fashion industry substantial opportunities to address its challenges and enhance operations, ethical practices, and consumer trust.

12.3 SUPPLY CHAIN CHALLENGES IN THE FASHION INDUSTRY

12.3.1 OVERVIEW OF THE FASHION SUPPLY CHAIN AND ITS COMPLEXITIES

The management and direction of the movement of materials, components, and finished goods to the consumer constitutes a supply chain (Stevens and Johnson, 2016). Supply chain management (SCM) is a multifaceted concept that can be interpreted as a management philosophy, a branch of management, a process of managing materials and products, or even material and product traffic (Kimani et al., 2020). The processes and undertakings associated with the production, distribution, and retail sale of fashion items comprise the fashion supply chain.

An intricate network is comprised of the participation of numerous stakeholders, including consumers, manufacturers, suppliers, distributors, and designers. The stages that comprise the fashion supply chain are as follows: In light of market trends, consumer inclinations, and imaginative stimulation, fashion designers generate novel designs and collections. Prototypes of designs are developed and refined in collaboration between designers, pattern makers, sample makers, and textile suppliers. Logistics constraints, rising shipping costs, material shortages, and manufacturing delays are a few of the difficulties the fashion industry is presently experiencing as a result of the pandemic (BoF-McKinsey and Company, 202).

Consequently, supply chain efficacy is under unprecedented pressure. Already, as a result of their manufacturing processes, the apparel and textile industries rank second in terms of pollution. (McFall-Johnsen, 2020; Nimkar, 2018). These procedures necessitate the application of an excessive quantity of hazardous chemicals, which produce unhealthy byproducts and residues. Fashion companies acquire essential components, fabrics, embellishments, and materials from vendors once the design has been finalized. To ensure timely delivery, cost-effectiveness, and quality, this requires identifying suitable suppliers, negotiating contracts, and overseeing the procurement procedure.

The manufacturing phase involves converting the design and sourced materials into completed fashion items. Typically, it involves multiple production stages, such as cutting, sewing, assembly, quality control, and finishing. Manufacturing can be performed in-house or outsourced to third-party factories, which are frequently located in various countries, and it involves:

1. **Distribution and Logistics:** Following production, fashion items must be transported from factories to distribution centers or retail locations. Distribution and logistics involve coordinating transportation, warehousing, inventory management, and order fulfillment in order to ensure the timely and efficient delivery of products to retailers or consumers directly. Fashion items are distributed through a variety of retail channels, including physical storefronts, e-commerce platforms, department stores, and boutiques. Retailers manage inventory, marketing, and sales in order to attract consumers, display and sell fashion products, and facilitate a seamless purchasing experience.

2. **Returns and Reverse Logistics:** The fashion supply chain also includes reverse logistics procedures for managing returns, exchanges, and product recalls. Retailers and manufacturers handle refunds, customer service, and stocking.

   The fashion supply chain is distinguished by the following complexities:

1. **Global Sourcing:** Fashion brands employ overseas facilities and suppliers for reduced production costs, more specialized labor, and more raw material options. This causes coordination, quality assurance, and worldwide regulation challenges.
2. **Seasonality and Fast Fashion:** The fashion industry is fast and trend-driven. To meet shifting consumer preferences, firms must release new collections swiftly. Effective supply chain management is needed to meet deadlines, manage supplies, and reduce lead times.
3. **Fragmentation and Multiple Intermediaries:** The fashion supply chain has multiple linkages, from raw material suppliers to completed product sellers. This dispersion hinders stakeholder coordination, communication, and information sharing.
4. **Demand Volatility and Forecasting:** Social media, consumer tastes, and market factors challenge demand and trend predictions. Demand changes that cause stockouts, surpluses, or inventory imbalances can reduce supply chain efficiency.
5. **Considerations Regarding Ethical and Environmental Sustainability:** The fashion industry is increasingly scrutinized regarding ethical sourcing, equitable labor practices, and environmental sustainability. Ensuring compliance with social and environmental standards throughout the supply chain is difficult and requires close monitoring and audits of suppliers and production processes.

To address these complexities in the fashion supply chain, effective communication, collaboration, and the adoption of technologies and practices that improve transparency, traceability, agility, and sustainability are required.

12.3.2 IDENTIFICATION AND ANALYSIS OF FASHION INDUSTRY SUPPLY CHAIN CHALLENGES

The fashion industry confronts obstacles when it comes to monitoring and tracing the origin of raw materials and components throughout the supply chain. Due to the complexity of procurement and manufacturing, it can be challenging to ensure transparency and traceability, which can lead to unethical practices, labor rights violations, and environmental degradation.

In accordance with industry-specific modifications and enhancements, ISO has defined traceability. In 2005, the ISO definition of it was introduced within the food industry; it entailed the processing history, distribution, and post-delivery location of the product, as well as the origin of materials and components. Nevertheless, a precise definition pertaining to the fashion and retail sectors remains elusive. The fashion and textile industries possess unique characteristics that render the current definitions and interpretations non-transferable (Saicheua et al., n.d.). The fashion supply chain's lack of transparency diminishes the ability to observe and assess supplier practices, working conditions, and production processes. The lack of clarity regarding this matter poses a challenge for fashion products' ethical and ecological implications for both consumers and brands. Furthermore, the identification and resolution of challenges pertaining to responsible sourcing and sustainability are further complicated.

Counterfeiting is a significant problem in the fashion industry, resulting in revenue loss and reputational harm for authentic brands. In addition to violating intellectual property rights, counterfeit products compromise product quality and safety. The proliferation of counterfeit products undermines consumer confidence and compromises the integrity of brands.

12.3.3 IMPLICATIONS FOR SUSTAINABILITY, CONSUMER CONFIDENCE, AND BUSINESS OPERATIONS

1. **Sustainability:** Sustainable development goal (SDG) #12 of the United Nations (2015) delineates the pertinent knowledge and consciousness regarding sustainable development and environmentally harmonious lifestyles. The fashion industry's dearth of traceability and transparency hinders efforts to promote sustainability. The lack of supply chain transparency makes sustainable

procurement, fair labor, and environmental principles harder to implement. This lack of transparency hinders the fashion industry's sustainability and accountability.

2. **Consumer Trust:** Supply chain issues, including traceability and counterfeiting, erode consumer trust in fashion. The ethical and environmental implications of consumer goods are becoming more significant to shoppers. Fashion companies risk losing clients' trust if they don't practice ethical business practices.
3. **Business Operations:** The fashion industry's supply chain issues affect daily business. Poor transparency and traceability can cause inefficiencies, interruptions, and greater costs. If not discovered and resolved, supply chain issues, including labor violations and unsustainable procedures, can damage a company's brand and legal position. Counterfeiting hurts brand sales and market share.
4. **Compliance with Legal and Regulatory Requirements:** Fashion industry success depends on restrictions. Supply chain issues, including poor traceability or labor law violations, may be illegal. Failure to follow sustainable and ethical standards can lead to negative publicity, customer boycotts, and legal action from advocacy groups or individuals.

The sustainability, reputation, and profitability of fashion brands require that these supply chain challenges be addressed. Implementing robust traceability systems, fostering transparency, and employing anti-counterfeiting measures can enhance sustainability efforts, rebuild consumer trust, and enhance business operations. Adopting technologies such as blockchain, which provide solutions for traceability, transparency, and product authentication, can play a crucial role in addressing these issues and fostering positive change in the fashion industry.

12.4 BLOCKCHAIN APPLICATIONS IN THE FASHION INDUSTRY

12.4.1 HOW THE FASHION INDUSTRY IS ADOPTING BLOCKCHAIN TECHNOLOGY?

In recent years, companies in the fashion industry have recognized the potential of blockchain technology to address supply chain challenges and improve industry-wide transparency, traceability, and authenticity. Here are some important factors regarding the fashion industry's blockchain adoption.

Blockchain technology provides a transparent and immutable distributed ledger that can monitor the movement of fashion products from the procurement of raw materials through manufacturing, distribution, and retail. By documenting each stage on the blockchain, brands can ensure transparency, verify product origins, and address counterfeiting and unethical issues.

1. **Product Authentication and Anti-Counterfeiting:** Blockchain enables fashion firms to develop a tamper-resistant system for confirming product authenticity. By preserving unique product identifiers and ownership information on the blockchain, brands can provide consumers with trustworthy proof of the product's origin, ensuring they are purchasing authentic items and reducing the circulation of counterfeit goods.
2. **Sustainable Sourcing and Ethical Practices:** Blockchain permits the transparent recording of information concerning sustainable sourcing, equitable trade, and compliance with labor and environmental standards. Brands can demonstrate their commitment to ethical business practices by providing verifiable data on the blockchain, thereby fostering consumer transparency and confidence.
3. **Supply Chain Efficiency and Cost Reductions:** The implementation of blockchain technology can streamline supply chain operations by reducing documentation, automating processes, and fostering stakeholder collaboration. Smart contracts on the blockchain can automate duties like payments, procurement, and inventory management, resulting in increased efficiency, decreased costs, and enhanced supply chain performance overall.
4. **Consumer Engagement and Brand Loyalty:** Blockchain-based systems that offer transparency and traceability can enable consumers to make informed purchasing decisions. By providing verifiable information on the blockchain, brands can establish credibility, increase consumer engagement, and cultivate brand loyalty among customers who value transparency and ethical business practices.

5. **Collaboration and Industry Standards:** In the fashion industry, brands, retailers, and technology providers have collaborated to develop industry standards and blockchain solutions. Consortia and partnerships have emerged to establish shared blockchain platforms that enable interoperability and data sharing throughout the supply chain, thereby facilitating the industry's adoption of blockchain technology.

Despite the fact that blockchain adoption in the fashion industry is still in its infancy, several pilot projects and initiatives have demonstrated the technology's prospective benefits. By leveraging the key characteristics of blockchain, fashion brands can increase supply chain transparency, improve product authenticity, promote sustainability efforts, and bolster consumer confidence in the industry. Fashion needs more research and collaboration before blockchain technology can benefit it.

Case studies of blockchain implementations in different areas of the supply chain (e.g., sourcing, manufacturing, distribution) are:

1. **Case Study 1:** The LVMH and Aura Blockchain Consortium Case Study: Luxury garment company LVMH formed the Aura Blockchain Consortium with ConsenSys. The cooperation seeks industry openness and traceability. LVMH uses the Aura blockchain network to verify Louis Vuitton and Christian Dior products' authenticity. Consumers can view the product's blockchain-stored manufacturing history, materials, and authenticity by scanning a unique code. This blockchain application lets luxury consumers verify their purchases with confidence.

2. **Case Study 2:** Zara and Blockchain for Transparency in the Supply Chain: Zara uses blockchain technology to improve supply chain transparency and traceability. Zara has partnered with ConsenSys and Microsoft to provide customers with accurate product data. The blockchain tracks every manufacturing and distribution stage. Scan the garment's label QR code for this data. Blockchain technology helps Zara demonstrate its commitment to transparency, responsible buying, and ethical business practices in fast fashion.

3. **Case Study 3:** Carrefour and Food Traceability: Carrefour, a worldwide retailer, used blockchain technology to track food supplies. IBM and the IBM Food Trust blockchain platform help Carrefour maintain food quality and transparency. The blockchain

system tracks food provenance, administration, and certifications. Scan the packaging's QR code to validate the product's authenticity, freshness, and ethical source. This blockchain solution boosts customer confidence in Carrefour's food safety and traceability.

4. **Case Study 4:** Adidas and Sustainable Supply Chain: Adidas is using blockchain technology to enhance supply chain transparency and sustainability. Adidas plans to create a digital platform that tracks its products' sustainability by working with Circular ID. This contains material, production, and environmental impact information. The blockchain implementation gives customers this data to make educated decisions and promote sustainable habits. Adidas promotes supply chain transparency and sustainability with blockchain technology.
5. **Case Study 5:** Collaboration of Provenance and Everledger for Sourcing: Everledger, a digital ledger provider, and Provenance, a blockchain technology firm, addressed ethical procurement in the fashion sector. They utilized blockchain to trace and authenticate the origins of diamond jewelry. Each diamond's mining location, cutting procedure, and certification were recorded by the system. The cooperation employed blockchain technology to promote transparency and confidence in the diamond supply chain, assuring ethical and conflict-free diamonds for jewelers.
6. **Case Study 6:** Manufacturing Smart Textile Platform by Loomia: Loomia's blockchain-powered smart textile platform improves product authenticity and transparency. Programmable, washable sensors were added to textiles during manufacturing. These sensors recorded the product's journey, including materials, production, and certifications. Consumers may authenticate the garment's authenticity and learn about its production process by scanning a QR code on the garment.
7. **Case Study 7:** Distribution Collaboration between VeChain and H&M: Blockchain platform VeChain worked with H&M to improve supply chain transparency and traceability. To educate consumers about the sustainability and authenticity of H&M's goods, they can scan the garment's NFC chip to access

VeChainThor blockchain data on design, manufacture, and shipment. H&M's blockchain adoption showed its commitment to sustainability and ethical buying, improving customer confidence.

These case studies show how fashion firms and retailers use blockchain to solve supply chain difficulties. These initiatives boost customer confidence, encourage responsible business practices, and provide reliable product information by improving transparency, traceability, authenticity verification, and sustainability. Blockchain technology could change fashion and set new transparency and accountability norms.

12.4.2 ANALYSIS OF THE BENEFITS AND LIMITATIONS OF BLOCKCHAIN IN ADDRESSING CHALLENGES IN THE FASHION SUPPLY CHAIN

12.4.2.1 ADVANTAGES OF BLOCKCHAIN FOR THE FASHION SUPPLY CHAIN

1. **Disclosure and Traceability:** Blockchain offers both transparency and traceability, enabling stakeholders to track and verify the movement of fashion items along the supply chain, from the procurement of raw materials to retail. This transparency contributes to solving issues linked with unethical business activities, counterfeit items, and inefficient supply chains.
2. **Product Authentication and Anti-Counterfeiting:** Blockchain provides an immutable record of goods' origin and travel, enabling product authenticity. It fights counterfeiting and boosts customer confidence by verifying trendy goods.
3. **Ethical Sourcing and Sustainability:** The unchangeable record of product origin and journey on blockchain ensures product authenticity. Verifying fashionable items prevents counterfeiting and promotes buyer confidence. It improves sustainability efforts by supplying verifiable data and promoting responsible practices throughout the entire fashion supply chain.

   By streamlining supply chain operations with blockchain technology, documentation can be reduced, processes can be automated, and stakeholder collaboration can be enhanced. Smart contracts on the blockchain automate tasks, eliminating the need for middlemen and optimizing inventory management, payments, and procurement.

4. **Consumer Engagement and Trust:** Blockchain promotes consumer trust by providing transparency and access to trustworthy information regarding product origins, manufacturing processes, and sustainability practices. It enables consumers to make informed, value-aligned purchasing decisions, thereby strengthening engagement and brand loyalty.

12.4.3 KEY CHALLENGES IN EMBRACING BLOCKCHAIN TECHNOLOGY IN THE FASHION INDUSTRY

1. **Implementation Complexity:** Integrating blockchain technology into existing supply chain systems requires substantial technical knowledge and resources. Comprehending interoperability obstacles and ensuring compatibility with legacy systems can be time-consuming and complicated.
2. **Scalability and Performance:** Blockchain networks, such as public blockchains, may encounter scalability and performance constraints when handling large transaction volumes. To meet the demands of large-scale fashion supply chains, the efficiency and scalability of blockchain technology need to be improved.
3. **Data Privacy and Compliance:** The inherent transparency of blockchain technology may conflict with data privacy regulations. Creating a balance between the need for transparency and the preservation of sensitive business and consumer information is difficult, especially in industries with stringent privacy requirements.
4. **Adoption and Collaboration:** The complete potential of blockchain technology can only be realized through widespread adoption and stakeholder collaboration. To establish shared standards and networks, the fashion industry requires extensive

participation, which can be difficult due to competitive dynamics and coordination efforts (Figure 12.3).

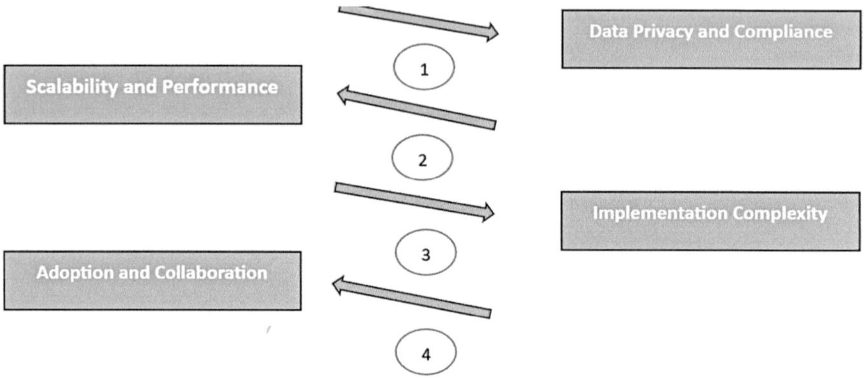

FIGURE 12.3 Key challenges in embracing blockchain technology in the fashion industry.

Source: Author's own compilation.

12.4.4 COMPARISON WITH TRADITIONAL SOLUTIONS AND EMERGING TECHNOLOGIES

Traditional supply chain systems frequently lack transparency and traceability, relying instead on manual processes and intermediaries. Blockchain provides a decentralized and immutable ledger, enhancing transparency, provenance, and productivity. It reduces reliance on intermediaries, optimizes processes, and increases the precision of data. While blockchain has unique benefits, other emerging technologies can complement or provide alternative solutions to fashion supply chain issues. Internet of Things (IoT) devices, for instance, can facilitate real-time product tracking and monitoring, while Artificial Intelligence (AI) and Machine Learning (ML) can improve demand forecasting and supply chain optimization. The combination of these technologies with blockchain can unlock additional benefits and comprehensive solutions.

It is essential to note that blockchain is not a one-size-fits-all solution and should be viewed as part of a larger technological ecosystem. Evaluating the fashion industry's unique requirements, obstacles, and resources will help determine the optimal combination of technologies to address supply chain challenges.

12.5 REPORT ON TRACEABILITY AND TRANSPARENCY IN THE FASHION INDUSTRY

12.5.1 TRACEABILITY AND TRANSPARENCY'S SIGNIFICANCE IN THE FASHION INDUSTRY

Traceability and transparency play a crucial role in the fashion industry, addressing critical challenges and meeting consumers' and stakeholders' evolving demands. The significance of traceability and openness may be summed up as follows.

The fashion and textile industries possess unique characteristics that render the current definitions and interpretations non-transferable (Olsen and Borit, 2013b). Traceability enables stakeholders to monitor the voyage of fashion products, ensuring ethical procurement practices, equitable labor conditions, and compliance with environmental standards. Transparency promotes sustainable practices and responsible procurement by providing visibility into supply chain processes.

1. **Consumer Confidence and Trust:** Traceability and transparency increase consumer confidence by providing verifiable information regarding product origins, manufacturing processes, and sustainability practices. Increasingly, consumers place a premium on openness and demand assurance that the products they purchase align with their values.
2. **Assurance of Product Quality and Authenticity:** Traceability aids in ensuring product quality and authenticity. In the fashion industry, counterfeit products pose a significant threat to brand reputation and consumer safety. Consumers are able to verify the authenticity and quality of their purchases when supply chains are transparent.

12.5.2 HOW BLOCKCHAIN IMPROVES SUPPLY CHAIN TRACEABILITY AND TRANSPARENCY?

In the fashion industry, blockchain technology offers unique capabilities for enhancing traceability and transparency. The main ways in which blockchain improves these elements are as follows:

1. **Transparent and Immutable Ledger:** Blockchain provides a transparent and immutable ledger in which transactions and data are recorded in a tamper-resistant manner. Each transaction is time-stamped and linked to previous transactions, thereby creating an auditable, participant-accessible information chain.
2. **Enhanced Supply Chain Visibility:** Blockchain enables real-time supply chain visibility and transparency. Participants can monitor and verify every stage of the product's voyage, from the procurement of raw materials to its manufacture, distribution, and retail sale.
3. **Secure Data Sharing:** Blockchain technology enables supply chain participants to securely share data. By employing cryptographic techniques, blockchain ensures that only authorized parties have access to sensitive information.
4. **Smart Contracts:** Blockchain-based smart contracts automate and enforce predetermined supply chain norms and conditions. They facilitate automated agreement verification and execution, eliminating the need for middlemen and streamlining processes.

12.5.3 EVALUATION OF BLOCKCHAIN'S EFFECTIVENESS IN ACHIEVING TRACEABILITY AND TRANSPARENCY

Garcia Torres et al. (2019) suggest that traceability and transparency may be indispensable components of a sustainable supply chain management system. In the fashion industry, blockchain technology contains great potential for attaining traceability and transparency objectives. Nonetheless, several factors must be regarded for a thorough evaluation:

1. **Adoption and Collaboration:** For blockchain implementations to be successful, widespread adoption and collaboration among fashion brands, retailers, and suppliers are essential. The participation of key stakeholders is necessary for blockchain to be effective in attaining traceability and transparency.
2. **Integration with Existing Systems:** For seamless data transfer and compatibility, blockchain must be integrated with existing supply chain systems and processes. To guarantee the efficacy of blockchain solutions, interoperability and system integration issues must be addressed.

3. **Accuracy and Verification of Data:** The veracity and precision of data recorded on the blockchain are essential for traceability and openness. Maintaining the system's integrity requires that the inputted data be accurate and verifiable by independent parties.

4. **Scalability and Performance:** For large-scale fashion supply chains, the scalability and performance of blockchain networks must be considered. Blockchain systems must be able to handle high transaction volumes and provide visibility in real-time without compromising efficacy.

5. **Privacy and Compliance:** It is essential to strike a balance between the transparency afforded by blockchain and privacy requirements. Traceability and transparency solutions based on blockchain technology must adhere to data protection regulations and be able to manage sensitive information securely.

The fashion and textile industries are facing increasing pressure from a growing consumer base to be more transparent (James and Montgomery, 2017). There is a growing expectation among influential consumer segments that their preferred brands adopt greater transparency and accountability (Griplas, 2021). Blockchain technology has immense potential to increase the transparency and traceability of the fashion industry. It offers solutions to critical problems by virtue of its capability to provide an immutable and transparent ledger, secure data exchange, and automation via smart contracts. But adoption, integration, data accuracy, scalability, privacy, and compliance are all factors that influence the effectiveness of blockchain implementations. In order to maximize the advantages of blockchain technology in achieving the goals of traceability and transparency within the fashion sector, it is imperative that stakeholders engage in meticulous assessment and cooperation. An industry that greenwashes its products takes advantage of consumers' ignorance regarding sustainable practices to sow confusion. As a consequence, purchasing decisions in sustainable fashion are influenced by a lack of confidence in environmentally friendly products and ethical manufacturing (Bhaduri and Ha-Brookshire, 2011). This gray area, which is frequently exploited by businesses to greenwash their products, would be eliminated by transparent supply chain operations, which would likely increase consumer interest in genuinely green products.

12.6 PRODUCT AUTHENTICITY AND ANTI-COUNTERFEITING

12.6.1 OBSTACLES RELATED TO AUTHENTICITY OF PRODUCTS IN THE FASHION INDUSTRY

The fashion industry confronts significant challenges regarding product authenticity; counterfeiting is a pervasive problem. The following difficulties are frequently encountered:

1. **Counterfeit Products:** Counterfeit products imitate authentic fashion items, violating intellectual property rights and misleading consumers. Not only do counterfeit products result in revenue loss for legitimate brands, but they also undermine consumer confidence and brand reputation.
2. **Lack of Transparency:** The lack of transparency in supply channels makes it difficult for consumers to confirm the authenticity of fashion products. Multiple intermediaries are frequently involved in supply chains, generating opportunities for counterfeit products to enter the market.
3. **Inconsistent Quality:** Genuine fashion goods are usually better made than fakes. Consumers may accidentally buy low-quality counterfeit goods, resulting in disappointment and brand damage.

12.6.2 THE FUNCTION OF BLOCKCHAIN IN GUARANTEEING PRODUCT AUTHENTICITY AND COMBATING COUNTERFEITING

Blockchain technology is crucial to fashion product authenticity and counterfeit prevention. Several blockchain features contribute to this:

1. **Product Details and Transactions:** They are immutable and tamper-proof using blockchain. Data recorded on the blockchain cannot be changed, assuring product validity.
2. **Unique Identifiers:** Blockchain lets fashion products have serial numbers or digital certifications. Blockchain-based IDs allow customers to verify product authenticity. Blockchain tracks a fashion product's complete journey, improving transparency and traceability. Buyers may trace the product's origin, manufacturing, and distribution to ensure authenticity.

3. **Consumer Engagement:** Blockchain-based technologies give customers fashion item authenticity verification. Scan a QR code or use a mobile app to verify product authenticity to boost consumer confidence.

12.6.2.1 CASE STUDIES AND INITIATIVES EMPLOYING BLOCKCHAIN FOR ANTI-COUNTERFEITING IN THE FASHION INDUSTRY

1. **VeChain and Authenticity Proofs:** VeChain, a blockchain network, worked with a luxury fashion firm to authenticate expensive handbags. NFC devices with blockchain-connected identifiers are in each bag. Customers may scan the chip to verify the bag's authenticity and learn about its manufacture and origins.
2. **Arianee and Digital Certificates:** Arianee's blockchain-based system creates fashion product digital certificates. Blockchain-stored digital certificates prove authenticity. Consumer-accessible and transferable certifications boost secondary market transparency and confidence.

12.6.3 EVALUATION OF THE EFFICACY AND LIMITATIONS OF BLOCKCHAIN FOR AUTHENTICATING PRODUCTS

Immutable records, unique identities, transparency, and traceability make blockchain effective in ensuring product authenticity. Blockchain-based technologies let customers authenticate fashion goods' authenticity, reducing counterfeit danger.

Blockchain has many benefits but also limitations. Blockchain-based anti-counterfeiting solutions require widespread acceptance and collaboration among companies, retailers, and customers to reach their full potential. Scalability, privacy, and interoperability must also be considered for large-scale application.

Blockchain technology may prevent fashion counterfeiting and ensure product authenticity. Blockchain improves customer confidence and fashion product verification with immutable data, unique IDs, transparency, and traceability. However, widespread acceptance, scalability, and privacy issues must be addressed to properly utilize blockchain for counterfeit prevention and product authentication.

12.7 ADOPTION AND IMPLEMENTATION CHALLENGES OF BLOCKCHAIN IN THE FASHION INDUSTRY

12.7.1 BARRIERS AND CHALLENGES TO BLOCKCHAIN ADOPTION IN THE FASHION INDUSTRY

1. **Dearth of Awareness and Comprehension:** Lack of knowledge and understanding of blockchain's potential benefits and uses in fashion is a major barrier to adoption. Many stakeholders may not comprehend how blockchain might address supply chain problems, preventing them from adopting and investing in it. Implementing blockchain technology requires extensive infrastructure, technical skills, and resources. Small and medium-sized fashion enterprises may lack the funds and resources to embrace blockchain, preventing widespread adoption.
2. **Coordination and Collaboration:** Blockchain implementation in the fashion industry requires coordination and collaboration among multiple stakeholders, including brands, manufacturers, suppliers, and retailers. In an industry characterized by competition and complex relationships, achieving consensus and cooperation among these parties can be difficult.

12.7.2 REGULATORY AND LEGAL CONCERNS REGARDING BLOCKCHAIN IMPLEMENTATION

1. **Compliance with Data Protection Regulations:** Blockchain technology involves the storage and distribution of data across multiple locations, which raises concerns about data protection and privacy. Blockchain solutions handling personal data must comply with the EU's General Data Protection Regulation (GDPR).

2. **Intellectual Property Rights:** In the fashion industry, intellectual property protection is a crucial factor. Blockchain-based solutions must consider how to safeguard intellectual property rights while still providing transparency and traceability. Determining ownership and access rights within a decentralized blockchain ecosystem can pose legal complications.

12.7.3 INTEGRATION DIFFICULTIES WITH EXISTING SYSTEMS AND TECHNOLOGIES

1. **Interoperability:** Integrating blockchain with existing supply chain technologies and systems can be difficult. For seamless data exchange and compatibility, it is essential to ensure interoperability between blockchain networks and legacy systems.
2. **Lack of Standard Protocols:** The absence of standard protocols and frameworks for blockchain integration in the fashion industry can hinder adoption. Common data formats, identifiers, and interfaces are necessary for interoperability and seamless integration with existing systems.

12.7.4 PRIVACY, DATA MANAGEMENT, AND CYBERSECURITY CONCERNS

1. **Privacy and Data Governance:** Blockchain's inherent openness can conflict with data privacy regulations. Challenges exist in achieving a balance between the need for openness and the preservation of sensitive business and consumer data. It is essential to implement privacy-protecting mechanisms, such as zero-knowledge proofs or selective disclosure, to resolve these concerns.
2. **Cybersecurity:** Blockchain networks are not immune to cybersecurity attacks. Ensuring the security of the blockchain infrastructure, including secure key management, protection against hijacking attempts, and prevention of unauthorized access, is

crucial for successful blockchain implementation in the fashion industry.

Addressing these adoption and implementation challenges calls for industry-wide collaboration, regulatory clarity, and technical proficiency. Overcoming these obstacles will allow the fashion industry to capitalize on the transformative potential of blockchain technology for enhanced supply chain transparency, traceability, and authenticity verification.

12.8 FUTURE TRENDS AND OPPORTUNITIES IN BLOCKCHAIN FOR THE FASHION INDUSTRY

12.8.1 EMERGING FASHION TRENDS AND INNOVATIONS IN BLOCKCHAIN

1. **Tokenization and Digital Assets:** Using blockchain, the fashion industry can explore the tokenization of fashion assets, such as limited-edition items or digital collectibles. This enhances the fashion ecosystem's liquidity, fractional ownership, and value exchange.
2. **Integrating Blockchain:** Emerging technologies such as IoT devices and sensors can improve real-time tracking, monitor product conditions, and automate supply chain processes. This integration can provide an exhaustive view of the product's journey, allowing for greater efficiency and transparency.
3. **Sustainability and Circular Economy:** Blockchain can help the fashion industry implement circular economy models. By recording and verifying information regarding the reuse, recycling, and proper disposal of fashion products on the blockchain, brands can demonstrate their commitment to sustainability and facilitate more sustainable business practices.

12.9 MANAGERIAL IMPLICATIONS OF BLOCKCHAIN TECHNOLOGY IN THE FASHION INDUSTRY

Using blockchain technology requires forethought and strategy. Supply chain efficiency, data security, and cooperation are essential for fashion executives. Standardization and sustainability will provide firms an edge, but strict compliance with regulations is essential (Figure 12.4).

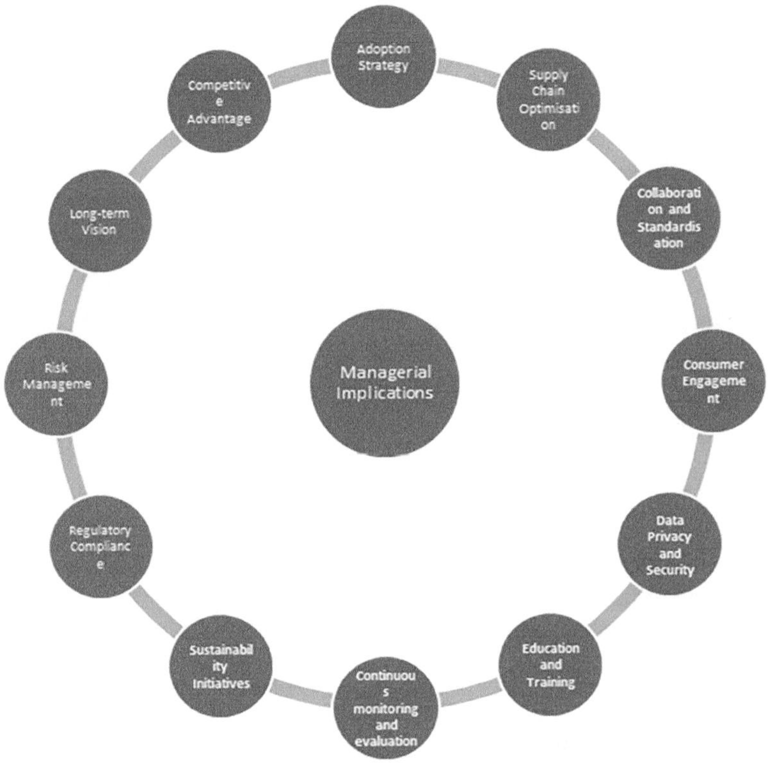

FIGURE 12.4 Managerial implications of blockchain technology in the fashion industry.
Source: Author's own compilation.

1. **Adoption Plan:** Fashion sector managers should integrate blockchain technology into their supply chain management plan. This includes funding blockchain-based traceability and transparency technologies.
2. **Supply Chain Optimization:** To maximize the benefits of blockchain technology, fashion firms should prioritize optimizing their supply networks. Streamlining processes, removing redundancies, and eliminating middlemen can boost blockchain deployments.
3. **Collaboration and Standardization:** Fashion brands and stakeholders should work on standardizing blockchain protocols and data formats. By doing so, the industry can improve supply chain transparency and interoperability.

4. **Consumer Engagement:** Managers should acknowledge the rising significance of authenticity and sustainability ideals. Blockchain may be used to market product origin and lifetime data to customers.
5. **Data Privacy and Security:** Prioritizing data privacy and security is essential in blockchain applications. Managers must implement strong cybersecurity measures and comply with GDPR.
6. **Education and Training:** Managers should invest in staff training for successful blockchain implementation. Successful adoption requires workforce blockchain knowledge.
7. **Continuous Monitoring and Evaluation:** Managers must evaluate the performance and efficacy of blockchain systems. Regular reviews will increase technology utilization and suggest opportunities for improvement.
8. **Sustainability Integration:** Fashion firms could incorporate blockchain into their sustainability efforts. They may significantly reduce their carbon footprint by publicly tracking product environmental impact.
9. **Regulatory Compliance:** Managers must remain updated on blockchain technology legislation. Avoiding legal and financial hazards requires compliance with local and international laws.
10. **Risk Management:** Understand that blockchain technology, like any other, has inherent risks and challenges. Management should establish risk mitigation strategies to address blockchain system vulnerabilities.
11. **Long-Term Vision:** Managers should envision blockchain integration in the long term. Blockchain technology is evolving, so remaining dedicated to its prospective benefits might yield significant returns.
12. **Competitive Advantage:** Fashion firms may gain a competitive edge by adopting blockchain for supply chain management early. Managers should focus on new blockchain applications to differentiate their brand.

12.10 CONCLUSION

This chapter examined blockchain technology's possibilities in fashion, focusing on supply chain traceability, transparency, and authenticity. It evaluated blockchain's pros and cons, fashion industry challenges, and their implications on sustainability, customer trust, and corporate operations. The study also included blockchain case studies from several fashion supply chain domains to demonstrate its usefulness in solving supply chain issues.

Many important facts and insights arose from the investigation. Blockchain technology might increase fashion product traceability, transparency, and authenticity. It helps supply chain stakeholders track fashion items, ensuring ethical sourcing, combating counterfeiting, and encouraging sustainability. Blockchain technologies boost customer trust, sustainability, and supply chain efficiency. We must still address implementation difficulties, scalability, data privacy, and uptake.

12.10.1 RECOMMENDATIONS TO PRACTITIONERS AND POLICYMAKERS

The findings suggest the following for fashion industry practitioners and policymakers:

1. **Working Together to Set Industry Standards:** Industry standards and protocols for blockchain technology in the fashion supply chain should be developed by stakeholders. This improves industry-wide data exchange, interoperability, and transparency.
2. **Accept Blockchain Pilots:** To assess blockchain's suitability for supply chain issues, fashion manufacturers and retailers could conduct pilot projects. These projects can provide data and insights to help spread the technology.
3. **Improve Consumer Engagement and Education:** Educating customers about blockchain and offering clear and verifiable information about product origins, sustainability, and authenticity may boost engagement and confidence.

4. **Address Implementation Challenges:** Fashion firms should allocate resources to address implementation challenges, such as integration with existing systems, concerns about data privacy, and scalability. Collaboration with technology providers and experts can aid in effectively navigating these obstacles.

12.10.2 IMPLICATIONS FOR FUTURE RESEARCH TO ADVANCE THE COMPREHENSION AND APPLICATION OF BLOCKCHAIN TECHNOLOGY IN THE FASHION INDUSTRY

Future research should concentrate on the following areas:

1. **Scalability and Performance Optimization:** Research should investigate solutions to address the scalability and performance limitations of blockchain networks, particularly when they are applied to large fashion supply chains.
2. **Privacy-Preserving Mechanisms:** Developing privacy-preserving mechanisms that strike a balance between the need for transparency and data privacy regulations is crucial. To safeguard sensitive information, research should concentrate on techniques such as zero-knowledge proofs and selective disclosure.
3. **Long-Term Effects:** Further research should evaluate the long-term effects of blockchain implementation on the fashion industry's sustainability, consumer trust, and business operations. Assessing the economic, social, and environmental consequences can provide decision-makers with valuable insight.
4. **Interdisciplinary Collaboration:** Collaboration between researchers, practitioners, and policymakers from various disciplines, such as fashion, technology, and law, can lead to comprehensive and holistic insights into the adoption of blockchain technology and its implications.

In conclusion, blockchain technology bears great promise for addressing the supply chain challenges of the fashion industry, particularly in terms of attaining product traceability, transparency, and authenticity. Collaboration, standardization, and overcoming implementation obstacles are required to implement blockchain solutions. By adopting blockchain technology, the fashion industry can improve its sustainability efforts, regain consumer trust, and promote positive industry change. Future research and innovation are required to realize blockchain's maximum potential in the fashion supply chain.

KEYWORDS

- **blockchain**
- **consumer engagement**
- **data exchange**
- **fashion industry**
- **fashion supply chain**
- **industry standards**
- **pilot projects**
- **stakeholders**
- **supply chain**
- **sustainability**
- **traceability**

REFERENCES

Agrawal, T. K., & Pal, R. (2019). Traceability in textile and clothing supply chains: Classifying implementation factors and information sets via Delphi study. *Sustainability, 11*(6), 1698. https://doi.org/10.3390/su11061698.

Arora, S., & Mittal, S. (2011). Intensifying export performance through planned capacity building: A study of the Indian apparel sector. *Journal of Global Fashion Marketing, 2*(1), 20–27. https://doi.org/10.1080/20932685.2011.10593079.

Bhaduri, G., & Ha-Brookshire, J. E. (2011). Do transparent business practices pay? Exploration of transparency and consumer purchase intention. *Clothing and Textiles Research Journal, 29*(2), 135–149. https://doi.org/10.1177/0887302x11407910.

Chauhan, R., & Kaul, V. (2021). Blockchain technology: A panacea for medical distribution ailments. In *Blockchain for Healthcare Systems* (pp. 43–60). CRC Press.

Chauhan, R., & Kaul, V. (2022). Blockchain technology: Unlocking the business model maze for evolving businesses and start-ups. In *Transformations through blockchain technology: The New Digital Revolution* (pp. 291–309). Cham: Springer International Publishing.

Entwistle, J. (2000). *The Fashioned Body: Fashion, Dress, and Modern Social Theory.* Malden, MA: Polity Press.

Garcia-Torres, S., Albareda, L., Rey-Garcia, M., & Seuring, S. (2019). Traceability for sustainability: Literature review and conceptual framework. *Supply Chain Management, 24*(1), 85–106. https://doi.org/10.1108/scm-04-2018-0152.

Griplas, L. (2021). Fibre: Unravelling the seams. *The Woolmark Company.* https://www.woolmark.com/fibre/unravelling-the-seams/ (accessed on 14 December 2024).

Igwe, P., & Kanyembo, F. (2020). The CAGE around internationalization of SMEs and the role of government.

Islam, S. (2021). Waste management strategies in fashion and textiles industry: Challenges in governance, materials culture, and design-centric approaches. In *Waste Management in the Fashion and Textile Industries* (pp. 275–293). https://doi.org/10.1016/b978-0-12-818758-6.00015-6.

James, A., & Montgomery, B. (2017). Engaging the fashion consumer in a transparent business model. *International Journal of Fashion Design, Technology and Education, 10*(3), 287–299. https://doi.org/10.1080/17543266.2017.1378730.

Kimani, D., Adams, K., Attah-Boakye, R., Ullah, S., Frecknall-Hughes, J., & Kim, J. (2020). Blockchain, business, and the fourth industrial revolution: Whence, whither, wherefore, and how? *Technological Forecasting and Social Change, 161,* 120254. https://doi.org/10.1016/j.techfore.2020.120254.

Kushwaha, S. S., & Joshi, S. (2021). An overview of blockchain-based smart contracts. *Lecture Notes on Data Engineering and Communications Technologies, 58,* 899–906. https://doi.org/10.1007/978-981-15-9647-6_70.

La Londe, B. J., & Masters, J. M. (1994). Emerging logistics strategies: Blueprints for the next century. *International Journal of Physical Distribution & Logistics Management, 24*(7), 35–47. https://doi.org/10.1108/09600039410070975.

Major, L. S., & Steele, V. (2019). Fashion industry. Illinois, United States: Encyclopedia Britannica, Inc.

McFall-Johnsen, M. (2020). These facts show how sustainable the fashion industry is. *World Economic Forum.* https://www.weforum.org/agenda/2020/01/fashion-industry-carbon-unsustainable-environment-pollution/ (accessed on 14 December 2024).

Mentzer, J. T., DeWitt, W., Keebler, J. S., Min, S., Nix, N. W., Smith, C. D., & Zacharia, Z. G. (2001). Defining supply chain management. *Journal of Business Logistics, 22*(2), 1–25. https://doi.org/10.1002/j.2158-1592.2001.tb00001.x.

Nimkar, U. (2018). Sustainable chemistry: A solution to the textile industry in a developing world. *Current Opinion in Green and Sustainable Chemistry, 9,* 13–17. https://doi.org/10.1016/j.cogsc.2017.11.002.

Olsen, P., & Borit, M. (2013). How to define traceability. *Trends in Food Science & Technology, 29*(2), 142–150. https://doi.org/10.1016/j.tifs.2012.10.003.

Ozer, A., Gurbuz, G., Calimli, A., & Korbahti, B. K. (2008). Investigation of nickel (II) biosorption on *Enteromorpha prolifera*: Optimization using response surface analysis. *Journal of Hazardous Materials, 152*(3), 778–788. https://doi.org/10.1016/j.jhazmat.2007.07.053.

Saicheua, V., Cooper, T., & Knox, A. (n.d.). Public understanding towards sustainable clothing and the supply chain.

Sanghi, R., & Bhattacharya, B. (2002). Review on decolorization of aqueous dye solutions by low-cost adsorbents. *Coloration Technology, 118*(5), 256–269.

Stevens, G. C., & Johnson, M. (2016). Integrating the supply chain … 25 years on. *International Journal of Physical Distribution & Logistics Management, 46*(1), 19–42. https://doi.org/10.1108/ijpdlm-07-2015-0175.

Tsay, A. A., Nahmias, S., & Agrawal, N. (1999). Modeling supply chain contracts: A review. In *Quantitative Models for Supply Chain Management* (pp. 299–336). https://doi.org/10.1007/978-1-4615-4949-9_10.

CHAPTER 13

Revolutionizing HR: Blockchain Unleashed for a Bright Future

LEENA1,* and SHOBHNA SIROHI[2]

[1]*Banasthali Vidyapith, Jaipur, Rajasthan, India*

[2]*Tata Consultancy Services, Deutschland GmbH, Germany*

*Corresponding author

ABSTRACT

Blockchain is it something that can add worth in terms of security and transparency in Human Resource functions? Does this technology have the potential to create an impact or incorporate a change in the way we work or do business? Many more questions flooded as soon as blockchain technology is linked with Human Resource Management. So, to answer all questions, we first need to understand the concept of Blockchain technology. Blockchain is a digitally encrypted ledger that stores information in groups called blocks. Blocks possess specific storage capacities, and once they reach full capacity, they are sealed and connected to the preceding filled block, creating a sequence of data referred to as the blockchain. Any recent data added after this new block is compiled into a newly created block, which, upon reaching full capacity, becomes part of the chain as well. It is a type of decentralized shared database system whose main purpose is to remove third-party validation, making it a trust protocol. The objective of blockchain is to enable the recording and distribution of digital information without allowing any edits, a concept also referred to as distributed ledger technology. Incorporating this technology into HR functions could be of great help in various domains such

Blockchain Technology: Unlocking the Value of 3Ts in the Business World.
Richa Chauhan & Vidhi Kaul (Eds.)

DOI: 10.1201/9781998511914-13

as payroll, recruitment, contract management, employee database, and so on. It may help to reduce fraud, reduce chapter work, and facilitate easier identity verification, which could help in smoother payment facilitation to employees, and so on.

Along with all benefits, there are many challenges in the adoption of blockchain in HR, but as of now, due to its infancy, it is still not a widespread technology to address the challenges being faced in different aspects of HR, especially in transparency and security of data. Although blockchains are enabled with various existing and emerging technologies like encryption and peer-to-peer networking, there is still a lack of a capable supporting system.

This chapter will unfold various functions of HR where blockchain technology can be used, explaining in detail how it could be of use to the organization and what contribution it could provide to the productivity and goals of organizations.

13.1 INTRODUCTION

Blockchain is still a nascent technology for HR, so the big challenge is where to begin with it. It is worth giving it a try with the help of IT leaders on how it can be incorporated into HR functions like recruitment, payroll, employee engagement, compensation, and benefits, and so on. Many organizations have already started working towards this goal of imbibing blockchain technology into HR domains for effective performance management and goal achievement. Still, a lot is to be known about it and to be worked on for the desired outcome based on accurate data free from all redundancies and protected from fraudulent activities. Blockchain has enormous potential in Human Resource Management functions; in the near future, it could be one of the greatest transformations in the way we do business and deal with employees (i.e., human resources).

This technology is a value addition that will benefit both employees and employers at both individual and organizational levels. It may be helpful to gain productivity in business with transparency and data security at every level. HR functions can be more convenient and cost-effective with the help of blockchain technology due to its distributed nature of database and security features (Ramachandran et al., 2023).

Blockchain technology can significantly enhance various functions within Human Resources (HR) departments, revolutionizing the way organizations manage their workforce and streamline operations. Let's explore how blockchain can be applied to different HR functions and the contributions it can make to the productivity and goals of organizations.

13.2 THE VARIOUS MANAGERIAL APPLICATIONS OF BLOCKCHAIN

1. **Secure Employee Data Management:** Blockchain's decentralized and immutable nature ensures that employee data, including personal information, certifications, and work history, is securely stored. By utilizing blockchain for data management, organizations can minimize the risk of data breaches and unauthorized access. Improved data security enhances employees' trust in the organization, leading to increased productivity and loyalty. Additionally, HR professionals can efficiently access and verify employee records, reducing administrative burdens and enhancing overall efficiency.
2. **Efficient Payroll Management:** Blockchain-powered smart contracts can automate payroll processes, ensuring timely and accurate salary payments to employees. Smart contracts for payroll can be coded to carry out payments according to predetermined conditions, like logged work hours or specific dates. This automation eradicates the necessity for manual payroll processing, thereby conserving time and resources for HR departments. Precise and timely salary payments additionally enhance the work environment and employee contentment, fostering increased productivity and motivation.
3. **Transparent Performance Evaluation:** By recording performance metrics on the blockchain, organizations can create a transparent and immutable performance evaluation system. Employees' achievements and contributions are securely recorded, providing an objective basis for performance reviews. This transparency fosters a culture of accountability and fairness, encouraging employees to strive for continuous improvement. Transparent performance evaluations lead to higher employee engagement and productivity, as employees understand their strengths and areas for development.

4. **Streamlined Recruitment Process:** Blockchain can streamline the recruitment process by creating a decentralized and tamper-proof repository of candidate qualifications and work history. Job seekers can upload their credentials and employment records onto the blockchain, enabling HR departments to access verified information directly. This eliminates the need for time-consuming and expensive third-party verification processes, reducing hiring timeframes and improving the efficiency of talent acquisition. Faster and more accurate recruitment processes lead to quicker onboarding and greater workforce productivity.
5. **Credential Verification and Compliance:** Blockchain technology can facilitate the verification of educational credentials, certifications, and licenses of job applicants and existing employees. Educational institutions and certification bodies can issue digital credentials directly onto the blockchain, ensuring their authenticity and validity. This streamlines the verification process for HR departments and reduces the risk of hiring candidates with fraudulent qualifications. Compliance with regulatory requirements is also simplified through blockchain, enhancing the organization's reputation and avoiding legal complications.
6. **Decentralized Employee Benefit Management:** Blockchain's decentralized architecture allows for efficient and transparent management of employee benefits. Smart contracts can be programmed to automatically distribute benefits, such as health insurance, retirement contributions, and bonuses, based on predefined rules and eligibility criteria. Employees can access real-time information about their benefits, promoting a sense of empowerment and reducing the administrative burden on HR personnel. A well-managed benefits system contributes to employee satisfaction and loyalty, increasing overall productivity.
7. **Enhanced Talent Development and Training:** Blockchain can facilitate the tracking and verification of employees' professional development and training activities. HR departments can record training completion on the blockchain, ensuring a verifiable and tamper-proof record of employees' skills and certifications. This information aids in talent development and succession planning, as HR can identify skill gaps and provide targeted training

opportunities. A skilled and competent workforce enhances organizational productivity and drives the achievement of business goals.

8. **Decentralized Identity and Access Management:** Blockchain's identity management capabilities can be utilized to enhance security in HR processes. Each employee can possess an individual digital identity recorded on the blockchain, managing entry to confidential company assets and information. This minimizes the likelihood of identity theft and unauthorized entry, guaranteeing that solely authorized staff can access designated information or systems. Improved data security and regulated access play a role in fostering a more efficient and productive work setting.

Blockchain technology has emerged as a transformative influence, challenging conventional business practices in diverse industries. Within the domain of Human Resource Management (HRM), there is growing attention to Blockchain's potential for reshaping and streamlining processes. HR professionals are recognizing the transformative power of blockchain in areas such as talent recruitment, employee management, payroll processing, credential verification, and talent development. By offering a decentralized, secure, and transparent infrastructure, blockchain has the potential to revolutionize the way HR departments operate, leading to enhanced efficiency, data integrity, and overall employee experience (Agarwal, 2023).

In this chapter, we will explore the myriad functions and applications of blockchain in HRM. We will delve into how blockchain technology can be harnessed to streamline recruitment processes, ensuring accurate and tamper-proof credential verification and compliance. Furthermore, we will discuss how blockchain can empower employees by giving them greater control over their digital identities and access to training and development opportunities tailored to their specific needs. From enhancing data security to enabling real-time payroll processing, the integration of blockchain in HRM promises to unlock a new era of transparency, efficiency, and trust in the workplace.

As the HR landscape continues to evolve, embracing blockchain technology becomes imperative for organizations seeking to stay ahead in the competition for top talent and create a more agile and employee-centric work environment. By understanding the potential and challenges of integrating blockchain in HRM, organizations can chart a course towards a future where talent management is optimized, compliance is streamlined, and employees are empowered to take charge of their professional growth. Let us embark on this exploration of blockchain in HRM and discover the innovative possibilities that lie ahead in transforming the way we manage and nurture our most valuable asset: our people (Mohammad Saif & Islam, 2022).

13.3 PRACTICAL IMPLICATIONS OF BLOCKCHAIN IN HR FUNCTIONS

1. **Secure Employee Data Management:** This is a critical aspect of HR operations in any organization. It involves the storage, handling, and protection of sensitive employee information, such as personal details, financial data, medical records, performance evaluations, and other confidential documents. Ensuring the security and integrity of employee data is essential to comply with data protection regulations, maintain trust among employees, and safeguard the organization from potential data breaches and legal consequences.

   Traditionally, HR departments have relied on centralized databases and file systems to manage employee data. However, these centralized systems are vulnerable to cyber-attacks and unauthorized access, making them a prime target for data breaches. As cyber threats continue to evolve, organizations must adopt more robust and secure methods to manage their employee data. This is where blockchain technology comes into play (Zahedi, 2023).

   i. **Decentralization and Immutability:** A fundamental characteristic of blockchain technology is its decentralized design. Rather than housing data in a singular central repository, blockchain disperses information across a network of nodes, with each node possessing a duplicate of the entire dataset. This decentralized configuration mitigates the threat of a single point of failure, heightening the complexity for malicious actors attempting to compromise the data.

      Moreover, the immutability inherent in blockchain means that once data is incorporated into the chain, it becomes unalterable and impervious to deletion. This feature prevents unauthorized modifications to employee records, preserving the data's integrity and establishing a reliable source of information.

ii. **Encryption and Data Privacy:** The information about employees stored on the blockchain undergoes encryption through cryptographic algorithms. This ensures that solely individuals with the proper private keys, who are authorized, can access and decipher the data. This robust encryption enhances the privacy of data and shields sensitive employee information from any unauthorized access.

iii. **Secure Access Control:** Blockchain technology facilitates the deployment of smart contracts—self-executing contracts with predetermined conditions. These contracts have the capability to regulate access to particular employee data, guaranteeing that only pertinent parties with appropriate authorization can retrieve specific information. For example, HR managers may have access to an employee's performance evaluations, while payroll administrators can access salary-related data. Smart contracts can be programmed to manage access rights dynamically, maintaining data privacy and security.

iv. **Consent Management:** In some cases, employees may need to provide consent for the use and sharing of their personal data. Blockchain can facilitate consent management by allowing employees to grant and revoke consent through smart contracts. This ensures that data usage is transparent and compliant with data protection regulations, such as the GDPR.

v. **Data Ownership and Portability:** With blockchain, employees can retain ownership of their personal data while granting permission to share it with relevant parties. This empowers employees with more control over their data and facilitates data portability when changing jobs or working in the gig economy.

vi. **Auditability and Transparency:** Every transaction and change made to employee data on the blockchain is recorded in a transparent and traceable manner. This audit trail enables

HR departments to track any modifications, ensuring accountability and transparency. In case of disputes or compliance audits, the blockchain provides a reliable source of truth for employee data records.

vii. **Integration with Existing Systems:** Blockchain technology can be integrated with existing HR systems, complementing and enhancing their capabilities. It can act as an additional layer of security, ensuring the integrity of critical data.

2. **Efficient Payroll Management:** It is a critical function within HR departments, as it involves the accurate and timely calculation and distribution of employee salaries, bonuses, benefits, and deductions. The process of payroll management can be complex and time-consuming, involving numerous calculations and compliance with various tax and labor regulations. Blockchain technology offers several advantages that can significantly streamline and enhance the efficiency of payroll management processes (Godavarthi et al., 2023).

   i. **Automated Payroll Processing:** The most impactful capability of blockchain for effective payroll management lies in its capacity to carry out smart contracts. These are contracts that self-execute based on predetermined conditions. In the context of payroll, smart contracts have the capability to be programmed for the automatic initiation of salary payments to employees upon the fulfillment of certain conditions, such as the conclusion of the pay period.

   Once the conditions are fulfilled, the smart contract automatically triggers the payment process, eliminating the need for manual intervention in payroll processing. This automation not only saves time and reduces administrative overhead but also ensures precise and timely payments to employees. This may:

   ii. **Reduced Human Errors:** With traditional payroll systems, human errors can occur during data entry, calculation, or record-keeping. Such errors can lead to incorrect salary calculations and cause dissatisfaction among employees. Blockchain-based payroll systems significantly reduce the risk of human errors due to their automated and immutable nature. After information is registered on the blockchain, it remains immutable and resistant to alterations or tampering, thereby guaranteeing the precision and integrity of payroll records.

iii. **Enhanced Security and Data Privacy:** Payroll information encompasses sensitive details like financial data, social security numbers, and other personal information. Security breaches in payroll systems can lead to serious repercussions, including financial losses and harm to the organization's reputation. The decentralized and encrypted nature of blockchain offers elevated security and data privacy. Employee information stored on the blockchain is encrypted, and access can be controlled via smart contracts, ensuring that only authorized individuals can access specific details. This substantially diminishes the likelihood of data breaches and unauthorized entry to payroll information.

iv. **Real-Time Tracking and Reporting:** Blockchain-based payroll systems provide real-time tracking of payroll transactions. HR departments can monitor salary payments as they are executed on the blockchain, allowing for instant verification of transactions and payroll disbursements. Real-time reporting capabilities enable HR professionals to access up-to-date payroll information, facilitating financial planning and budgeting for the organization.

v. **Compliance with Tax and Labor Regulations:** Compliance with tax regulations and labor laws is a critical aspect of payroll management. Failure to comply with these regulations can lead to legal penalties and financial losses. Blockchain can integrate smart contracts with tax and labor regulations, ensuring that payroll calculations are automatically adjusted based on the latest rules and tax rates. This reduces the likelihood of non-compliance and eases the burden on HR departments to stay updated with changing regulations.

vi. **Streamlined Cross-border Payments:** For organizations with a global workforce or international operations, processing cross-border payments can be complex and costly. Blockchain technology can facilitate streamlined cross-border payroll transactions by eliminating intermediaries and currency conversion fees. Additionally, the transparency of blockchain transactions enables both employees and the organization to track payments through each stage of the process, promoting trust and transparency in cross-border payroll management.

vii. **Cost Savings:** Blockchain-powered payroll systems can lead to cost savings for organizations. Automating payroll procedures diminishes the reliance on manual labor and administrative duties, allowing HR personnel to dedicate their time to other strategic tasks. Furthermore, the heightened security and precision offered by blockchain-based systems decrease expenses related to error correction and the handling of data breaches.

3. **Transparent Performance Evaluation:** It is an approach to assessing employee performance that promotes openness, fairness, and accountability within an organization? It involves creating a system where the performance evaluation process and its outcomes are clear and visible to both employees and management. Implementing transparent performance evaluation can have numerous benefits for an organization, including improved employee engagement, increased productivity, and a stronger organizational culture (Deepa, 2023).

i. **Clear Performance Metrics and Criteria:** Transparent performance evaluation begins with clearly defined performance metrics and criteria. Employees should have a clear understanding of the key performance indicators (KPIs) against which their performance will be evaluated. This clarity ensures that employees know what is expected of them, enabling them to align their efforts with organizational goals and objectives.

ii. **Access to Performance Data:** The utilization of blockchain technology can have a notable impact on facilitating transparent performance assessments. When performance metrics and feedback are documented on the blockchain, the information becomes available to pertinent parties, including both employees and managers. This transparency guarantees that the evaluation process relies on objective and verifiable data rather than subjective opinions.

iii. **Real-Time Feedback and Tracking:** Blockchain-based systems can provide real-time feedback on employee performance. As employees achieve milestones or complete tasks, the progress can be recorded on the blockchain. This allows employees to track their performance and see how their efforts contribute to the organization's overall goals. Regular feedback helps employees stay motivated, identify areas for improvement, and make necessary adjustments to meet performance expectations.

iv. **Fair and Objective Evaluations:** The transparency of performance evaluations aids in eradicating biases and favoritism, common pitfalls in conventional evaluation procedures. Since blockchain records are unchangeable, once performance data is documented, it remains unaltered and resistant to tampering. This guarantees that assessments rely on precise and impartial data, resulting in fair and objective evaluations.

v. **Employee Development and Recognition:** Transparent performance evaluation provides valuable insights into employee strengths and areas for improvement. Managers can use this data to offer targeted coaching and development opportunities to help employees grow professionally. Similarly, when employees consistently perform well, transparent evaluation systems allow for timely recognition and rewards, promoting a culture of appreciation and motivation.

vi. **Building Trust and Engagement:** When employees have visibility into the evaluation process and the criteria used to assess their performance, they feel more involved and empowered. Transparent evaluation fosters trust between employees and management, as employees can see that their performance is evaluated based on measurable results. Trust and engagement are essential for a positive work environment, which, in turn, leads to higher job satisfaction and increased productivity.

vii. **Supporting Performance-based Compensation:** Transparent performance evaluation enables organizations to implement performance-based compensation structures. With clear performance data available on the blockchain, organizations can link pay and rewards directly to employee achievements and contributions. This encourages employees to excel in their performance and aligns their efforts with the goals of the organization.

viii. **Facilitating Performance Reviews and Talent Management:** With a transparent performance evaluation system, performance reviews become more structured and data driven. Managers can use blockchain-based data to conduct meaningful performance discussions and provide evidence-based feedback. Additionally, transparent evaluation data aids in talent management processes, such as identifying high-potential employees and making informed decisions about promotions and career development (Kral et al., 2022).

4. **A Streamlined Recruitment Process:** It is crucial for attracting and hiring top talent efficiently. It involves optimizing each step of the recruitment journey to reduce time-to-hire, attract qualified candidates, and enhance the overall candidate experience. By leveraging blockchain technology, organizations can revolutionize their recruitment process, making it more transparent, secure, and cost-effective (Fachrunnisa & Hussain, 2020).

   i. **Decentralized Candidate Database:** Blockchain allows organizations to create a decentralized candidate database where job seekers can securely upload their qualifications, work history, and other relevant information. The distribution of candidate data across the network, facilitated by its decentralized nature, diminishes the likelihood of data breaches and unauthorized access. HR professionals can access this database to find potential candidates for open positions, eliminating the need for multiple job boards or recruitment agencies.

ii. **Verified Credentials and Background Checks:** Candidates can upload their educational credentials, certifications, and work history directly onto the blockchain. These records are time-stamped and cannot be altered, ensuring their authenticity. This removes the necessity for manual verification and background checks, leading to time savings and diminished administrative responsibilities for HR departments. Hiring managers can confidently rely on blockchain-verified credentials, streamlining the candidate selection process.

iii. **Automated Candidate Matching:** Using smart contracts, organizations can create automated candidate matching systems. Smart contracts can be programmed to match candidate qualifications with the requirements of specific job openings. When a job is posted, the smart contract can automatically identify and notify potential candidates whose skills and experience match the position. This automated process accelerates the recruitment process and improves the chances of finding the right candidates quickly.

iv. **Efficient Candidate Communication:** Blockchain technology allows for secure and encrypted communication between candidates and HR professionals. Candidates can receive notifications and updates on their application status through the blockchain platform, ensuring that they stay informed throughout the recruitment process. Additionally, smart contracts can automate responses to common candidate inquiries, providing a seamless communication experience.

v. **Elimination of Duplicate Applications:** With a decentralized candidate database, candidates' unique digital identities are recorded on the blockchain. This prevents individuals from submitting duplicate applications or misrepresenting themselves under multiple identities. Duplicate applications are common in traditional recruitment systems and can be time-consuming to manage. Blockchain mitigates this issue, streamlining the candidate pool and ensuring each candidate is represented accurately.

vi. **Accelerated Reference Checking:** Reference checking is an essential part of the recruitment process but can be time-consuming and resource-intensive. Blockchain technology simplifies reference checking by providing a secure and verifiable record of candidate work history and qualifications. HR professionals can access this information quickly, reducing the time required for reference checks and accelerating the decision-making process.

vii. **Candidate Feedback and Ratings:** Decentralized candidate databases can also facilitate candidate feedback and ratings systems. Organizations can implement a transparent feedback mechanism where previous employers and colleagues can provide ratings and reviews of candidates they have worked with in the past. These ratings can help HR professionals make informed decisions and assess the cultural fit of potential candidates.

viii. **Enhanced Employer Branding:** Implementing a streamlined and technologically advanced recruitment process, such as one powered by blockchain, enhances an organization's employer branding. Candidates are more likely to perceive the organization as innovative, efficient, and employee-focused, which can attract top talent. Positive employer branding can also lead to increased application rates and higher candidate quality.

5. **Credential Verification and Compliance:** These are essential components of the hiring and talent management process within organizations. Credential verification ensures that the qualifications, certifications, and licenses presented by job applicants and employees are legitimate and accurate. Compliance, on the other hand, refers to adhering to regulatory requirements and industry standards related to employee qualifications and certifications. Blockchain technology can play a significant role in streamlining and enhancing the credential verification and compliance processes (Zhu et al., 2023).

   i. **Immutable and Verified Records:** The immutable quality of blockchain guarantees that credentials, once documented, remain impervious to alteration or tampering. This characteristic establishes a reliable and tamper-proof repository for verified credentials. Educational institutions, certification bodies, and licensing authorities have the capability to issue digital certificates and credentials directly onto the blockchain.

Job applicants and employees can then present their blockchain-verified credentials, providing HR departments with a reliable and verified source of information.

ii. **Reduction of Fraudulent Credentials:** Fraudulent credentials are a significant concern in the hiring process. Traditional paper-based certificates and documents can be easily forged or manipulated. By storing credentials on the blockchain, organizations can significantly reduce the risk of hiring candidates with fake qualifications. HR professionals can access the blockchain to verify the authenticity of the presented credentials, ensuring that only qualified candidates are considered for positions.

iii. **Efficient Verification Process:** The traditional process of credential verification involves contacting educational institutions, certification bodies, and licensing authorities to verify the authenticity of candidates' qualifications. This process can be time-consuming and require extensive communication between different parties. Blockchain streamlines this process by providing a single source of truth for verified credentials. HR departments can access the blockchain to obtain instant and verifiable proof of candidates' qualifications, reducing the time and effort required for verification.

iv. **Compliance with Industry Standards and Regulations:** Many industries have strict compliance requirements related to employee qualifications and certifications. For example, healthcare organizations must ensure that medical professionals possess valid licenses and certifications. Blockchain can be integrated with industry-specific regulatory frameworks, automating compliance checks based on predefined rules. This ensures that employees meet the necessary qualifications and comply with relevant regulations, reducing the risk of legal and compliance issues.

v. **Decentralized Access to Credentials:** With blockchain, candidates have greater control over their credentials. They can choose to share specific credentials with employers or other parties while keeping the rest of their information private. This decentralized access control allows candidates to maintain ownership of their credentials, promoting privacy and data ownership rights.

vi. **Real-Time Updates and Renewals:** Credentials on the blockchain can be linked to expiration dates and renewal requirements. HR departments can receive real-time updates about the validity of employee certifications, ensuring that employees' qualifications are up-to-date and compliant. This feature is particularly valuable in industries where certifications must be periodically renewed.

vii. **Global Talent Management:** For organizations with a global workforce, blockchain-enabled credential verification facilitates the recognition of international qualifications. Educational credentials and certifications from different countries can be verified through a single decentralized platform, streamlining the process of hiring and managing talent from diverse geographical locations.

viii. **Auditable Records for Compliance Checks:** Blockchain maintains an immutable audit trail of all credential verification processes. This feature is valuable during compliance audits, as HR departments can provide auditors with a transparent and traceable record of credential verification activities, ensuring compliance with relevant regulations.

6. **Decentralized Employee Benefit Management:** Leverages blockchain technology to create a transparent, secure, and efficient system for administering employee benefits. Employee benefits are an integral part of compensation packages, including health insurance, retirement plans, bonuses, and other perks. By utilizing blockchain, organizations can streamline benefit distribution,

enhance data security, and provide employees with greater control over their benefits (Dash, 2023).

i. **Transparent Benefit Allocation:** Blockchain's decentralized nature enables a transparent system for benefit allocation. HR departments can use smart contracts to define the rules and conditions for benefit distribution. For instance, once an employee becomes eligible for specific benefits, such as health insurance coverage or retirement contributions, the smart contract automatically executes the allocation without the need for manual intervention. This transparency ensures that eligible employees receive their entitled benefits, promoting trust and employee satisfaction.

ii. **Secure Benefit Data Storage:** Employee benefit data contains sensitive and confidential information, such as medical records and financial details. Storing this data on a centralized server poses security risks, making it susceptible to data breaches and unauthorized access. Blockchain's decentralized and encrypted structure enhances data security, ensuring that employee benefit information is distributed across the network and accessible only to authorized personnel. This heightened security mitigates the risk of data breaches and protects employee privacy.

iii. **Efficient Benefit Enrollment and Management:** Blockchain-based systems facilitate streamlined benefit enrollment and management processes. Employees can access their benefits and enrollment status on the blockchain platform, enabling self-service functionalities. HR departments can easily track benefit enrollments, changes, and updates in real-time, eliminating the need for manual paperwork and reducing administrative burdens. This efficiency leads to faster and more accurate benefit management for both HR professionals and employees.

iv. **Automated Benefit Distribution:** Blockchain-powered smart contracts automate benefit distribution based on predefined criteria and eligibility rules. For instance, when an employee reaches a certain tenure milestone, the smart contract may trigger a bonus or performance-based incentive. This automation ensures that employees receive their benefits promptly and accurately, contributing to employee satisfaction and loyalty.

v. **Employee Empowerment and Choice:** Decentralized employee benefit management empowers employees with more control over their benefits. Using blockchain, employees can access and manage their benefit information directly, making decisions that best suit their individual needs. Employees can choose from various benefit options, modify their coverage, or enroll in specific plans according to their preferences. This level of choice and autonomy improves employee engagement and satisfaction.

vi. **Accurate and Transparent Benefit Records:** Blockchain maintains a tamper-proof and immutable record of all benefit transactions and changes. This audit trail ensures the accuracy and integrity of benefit records, making it easier to track and verify historical benefit data. This transparency also facilitates compliance with regulatory requirements and provides a reliable source of information during audits.

vii. **Global Workforce Management:** For organizations with a global workforce, decentralized benefit management ensures a standardized and consistent approach to benefit distribution across various geographical locations. Blockchain eliminates the complexities associated with managing benefits in different regions and currencies, providing a unified and efficient system for HR departments to handle benefits for all employees.

viii. **Automated Compliance Checks:** Blockchain can be integrated with compliance regulations, ensuring that benefit distribution adheres to legal requirements. Smart contracts can be programmed to verify employees' eligibility for specific benefits based on regulatory guidelines. This automation streamlines compliance checks and reduces the risk of errors or non-compliance.

7. **Enhanced Talent Development and Training:** Are essential components of a successful talent management strategy. It involves providing employees with opportunities to develop their skills, acquire new knowledge, and enhance their competencies to contribute effectively to the organization's success. Blockchain technology can have a substantial impact on enhancing the effectiveness of talent development and training programs, rendering them more individualized, accessible, and transparent (Jain et al., 2021).

   i. **Personalized Learning Paths:** Blockchain technology empowers HR departments to devise customized learning journeys for employees, considering their unique skills, interests, and career aspirations. By documenting employees' training records, certifications, and performance data on the blockchain, HR professionals can pinpoint skill gaps and suggest focused training initiatives. This individualized strategy guarantees that employees undergo training pertinent to their roles and future career ambitions, thereby optimizing the effectiveness of talent development initiatives.

   ii. **Immutable Learning Records:** Blockchain's immutability ensures that learning records, certifications, and achievements are securely stored and cannot be altered or manipulated. This feature provides a verifiable and trustworthy source of employees' training accomplishments, making it easier to track their progress and share credentials with stakeholders. Immutable learning records also simplify compliance audits and regulatory requirements, ensuring that the organization maintains accurate and transparent training records.

   iii. **Competency-based Learning and Certification:** Blockchain can be leveraged to implement competency-based learning and certification systems. Instead of relying solely on traditional academic qualifications, organizations can record and verify employees' practical skills and achievements on the blockchain. This approach enables HR departments to identify high-performing employees and provide them with advancement opportunities based on their demonstrated competencies.

iv. **Micro-Credentials and Badges:** Blockchain allows for the issuance of micro-credentials and badges, which are smaller, focused certifications that validate specific skills or achievements. Employees can earn micro-credentials for completing short courses, mastering particular competencies, or contributing to specific projects. These badges are recorded on the blockchain, enabling employees to showcase their skills and accomplishments more effectively.

v. **Learning and Collaboration Incentives:** Blockchain technology can be integrated into employee incentive programs, rewarding employees for their engagement in learning and collaborative activities. Smart contracts can be programmed to trigger incentives automatically when employees complete training modules, share knowledge with peers, or contribute to collaborative projects. These incentives create a culture of continuous learning and knowledge-sharing, fostering employee growth and development.

vi. **Global Talent Development and Accessibility:** For organizations with a geographically dispersed workforce, blockchain-based talent development platforms provide a standardized and accessible training environment. Employees from different locations can access the same training materials and resources on the blockchain, promoting consistent learning experiences and eliminating geographical barriers to talent development.

vii. **Enhanced Learning Analytics:** Blockchain-based learning platforms provide enhanced learning analytics and insights. HR departments can track employees' learning progress, identify patterns in training engagement, and evaluate the effectiveness of training programs. These data-driven insights enable continuous improvement in talent development initiatives and help HR professionals tailor training content to meet employees' evolving needs.

viii. **Peer-to-Peer Learning and Collaboration:** Blockchain facilitates peer-to-peer learning and collaboration through decentralized networks. Employees can share knowledge, experiences, and best practices with their peers on the blockchain platform. This decentralized sharing of knowledge enhances organizational learning and promotes a culture of knowledge exchange and innovation.

8. **Decentralized Identity and Access Management (IAM):** Is an innovative approach to managing user identities and controlling access to resources and data within an organization. Traditional IAM systems typically rely on centralized databases or third-party identity providers, which can present security and privacy challenges. Decentralized IAM, powered by blockchain technology, offers a more secure, transparent, and user-centric solution. It gives individuals greater control over their digital identities while ensuring data privacy and reducing the risk of identity-related cyber-attacks (Sharif & Ghodoosi, 2022).

   i. **Self-Sovereign Identity:** Decentralized IAM enables the concept of self-sovereign identity, where individuals have full control over their digital identities. With blockchain, users can create a unique and tamper-proof digital identity that is under their ownership and control. This identity is cryptographically secured, ensuring that only the user possesses the private keys needed to access and manage their identity information.

   ii. **Elimination of Centralized Authorities:** Decentralized IAM eliminates the need for centralized identity providers, reducing the dependency on single points of failure. Each user's identity information is stored and managed on the blockchain in a distributed and decentralized manner. This ensures that no single entity has complete control over user identities, reducing the risk of data breaches and identity theft.

   iii. **Enhanced Security and Privacy:** Blockchain's encryption and consensus mechanisms enhance the security and privacy of user identities. Identity information stored on the blockchain is encrypted, and access is controlled by the individual through their private keys. The consensus mechanism ensures that any changes to identity records must be agreed upon by multiple nodes on the network, preventing unauthorized modifications.

iv. **User-controlled Access Permissions:** With decentralized IAM, users have granular control over access permissions. They can provide consent for specific applications or services to access certain aspects of their identity data. By adopting this approach, user information is made accessible only to authorized parties, thereby improving data privacy and reducing the potential for data misuse.

v. **Decentralized Single Sign-On (SSO):** Blockchain-powered IAM enables decentralized SSO capabilities. Once a user's identity is authenticated and verified on the blockchain, they can use their digital identity to access multiple services and applications without the need for separate logins. This simplifies the authentication process for users and reduces the burden of managing multiple credentials.

vi. **Identity Portability and Interoperability:** Decentralized IAM promotes identity portability and interoperability across different platforms and organizations. Users can use their blockchain-based identity to access services from various providers seamlessly. This fosters a more user-centric experience and allows users to retain control of their identity information across different contexts.

vii. **Auditability and Compliance:** The transparency and unalterable characteristics of blockchain guarantee the recording of all identity-related transactions and modifications on the blockchain. This establishes a dependable audit trail for identity management activities, simplifying compliance with regulatory mandates and showcasing adherence to data protection standards.

viii. **Revocation and Recovery of Identity:** In case of compromised credentials or lost private keys, decentralized IAM provides mechanisms for identity revocation and recovery. Users can revoke access permissions from compromised devices or recover access to their identity by using predefined recovery mechanisms. This enhances the overall security and resilience of the IAM system (Zeadally & Abdo, 2019).

13.4 DISCUSSIONS AND RESULTS

Integrating blockchain technology into HR functions has the potential to revolutionize how organizations manage their workforce. From secure data management to efficient payroll processing and talent development, blockchain brings transparency, security, and efficiency to HR processes, ultimately contributing to increased productivity and the achievement of organizational goals. However, while the benefits of blockchain in HR are promising, it is essential for organizations to carefully evaluate the technology's implementation and address any challenges related to adoption and integration. With strategic planning and consideration, blockchain can undoubtedly play a transformative role in shaping the future of HR management.

Blockchain technology offers a wide range of applications in HR, presenting numerous benefits for organizations seeking to improve their workforce management and overall productivity, such as:

1. **Data Security and Trust:** One of the primary advantages of using blockchain in HR is the enhanced data security and trust it provides. The decentralized and immutable nature of blockchain ensures that sensitive employee data remains secure and tamper-proof. This heightened security builds trust among employees, leading to increased productivity and loyalty.

2. **Automation and Efficiency:** The integration of smart contracts in HR processes enables automation, reducing manual administrative tasks and saving time and resources. Automated payroll management and benefit distribution, for example, streamline HR operations, leading to increased efficiency within the department.

3. **Transparency and Fairness:** Transparent performance evaluations and recruitment processes based on blockchain records promote fairness and accountability. Employees appreciate knowing that their performance evaluations are based on objective data, leading to higher engagement and motivation.

4. **Cost Savings and Resource Optimization:** By leveraging blockchain for recruitment and credential verification, organizations can save costs associated with third-party verification services and reduce the time spent on manual verification processes. This

allows HR professionals to focus on strategic talent management initiatives.

5. **Employee Empowerment:** Blockchain's self-sovereign identity concept empowers employees to have greater control over their digital identities and data. This empowerment enhances the overall employee experience and fosters a sense of ownership and responsibility within the workforce.
6. **Compliance and Reputation:** Blockchain simplifies compliance efforts by providing a transparent and auditable record of HR activities. This ensures that organizations meet regulatory requirements, contributing to their reputation as responsible employers.
7. **Data-Driven Talent Development:** The ability to track and verify employees' training and development on the blockchain enables data-driven talent development strategies. HR departments can identify skill gaps and offer targeted training, aligning employees' skills with organizational goals.
8. **Streamlined Onboarding:** Blockchain-based recruitment and credential verification processes lead to faster onboarding of new employees, reducing time-to-hire and enabling them to contribute to the organization's productivity sooner.
9. **Enhanced Collaboration:** The decentralized and transparent qualities of blockchain can support collaborative interactions and the exchange of knowledge among peers within the organization. This fosters a culture of ongoing learning and innovation, ultimately leading to increased productivity.
10. **Global Workforce Management:** Blockchain-enabled HR systems simplify talent management for organizations with a global workforce. The unified and standardized approach to identity and access management enables consistent HR practices across different regions.

13.5 CONCLUSION

Incorporating blockchain technology into HR holds the promise of transforming how organizations oversee their workforce and enhance

HR procedures. Blockchain offers a myriad of benefits that can significantly contribute to achieving organizational goals and enhancing overall productivity.

The enhanced data security and trust provided by blockchain's decentralized and immutable nature create a safe and tamper-proof environment for storing sensitive employee data. This fosters trust among employees, leading to increased loyalty and productivity within the organization.

By leveraging smart contracts, HR departments can automate repetitive tasks, such as payroll management and benefit distribution, resulting in increased efficiency and resource optimization. Employees benefit from transparent performance evaluations and recruitment processes based on objective data, promoting fairness and accountability, thereby boosting engagement and motivation.

Moreover, organizations can save costs and streamline processes by utilizing blockchain for recruitment and credential verification, reducing reliance on third-party verification services and manual administrative tasks.

The concept of self-sovereign identity empowers employees to have greater control over their digital identities and data, enhancing the overall employee experience and instilling a sense of ownership and responsibility.

Blockchain's transparent and auditable nature ensures compliance with regulatory requirements, contributing to the organization's reputation as a responsible employer.

HR departments can harness the data-driven talent development opportunities offered by blockchain, enabling targeted training initiatives that address skill gaps and align employees' competencies with organizational objectives.

Blockchain-based recruitment and verification processes lead to faster onboarding of new employees, allowing them to contribute to the organization's productivity more rapidly. The decentralized and transparent nature of blockchain fosters a collaborative work environment, promoting continuous learning and innovation through peer-to-peer knowledge sharing.

For organizations with a global workforce, blockchain-enabled HR systems simplify talent management by providing a unified and standardized approach to identity and access management across different regions.

Blockchain technology represents a game-changing opportunity for HR departments to elevate their practices, streamline operations, and empower employees. As organizations continue to explore the potential of blockchain in HR, careful planning, integration, and consideration of regulatory frameworks will be crucial in fully realizing the transformative benefits that this technology has to offer. Embracing blockchain in HR will undoubtedly shape the future of talent management and contribute to the success and growth of organizations in an increasingly competitive and dynamic business landscape.

KEYWORDS

- **automation**
- **blockchain**
- **data security**
- **global work force management**
- **human resources**
- **payroll management**
- **resource optimization**
- **smart contracts**
- **workforce management**

REFERENCES

Agarwal, A. (2023). Use cases of blockchain technology for sustainable global HR operations in industry 4.0. *Blockchain in a Volatile-Uncertain-Complex-Ambiguous World*, 149–161.

Dash, S. P. (2023). HR digital transformation: Blockchain for business. In *Recent Advances in Blockchain Technology: Real-World Applications* (pp. 59–87). Cham: Springer International Publishing.

Deepa, R. (2023). The application of blockchain in talent supply chain management. In *Blockchain in a Volatile-Uncertain-Complex-Ambiguous World* (pp. 121–139). Elsevier.

Fachrunnisa, O., & Hussain, F. K. (2020). Blockchain-based human resource management practices for mitigating skills and competencies gap in workforce. *International Journal of Engineering Business Management, 12*, 1847979020966400.

Godavarthi, B., Dhar, M., Devi, S. A., Raju, S. S., Balaram, A., & Srilakshmi, G. (2023). Blockchain integration with the internet of things for the employee performance management. *The Journal of High Technology Management Research, 34*(2), 100468.

Jain, G., Sharma, N., & Shrivastava, A. (2021). Enhancing training effectiveness for organizations through blockchain-enabled training effectiveness measurement (BETEM). *Journal of Organizational Change Management, 34*(2), 439–461.

Kral, P., Janoskova, K., & Dawson, A. (2022). Virtual skill acquisition, remote working tools, and employee engagement and retention on blockchain-based metaverse platforms. *Psychosociological Issues in Human Resource Management, 10*(1), 92–105.

Mohammad Saif, A. N., & Islam, M. A. (2022). Blockchain in human resource management: A systematic review and bibliometric analysis. *Technology Analysis & Strategic Management*, 1–16.

Ramachandran, R., Babu, V., & Murugesan, V. P. (2023). The role of blockchain technology in the process of decision-making in human resource management: A review and future research agenda. *Business Process Management Journal, 29*(1), 116–139.

Sharif, M. M., & Ghodoosi, F. (2022). The ethics of blockchain in organizations. *Journal of Business Ethics, 178*(4), 1009–1025.

Zahedi, M. R. (2023). The use of blockchain to improve human resource management in organizations, a case study of one of the country's universities. *Quarterly Journal of Industrial Technology Development.*

Zeadally, S., & Abdo, J. B. (2019). Blockchain: Trends and future opportunities. *Internet Technology Letters, 2*(6), e130.

Zhu, G., Gu, Z., & Dai, Y. (2023). Construction of a human resource sharing system based on blockchain technology. *Information, 14*(3), 177.

CHAPTER 14

A Systematic Literature Review of Determinants Affecting the Adoption of Blockchain Technology in Organizations: TOE Framework Perspective

DIVYA MISHRA[1] and NIDHI MAHESHWARI[2]

[1]*Assistant Professor, Vivekananda School of Business Studies, Vivekananda Institute of Professional Studies-Technical Campus, AU-Block (Outer Ring Road), Pitampura, Delhi, India*

[2]*Delhi Technological University, University School of Management and Entrepreneurship, New Delhi, India*

*Corresponding author

ABSTRACT

In the dynamic landscape of technological evolution, blockchain technology has emerged as a potent force with transformative potential, poised to reshape organizational foundations and redefine business transactions. This systematic literature review, employing the technological, organizational, and environmental (TOE) framework, meticulously examines the determinants influencing blockchain adoption within organizational contexts. The study refines the TOE framework for nuanced insights into the unique challenges and opportunities presented by blockchain. Emphasizing the intricate interplay among technological, organizational, and environmental factors, the research advocates a holistic approach to adoption success. Theoretical implications contribute to the evolving

Blockchain Technology: Unlocking the Value of 3Ts in the Business World.
Richa Chauhan & Vidhi Kaul (Eds.)

DOI: 10.1201/9781998511914-14

understanding of technology adoption, while practical insights guide decision-makers in strategic planning, resource allocation, and regulatory engagement. As organizations navigate the complexities of blockchain integration, this study provides a valuable foundation for informed decision-making and optimized implementation.

14.1 INTRODUCTION

In the rapidly evolving landscape of technological innovation, blockchain technology has emerged as a transformative force with the potential to reshape the foundations of organizations and redefine the dynamics of business transactions (Behnke and Janssen, 2019). The pervasive influence of blockchain has led to considerable discourse among industry experts and researchers, prompting a critical examination of the challenges faced by early adopters during this nascent phase of its evolution.

At its essence, a blockchain is a sequential arrangement of secure blocks that encapsulates data through hash functions, incorporating timestamps and links to previous blocks. This structure forms a distributed ledger, stored across multiple nodes, thereby mitigating centralized vulnerabilities exploited by cybercriminals. Beyond its fundamental role in data security, blockchain technology facilitates trustless transactions through tokenization and the automated execution of “smart contracts” under specified conditions (Marsal-Llacuna, 2018).

While the adoption of blockchain spans diverse industry sectors, its most pronounced impact has been witnessed in financial services (Hughes et al., 2019). Often synonymous with cryptocurrencies, blockchain introduces crypto tokens, specialized virtual currency units residing on dedicated blockchains, representing assets or utilities. These tokens serve a spectrum of purposes, ranging from cryptocurrencies to utility tokens and tokenized securities (Wood, 2014). The concept of a distributed ledger forms the backbone of blockchain applications, enabling information exchange among users beyond traditional database custodianship. Central to this paradigm are “smart contracts,” digital agreements that automate and enforce contractual obligations based on predefined conditions (Buterin, 2014).

In the dynamic landscape of blockchain development, various architectures have emerged for diverse applications. Initiatives such as Block.one’s EOS.io and the utilization of Hyperledger’s technology by industry leaders like American Express and IBM exemplify the vast potential of blockchain (Androulaki et al., 2018).

Introducing disruptive technologies like blockchain to any sector inevitably presents multifaceted challenges, spanning technical, regulatory, social, and adoption-related realms (Hughes et al., 2019). Despite assertions that blockchain inherently obviates the need for intermediaries, governance remains paramount for designing, operating, and maintaining these systems (Ølnes et al., 2017). Recognizing the nuanced landscape of blockchain technology, this chapter embarks on a systematic literature review, aiming to comprehensively analyze and synthesize the existing body of knowledge concerning the determinants influencing the adoption of blockchain technology within organizational contexts.

To structure our exploration, we adopt the Technological, Organizational, and Environmental (TOE) framework as a conceptual lens. This well-established framework provides a systematic approach to categorize and analyze the determinants influencing technology adoption within organizational settings. As we delve into the rich tapestry of factors shaping the adoption of blockchain technology, this systematic literature review aims to provide valuable insights for both academia and practitioners. The subsequent sections will expound upon the TOE framework's relevance, delineate the methodology employed, and present a synthesis of the existing literature, culminating in a nuanced understanding of the determinants influencing the adoption of blockchain technology in organizational contexts.

14.2 RESEARCH METHODOLOGY

To identify the determinants affecting the adoption of blockchain technology within organizations, a methodical research approach was undertaken, guided by the TOE framework. The initial phase involved an in-depth exploration of existing literature in the realms of technology and organizational studies. This literature review revealed a diverse set of factors crucial for organizations contemplating the adoption of blockchain, prompting an investigation into how the adoption of this disruptive technology can be optimally managed. The identified factors were then systematically categorized within the TOE framework, which offers a three-dimensional lens incorporating technological, organizational, and environmental dimensions. This framework proved invaluable for understanding the intricate interplay of factors within the context of blockchain adoption, given the technology's disruptive nature, the involvement of various stakeholders, and the influence of market forces. To refine the focus, a targeted keyword search using "blockchain" and "adoption" was conducted across reputable databases, resulting in an initial pool of over 800 results. The abstracts of each chapter were scrutinized, leading to the identification of 35 chapters specifically addressing blockchain adoption as their primary objective. Thorough reviews of these selected chapters were conducted to extract and analyze relevant factors associated with blockchain adoption.

The culmination of this research effort resulted in the identification of 12 determinants, which were then categorized within the TOE framework. In the technological dimension, factors related to the disruptive nature of blockchain, its malleability by various actors and markets, and its potential applications were delineated. In the organizational dimension, the focus was on governance, trust, and the need for intermediaries despite blockchain's trust mechanisms. Finally, within the environmental dimension, considerations included challenges and complexities arising across technical, regulatory, social, and adoption-related domains. This systematic identification and categorization process within the TOE framework contribute to a comprehensive understanding of the determinants influencing the adoption of blockchain technology in organizations. The subsequent sections of the chapter will delve into these factors within the technological, organizational, and environmental dimensions, offering nuanced insights into the complexities of blockchain adoption.

14.3 RESULTS: FINDINGS AND ANALYSIS

This section delves into the determinants influencing the adoption of blockchain technology in organizations, categorized within the technological, organizational, and environmental contexts, drawing inspiration from the TOE framework (Tornatzky and Fleischer, 1990). These determinants can either act as enablers (+), positively influencing adoption, or inhibitors (–), posing challenges to adoption.

14.3.1 TECHNOLOGICAL DETERMINANTS OF BLOCKCHAIN ADOPTION

In the dynamic landscape of blockchain adoption within organizational frameworks, delving into the nuances of technological determinants is essential for comprehending the intricacies of adoption patterns. These determinants, namely technological compatibility, technological complexity, and relative advantage, play distinctive roles, acting as either enablers or inhibitors throughout the adoption journey, defined as:

1. **Technological Compatibility:** This emerges as a fundamental enabler when assessing the seamless integration of blockchain within existing organizational practices. High compatibility signifies a smooth alignment, allowing blockchain to enhance operational processes and meet organizational needs (Guo and Liang, 2016; Lansiti and Lakhani, 2017). This enabler facilitates a positive reception of blockchain technology within the organization. However, it can transform into an inhibitor when there is a misalignment, introducing concerns related to data security and privacy. Instances where blockchain features do not align with organizational practices may impede its adoption, rendering compatibility a decisive factor in the adoption decision.

2. **Technological Complexity:** The determinant of technological complexity presents a dual role in influencing blockchain adoption. As an enabler, lower technological complexity contributes to user-friendliness and expedites the adoption process (Morabito, 2017; Shrier et al., 2016). Intuitive blockchain platforms that are easy to understand and implement garner positive responses, facilitating the technology's integration into organizational workflows. Conversely, technological complexity acts as an inhibitor when it surpasses acceptable thresholds. Features that are overly intricate, demanding extensive training or presenting steep learning curves can hinder the adoption journey, deterring organizations from embracing blockchain technology.

3. **Relative Advantage:** The concept of relative advantage serves as a paramount enabler throughout the blockchain adoption process. A clear and tangible relative advantage, such as improved efficiencies, competitive edges, or innovative opportunities, propels organizations toward embracing blockchain (Woodside et al., 2018; Zamani and Giaglis, 2018). When organizations perceive a significant benefit in adopting blockchain technology, it becomes a driving force, expediting the adoption journey. On the contrary, the absence of evident advantages acts as a notable inhibitor. If blockchain fails to demonstrate a clear superiority over existing technologies, organizations may hesitate to invest time and resources in its adoption, slowing down the integration process.

Therefore, the technological determinants play a nuanced role, acting as either enablers or inhibitors depending on the degree to which they align with organizational needs and expectations. A strategic understanding and management of technological compatibility, complexity, and relative advantage are imperative for organizations seeking to successfully navigate the landscape of blockchain adoption.

14.3.2 ORGANIZATIONAL DETERMINANTS OF BLOCKCHAIN TECHNOLOGY ADOPTION

In the intricate landscape of blockchain technology adoption within organizational settings, a profound exploration of organizational determinants is crucial for a comprehensive understanding of the adoption dynamics. These organizational determinants encompass internal factors that either enable or inhibit the assimilation of blockchain technology into the fabric of an organization, such as:

1. **Organizational Culture:** This emerges as a pivotal determinant influencing the adoption of blockchain technology. As an enabler, a culture that embraces innovation, transparency, and collaboration facilitates the acceptance of blockchain (Chen et al., 2018; Shrier et al., 2016). When organizational culture aligns with the principles of decentralization and trust inherent in blockchain, it fosters a conducive environment for adoption. Conversely, a conservative or resistant organizational culture can act as an inhibitor, impeding the acceptance of blockchain's transformative potential.
2. **Leadership Support and Vision:** The support and vision provided by organizational leadership play a critical role in shaping blockchain adoption. Leadership endorsement serves as a significant enabler, signaling a commitment to innovation and technological advancement (Moabito, 2017). When leaders actively champion blockchain initiatives and communicate a clear vision for its integration, it encourages buy-in from the broader organizational ecosystem (Chen et al., 2018). In contrast, a lack of leadership support or a failure to articulate a compelling vision can hinder adoption efforts, leading to skepticism and reluctance among organizational members.

3. **Resource Allocation and Budget Constraints:** Resource allocation and budget considerations are organizational determinants that significantly impact blockchain adoption. Adequate resource allocation, including financial investments, human capital, and technological infrastructure, serves as an enabler (Lansiti and Lakhani, 2017). Organizations willing to allocate resources for blockchain development, implementation, and training are better positioned for successful adoption (Zamani and Giaglis, 2018). Conversely, budget constraints can act as inhibitors, limiting the ability to invest in blockchain initiatives and slowing down adoption efforts.
4. **Organizational Structure and Processes:** The existing organizational structure and processes play a crucial role in determining the feasibility of blockchain adoption. A flexible and adaptable organizational structure serves as an enabler, allowing for smoother integration of blockchain into existing workflows (Chen et al., 2018). When organizational processes are streamlined and conducive to decentralized and transparent frameworks, blockchain adoption is facilitated. Conversely, rigid organizational structures and cumbersome processes can act as inhibitors, posing challenges to the seamless assimilation of blockchain.
5. **Employee Training and Change Management:** The readiness of employees and effective change management strategies are instrumental in determining the success of blockchain adoption. Adequate training programs and change management initiatives act as enablers, equipping employees with the necessary skills and fostering a positive attitude toward the technology (Lindman et al., 2017). Organizations that invest in comprehensive training and change management are more likely to overcome resistance and facilitate a smooth transition (Seebacher and Schuritz, 2017). Conversely, inadequate training and a lack of change management strategies can act as inhibitors, leading to employee apprehension and potential roadblocks in the adoption journey.

Therefore, organizational determinants serve as key influencers in the complex process of blockchain technology adoption. Navigating organizational culture, securing leadership support, resource allocation, aligning with organizational structures, and prioritizing employee readiness are pivotal considerations for organizations seeking to embrace the transformative potential of blockchain technology. A strategic and holistic approach to these determinants is essential for successful and sustainable adoption within the organizational ecosystem.

14.3.3 ENVIRONMENTAL DETERMINANTS OF BLOCKCHAIN TECHNOLOGY ADOPTION

Understanding the environmental determinants is imperative for organizations navigating the landscape of blockchain technology adoption. These external factors, shaped by the broader environment in which organizations operate, play a crucial role in either enabling or inhibiting the assimilation of blockchain technology, such as:

1. **Regulatory Landscape:** The regulatory environment significantly influences blockchain adoption. Clarity and favorable regulatory frameworks act as enablers, providing organizations with a stable and supportive context for adopting blockchain (Morabito, 2017). When regulations align with the decentralized and transparent nature of blockchain, organizations are more likely to explore and adopt the technology (Zamani and Giaglis, 2018). Conversely, ambiguous or restrictive regulations can act as inhibitors, introducing uncertainties and legal challenges that may impede adoption efforts.
2. **Industry Standards and Collaboration:** The existence of industry standards and collaborative ecosystems is a determinant that shapes the adoption of blockchain technology. Well-defined standards and collaborative initiatives act as enablers, fostering interoperability and seamless integration of blockchain across industries (Zamani and Giaglis, 2018). Organizations operating in industries with established standards are more likely to benefit from the positive externalities of blockchain adoption (Tapscott and Tapscott, 2016). Conversely, the absence of standards and a lack of collaboration can act as inhibitors, hindering the widespread acceptance and adoption of blockchain solutions.
3. **Market Demand and Competitor Adoption:** Market dynamics and the adoption landscape among competitors play a pivotal role

in influencing blockchain adoption. High market demand and a competitive environment where peers adopt blockchain act as enablers, motivating organizations to stay innovative and relevant (Mendling et al., 2017). When organizations perceive a growing demand for blockchain-based solutions and witness competitors reaping benefits, it becomes a driving force for adoption (Woodside et al., 2017). Conversely, a lack of market demand or a slow adoption pace among competitors can act as inhibitors, leading organizations to question the urgency and relevance of blockchain adoption.

4. **Global Economic and Geopolitical Factors:** Global economic conditions and geopolitical factors contribute to the environmental determinants of blockchain adoption. Economic stability and geopolitical harmony act as enablers, providing a conducive environment for organizations to explore and invest in blockchain solutions (Pilkington, 2016). In times of uncertainty or geopolitical tensions, organizations may exhibit hesitancy in adopting transformative technologies (Tapscott and Tapscott, 2016). These external factors, therefore, play a pivotal role in influencing the risk appetite and strategic decisions of organizations regarding blockchain adoption.

Therefore, environmental determinants constitute a complex interplay of external factors that shape the adoption journey of blockchain technology. Navigating the regulatory landscape, aligning with industry standards, responding to market dynamics, and considering global economic and geopolitical factors are critical considerations for organizations seeking to harness the transformative potential of blockchain technology within their environmental context.

Figure 14.1 presents a proposed framework that categorizes the potential blockchain adoption determinants into three different contexts of the TOE framework (Tornatzky and Fleischer, 1990).

14.4 DISCUSSION

The systematic literature review, conducted through the TOE framework, provides a comprehensive understanding of the determinants influencing the adoption of blockchain technology within organizational contexts. The

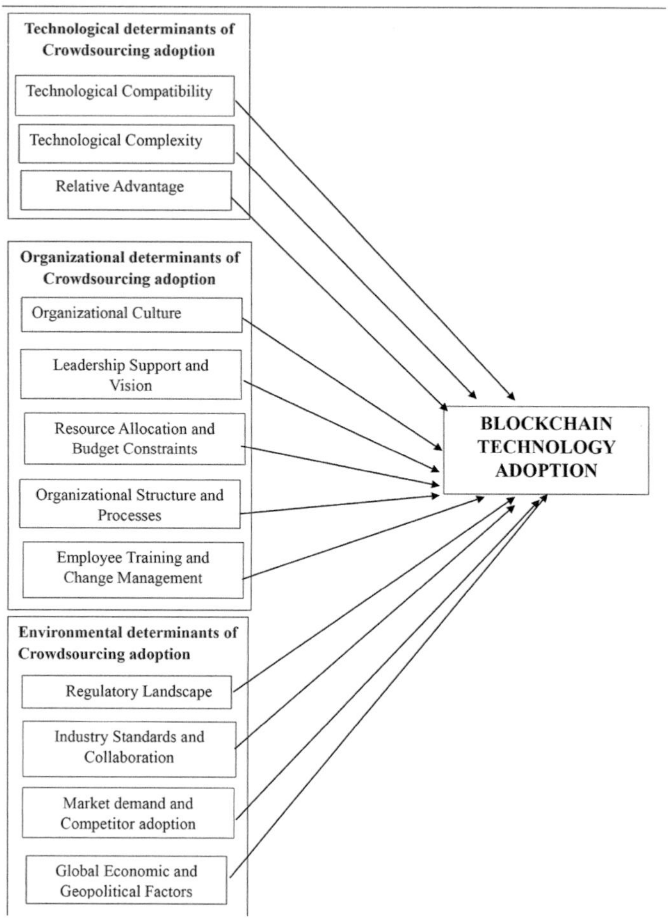

FIGURE 14.1 Proposed framework of blockchain adoption in organizations.

discussion delves into the nuanced interplay of technological, organizational, and environmental factors, shedding light on the complexities of blockchain adoption.

1. **Technological Determinants:**

   i. **Technological Compatibility:** The analysis highlights that technological compatibility serves as a pivotal enabler when seamlessly integrating blockchain into existing organizational practices. However, the potential misalignment introduces inhibitions related to data security and privacy. Organizations must meticulously evaluate the compatibility of blockchain features with their established practices to harness the benefits without compromising security.

ii. **Technological Complexity:** The dual role of technological complexity is evident, with lower complexity expediting adoption but excessive complexity acting as a hindrance. Striking a balance is crucial, emphasizing the need for intuitive blockchain platforms that align with organizational workflows. Overly intricate features may deter organizations, emphasizing the importance of user-friendly solutions in driving successful adoption.

iii. **Relative Advantage:** The concept of relative advantage emerges as a paramount enabler, aligning with the innovation's expected benefits to the organization. Blockchain adoption gains momentum when perceived advantages, such as improved efficiencies and competitive edges, resonate with organizational goals. Clear communication of these advantages is essential to drive organizational commitment toward blockchain adoption.

2. **Organizational Determinants:**

i. **Organizational Culture:** It is identified as a pivotal determinant, acting as an enabler when embracing innovation, transparency, and collaboration. Aligning culture with the principles of decentralization and trust inherent in blockchain fosters an environment conducive to adoption. Organizations should actively cultivate a culture that values innovation to facilitate a smoother transition to blockchain.

ii. **Leadership Support and Vision:** The critical role of leadership support is evident, serving as a significant enabler when actively championing blockchain initiatives. A clear vision communicated by organizational leaders propels buy-in from the broader ecosystem. Conversely, a lack of leadership support hinders adoption efforts, emphasizing the need for strong leadership endorsement to navigate the complexities of blockchain integration.

iii. **Resource Allocation and Budget Constraints:** Resource allocation emerges as a decisive enabler when organizations invest in blockchain development, implementation, and training. Adequate resources, including financial investments and human capital, position organizations for successful adoption. Budget constraints, however, pose a significant inhibitor, emphasizing the importance of strategic resource allocation to ensure sustained adoption efforts.

iv. **Organizational Structure and Processes:** The flexibility of organizational structure serves as an enabler, allowing for the smoother integration of blockchain into existing workflows. Streamlined and decentralized processes align well with blockchain frameworks. Rigid structures, on the other hand, present challenges, necessitating organizations to assess and adapt their structures to accommodate the transformative nature of blockchain.

v. **Employee Training and Change Management:** The readiness of employees and effective change management strategies emerge as instrumental enablers. Comprehensive training programs equip employees with the necessary skills, fostering a positive attitude toward blockchain. Conversely, inadequate training and a lack of change management strategies pose inhibitors, emphasizing the need for organizations to prioritize employee readiness.

3. **Environmental Determinants:**

i. **Regulatory Landscape:** The regulatory environment's influence on blockchain adoption is evident, with clarity and favorable frameworks serving as enablers. Regulatory alignment with the decentralized and transparent nature of blockchain encourages exploration and adoption. Ambiguities and restrictive regulations, however, act as inhibitors, highlighting the need for clear and supportive regulatory environments to facilitate blockchain integration.

ii. **Industry Standards and Collaboration:** Well-defined industry standards and collaborative ecosystems act as enablers, fostering interoperability and seamless integration of blockchain. Organizations operating in industries with established standards are better positioned for positive externalities. The absence of standards and collaboration, conversely, poses inhibitors, emphasizing the need for industry-wide cooperation to enhance the acceptance and adoption of blockchain solutions.

iii. **Market Demand and Competitor Adoption:** High market demand and a competitive landscape act as enablers, motivating organizations to stay innovative and relevant through blockchain adoption. Perceiving a growing demand and witnessing competitors reaping benefits become driving forces for adoption. Lack of market demand and slow adoption among competitors act as inhibitors, urging organizations to carefully assess market dynamics before embarking on blockchain integration.

iv. **Global Economic and Geopolitical Factors:** Global economic conditions and geopolitical factors emerge as influential determinants. Economic stability and geopolitical harmony act as enablers, providing a conducive environment for organizations to explore and invest in blockchain solutions. Uncertainty and geopolitical tensions, however, introduce hesitancy, emphasizing the need for organizations to consider external factors in their strategic decisions regarding blockchain adoption.

v. **Integration of Determinants:** The discussion underscores the interconnectedness of these determinants, emphasizing the need for a holistic approach. Organizations must navigate the complex landscape by aligning technological capabilities with organizational culture, ensuring leadership endorsement, strategic resource allocation, and addressing external environmental factors. The successful adoption of blockchain hinges on a delicate balance, where enablers are maximized, and inhibitors are mitigated through strategic planning and proactive management.

14.5 IMPLICATIONS

14.5.1 THEORETICAL IMPLICATIONS

This systematic literature review on the determinants affecting the adoption of blockchain technology within organizational contexts contributes significantly to theoretical advancements. First and foremost, the study refines the established TOE framework by identifying and categorizing determinants specific to blockchain adoption. This refinement enhances the framework's applicability to disruptive technologies, providing a nuanced lens for future studies examining the adoption dynamics of emerging innovations. Additionally, the analysis highlights the intricate interconnectedness of technological, organizational, and environmental determinants, advancing our understanding of how these dimensions influence each other. The study emphasizes the need for a holistic approach in theoretical models exploring technology adoption within organizations. By contextualizing blockchain adoption within the TOE framework, this research contributes to the theoretical foundation of technology adoption theories, showcasing the unique challenges and opportunities presented by blockchain and emphasizing the importance of tailoring theoretical models to the distinct characteristics of disruptive technologies.

14.5.2 PRACTICAL IMPLICATIONS

The findings of this systematic literature review carry crucial practical implications that can guide decision-makers and practitioners in successfully adopting blockchain technology within organizational settings. Organizations can leverage the refined TOE framework to inform strategic decision-making processes, assessing their readiness, identifying potential challenges, and developing targeted strategies for blockchain adoption. Practical insights into the significance of leadership support and organizational culture provide guidance for fostering an environment conducive to blockchain adoption. Decision-makers can strategically allocate resources and implement comprehensive training programs to facilitate a seamless adoption process. Furthermore, the study emphasizes the importance of regulatory engagement, market awareness, and considering global economic and geopolitical factors in the adoption strategy. The integration of theoretical advancements and practical implications provides organizations with a comprehensive foundation for informed decision-making, strategic planning, and successful implementation of blockchain technology in diverse organizational contexts.

14.6 CONCLUSION

The systematic literature review within the TOE framework provides a nuanced understanding of the determinants influencing the adoption of blockchain technology in organizational contexts. The interplay of technological, organizational, and environmental factors reveals a multifaceted landscape, necessitating a strategic and holistic approach for successful adoption. As organizations continue to explore the transformative potential of blockchain, recognizing and navigating these determinants becomes imperative. Technological considerations must align with organizational practices, leadership must champion blockchain initiatives, and external factors such as regulatory environments and market dynamics must be carefully assessed. The proposed framework serves as a valuable guide for organizations embarking on the blockchain adoption journey, offering insights to inform decision-making and optimize the integration process.

In the ever-evolving landscape of technological innovation, blockchain stands as a powerful disruptor. The success of its adoption within organizations rests on the ability to navigate the intricate web of determinants, transforming challenges into opportunities and leveraging the transformative potential of blockchain to reshape the foundations of modern business transactions.

KEYWORDS

- **adoption dynamics**
- **blockchain**
- **determinants**
- **environmental determinants**
- **organizational determinants**
- **theoretical advancements**
- **TOE framework**

REFERENCES

Androulaki, E., Barger, A., Bortnikov, V., Cachin, C., Christidis, K., De Caro, A., et al. (2018). Hyperledger fabric: A distributed operating system for permissioned blockchains. In *Proceedings of the Thirteenth EuroSys Conference* (p. 30). https://doi.org/10.1145/3190508.3190538.

Behnke, K., & Janssen, M. F. W. H. A. (2019). Boundary conditions for traceability in food supply chains using blockchain technology. *International Journal of Information Management, 52, 101969.* https://doi.org/10.1016/j.ijinfomgt.2019.05.025.

Buterin, V., et al. (2013). *Ethereum White Paper*. GitHub repository. Retrieved from: https://github.com/ethereum/wiki/wiki/White-Paper (accessed on 14 December 2024).

Chen, G., Xu, B., Lu, M., & Chen, N. S. (2018). Exploring blockchain technology and its potential applications for education. *Smart Learning Environments, 5*(1), 1–10. https://doi.org/10.1186/s40561-017-0050-x.

Guo, Y., & Liang, C. (2016). Blockchain application and outlook in the banking industry. *Financial Innovation, 2*(1), 1–12. https://doi.org/10.1186/s40854-016-0034-9.

Hughes, L., Dwivedi, Y. K., Misra, S. K., Rana, N. P., Raghavan, V., & Akella, V. (2019). Blockchain research, practice, and policy: Applications, benefits, limitations, emerging research themes, and research agenda. *International Journal of Information Management, 49,* 114–129. https://doi.org/10.1016/j.ijinfomgt.2019.06.002.

Iansiti, M., & Lakhani, K. R. (2017). The truth about blockchain. *Harvard Business Review, 95*(1), 118–127.

Lindman, J., Tuunainen, V. K., & Rossi, M. (2017). Opportunities and risks of blockchain technologies: A research agenda. In *Proceedings of the 50th Hawaii International Conference on System Sciences* (pp. 1533–1542). https://doi.org/10.24251/HICSS.2017.186.

Marsal-Llacuna, M. L. (2018). Future living framework: Is blockchain the next enabling network? *Technological Forecasting and Social Change, 128,* 226–234. https://doi.org/10.1016/j.techfore.2017.12.005.

Mendling, J., Weber, I., van der Aalst, W., Brocke, J. V., Cabanillas, C., Daniel, F., & Gal, A. (2017). Blockchains for business process management: Challenges and opportunities. *ACM Transactions on Management Information Systems, 9*(1), 4. https://doi.org/10.1145/3183367.

Morabito, V. (Ed.). (2017). Blockchain value system. In *Business Innovation through Blockchain* (pp. 21–39). Springer. https://doi.org/10.1007/978-3-319-48478-5_2.

Ølnes, S., Ubacht, J., & Janssen, M. (2017). Blockchain in government: Benefits and implications of distributed ledger technology for information sharing. *Government Information Quarterly, 34*(3), 355–364. https://doi.org/10.1016/j.giq.2017.09.007.

Pilkington, M. (2016). Blockchain technology: Principles and applications. In F. X. Olleros & M. Zhegu (Eds.), *Research Handbook on Digital Transformations* (pp. 225–254). Edward Elgar. https://doi.org/10.4337/9781784717766.00015.

Seebacher, S., & Schüritz, R. (2017). Blockchain technology as an enabler of service systems: A structured literature review. In *International Conference on Exploring Services Science* (pp. 12–23). Springer. https://doi.org/10.1007/978-3-319-56925-3_2.

Shrier, D., Sharma, D., & Pentland, A. (2016). Blockchain & financial services: The fifth horizon of networked innovation. *MIT Science Connection,* 1–10.

Tapscott, D., & Tapscott, A. (2016). *Blockchain Revolution.* Portfolio/Penguin.

Tornatsky, L., & Fleischer, M. (1990). *The Process of Technology Innovation.* Lexington Books.

Wood, G. (2014). Ethereum: A secure decentralized generalized transaction ledger. *Ethereum Project Yellow Paper, 151,* 1–32.

Woodside, J. M., Augustine, F. K., & Giberson, W. (2017). Blockchain technology adoption status and strategies. *Journal of International Technology and Information Management, 26*(2), 65–93.

Zamani, E. D., & Giaglis, G. M. (2018). With a little help from the miners: Distributed ledger technology and market disintermediation. *Industrial Management & Data Systems, 118*(3), 637–652. https://doi.org/10.1108/IMDS-05-2017-0231.

Index

D

M

N

O

P

T

U

V

For Product Safety Concerns and Information please contact our EU representative GPSR@taylorandfrancis.com Taylor & Francis Verlag GmbH, Kaufingerstraße 24, 80331 München, Germany